AF560299

ENCYCLOPEDIA OF AGRICULTURE

INTRODUCTION

There is growing international *consensus* on the urgency of slowing the human-induced deterioration of biodiversity, a deterioration that may be coming at high costs to present and future generations. Indeed, within the United Nations System, the adoption of the International Undertaking (IU) on Plant Genetic *Resources* in 1983, and of the Convention on Biological Diversity (CBD) at the Rio Earth Summit in 1992, was motivated by the goal of *maintaining sustainability* and diversity of *species* and ecosystems.

In addition to issues regarding wild species diversity, the *Convention* also recognizes the *particular importance* of biodiversity of relevance for food and agriculture. In 1993 the Food and *Agriculture* Organization of the *United* Nations (FAO) adopted a resolution requesting member countries to negotiate-through the FAO inter-governmental *Commission* on Genetic *Resources* for Food and Agriculture (CGRFA-the revision of the IU in harmony with the CBD.

The Third *Conference* of the Parties (COP) to the Convention also decided to establish a multi-year *programme* of activities on agricultural biological diversity with the aims of:

(1) promoting the positive effects and mitigating the negative impacts of *agricultural* practices on biological diversity in *agroecosystems* and their interface with other ecosystems;

(2) promoting the *conservation* and sustainable use of genetic resources of actual or potential value for food and agriculture; and

(3) promoting the fair and equitable sharing of benefits arising out of the *utilization* of genetic resources.

Benefit sharing is also called for under the IU's *endorsement* of the concept of Farmers' Rights, which aims to, inter alia, "allow farmers, their *communities,* and countries in all regions, to *participate* fully in the benefits derived, at present and in the future, from the improved use of plant genetic resources."

A major observation underlying the negotiations is that *agricultural* biodiversity "hotspots" tend to be in the developing world, while modern commercial varieties based on plant and genetic resources for food and *agriculture* (PGRFAs) from these *hotspots* tend to be *developed* and *marketed* by developed countries.

As such, many promoters of the *Undertaking* assert that developed countries are *benefiting* more from the utilization of PGRFAs from developing countries than do the developing countries *themselves,* and that these developing countries are not being compensated in return for use of these resources. As *touched* upon already in Chapter 1, enough concern has developed internationally over the need to conserve *agricultural* genetic resources that in April 1999, the 161-member nations of the UN-based CGRFA agreed that a *multilateral* system of access and benefit sharing should be established for key crops, with proposals for payment for conservation of agricultural genetic resources in developing countries. This proposal falls under the auspices of the IU on Plant Genetic Resources for Food and Agriculture, which is the first comprehensive *international* agreement dealing with PGFRAs.

The IU is an evolving international agreement related to the Convention of Biodiversity (COB), yet a separate agreement in its own right (e.g., some countries that are a party to the CGRFA are not a party to the COB). *According* t o the proposal, financing of the Global Plan of Action for the conservation and sustainable development of plant *genetic* resources will cost the international *community* an estimated US$155 to $455 million annually.

In November 2001, the 161-member nations of CGRFA agreed that a multilateral system of access and benefit-sharing should be established for key crops and approved the legally binding. International Treaty on Plant Genetic Resources for Food and Agriculture. This international treaty-hereafter denoted by *IT-entered* into force on June 29, 2004.

The IT's objectives "are the conservation and sustainable use of plant genetic resources for food and agriculture and the fair and equitable sharing of the benefits arising out of their use, in harmony with the CBD, for sustainable agriculture and food security" (CGRFA, undated). Under the IT, countries agree to *establish* a multilateral system for *facilitating* access to PGRFAs, and to share the benefits in a fair and equitable way.

The *multilateral* system applies to *approximately* 60 major crops and forages as listed in the IT's annex. The governing body of the IT, which is composed of the countries that have ratified it, will set out the conditions for access and benefit-sharing in a "*Material Transfer* Agreement" (MTA). PGRFAs may be obtained from the multilateral system for utilization and conservation in research, breeding, and training.

When a commercial product is developed using these resources, the IT provides for payment of an "equitable share" of the *resulting* monetary benefits, if this product may not be used without restriction by others for further research and breeding. If others may use it, payment is *voluntary.* The Treaty provides for sharing the benefits of using PGRFA through information-exchange, access to and the *transfer* of technology, capacity-building, through the sharing of monetary *benefits* as mentioned above, and through financial *contributions*. The IT foresees that funds received under the multilateral system "should flow primarily, directly and indirectly, to farmers in all countries, *especially* in developing countries, and countries with *economies* in transition, who conserve and *sustainably* utilize plant genetic *resources* for food and agriculture". While the general funding principles are laid out in the treaty, specifics of who pays, and how much, are not laid out.

This chapter discusses the concept of the *economic* value of the contribution of PGRFAs to *commercial* and other uses of plant genetic resources, identifies *proxies* for this measure that can be used to determine the relative contribution of each country to the benefitssharing fund, and evaluates the *suitability* of each proxy to this task. Given political realities and the lack of existing data on benefits, the level of total annual contributions to this fund will be determined through a *multilateral* negotiation process that is *independent* of any estimates of the value of PGRFAs.

Hence, a goal of this chapter is to discuss how, given the existing data, each country's relative contribution to this fund can be as highly (and positively) correlated as possible with each country's benefits from *utilizing* a defined set of PGRFAs as well as satisfy *equity considerations*. Since country contributions will be of a *monetary* form, the focus of this chapter will be on the economic *value* of PGRFAs.

ECONOMIC INDICATOR FOR BENEFIT SHARING

If, for the sake of argument, the idealized goal is to set total country contributions to the *benefit-sharing arrangement* proportional to the benefits derived from the *commercial* and other uses of plant genetic resources obtained from outside *sources*, then the question is how to derive this value.

Figure elsewhere in this chapter is a stylized graphical representation of the economics benefits *associated* with an agricultural *output-enhancing* set of PGRFAs, denoted as I (possible definitions of this set are discussed below).[3] In the figure, *equilibrium* crop *quantity* produced and supplied *without availability* of *I* (denoted as " \V) is denoted by Q_1 and $P_{1'}$, respectively.

Assuming that set I served as an input to a crop breeding program that increased the crop supply by a fixed amount at any given price, and that in this case, the *equilibrium* crop quantity produced and supplied with availability of *I* is denoted by Q_0 and $P_{0'}$ respectively. The shaded area in the figure is the loss in *economic* benefits (defined in terms of *consumer* and producer surplus) of researchers and crop breeders not having access to this set I, say, due to loss of these crop *varieties* over time.

The basic difficulty in arriving at the economic value of PGRFAs, such as that depicted in Figure elsewhere in this chapter, is largely due to the structure of the market for PGRFAs. One can assume that PGRFAs are *valuable*: Breeders need them as input to *producing* new *varieties* and, furthermore, to the extent that *downstream* producers as well as consumers benefit from these new *varieties*, they benefit from PGRFAs as well.

However, the number of suppliers of PGRFAs is high enough that the market price for PGRFAs is *essentially* driven to zero. Because many farmers may grow the same PGRFA, they are all potential suppliers. Plus, the number of *suppliers* may cut across country boundaries. *Furthermore*, for most PGRFAs, the level of substitutability by other PGRFAs appears to be quite high.

Given the divergence between the social and the market value of PGRFAs, a related difficulty in *establishing* the value of genetic resources is that it is *extremely* difficult to separate the value of the raw PGRFA input from the value of the research used to produce a new variety. With enough *information* on the plant breeders' PGRFA choice set and their process of selecting between PGRFAs, it may be possible under certain conditions to separate these values, perhaps in some manner *analogous* to *bioprospecting* models for *pharmaceuticals,* at least in some narrowly focused studies.

However, the necessary information is largely proprietary and/or costly to collect. The contingent valuation and hedonic methods may come to mind as shortcuts to estimating values of PGRFAs, but have some serious *shortcomings.* The general concept of economic value in period *j* of a set of PGRFAs, can be formally stated in a simple *manner.*

Ignoring for ease of *exposition* the lags in the development of new products that use I as input, the value of I summed across all stakeholders in period *j* is

$$V_i(I)=\left[(W_i)-(W_i \setminus I)\right]$$

where W_i is the value of the agricultural sector in the PGRFA-utilizing countries. This value can be defined as consumer plus producer surplus at the retail level, and $W_i \setminus I$ *(i.e.,* W_i without 1) is the value of the agricultural sector *assuming* that it could not obtain PGRFA set *I.* If suppliers of *I* had some form of market power of the supply of set *I* in period *j,* then they would *capture* a portion of *V(I).*

For an *illustration* of how *V(I)* can be formally represented, take the seed industry as an example and assume for simplicity that $_{vj}(I)$ represents only *benefits* to that stakeholder. In a highly stylised form that considers as beneficial only the number, and not the quality, of varieties in set *I,* the seed industry *maximises* the following profit function:

$$\pi_{I.}=p(S)S-c(S,N)$$

where *S* is seed sales, *N* is the number of varieties in set *I, p (S)* is the demand *function* for S, and *c(S, N)* is the cost function, and *S = S(N).* While plant breeders *usually* do not pay for *N,* because research costs are a function of *N, N* enters into the cost function directly as well as indirectly. Totally *differentiating* n with respect to *N* yields

$$\frac{d\pi}{dN}=\left(P+\frac{\delta p}{\delta S}S-\frac{\delta c}{\delta S}\right)\left(\frac{\delta S}{\delta N}\right)-\frac{\delta c}{\delta N}$$

If the industry is producing *S* at profit-maximising levels, then marginal cost equals marginal revenue and the difference in the first set of *brackets* is zero, yielding

$$\frac{d\pi}{dN}=-\frac{\delta c}{\delta N}$$

says that the increase in profits for a marginal increase in *N* equals the *negative* of the decrease in costs associated with a marginal increase in *N.* If we assume that research costs fall with increases in *N,* then $d\pi/dN > 0$ and profits increase with increases in *N.* If the seed industry were to be taxed for *contributions* to the benefit-sharing fund in the amount t(N), so that $\pi =$

$p(S)S - C(N, S) - t(N)$, where $\delta t/\delta N > 0$, then di/dN would be positive only if $-\delta c/\delta N > \delta t/\delta N$. In other words, the marginal tax rate on *N* must be less than the decrease in marginal cost of *N* for breeding firms to benefit from increases in *N. Similar* analysis can be used evaluate the impacts of *N* and of *t(N)* on consumer and producer surplus. Of course, *V(I)* is a function of the scope of PGRFAs included in *I* as well as the time period over which *benefits* are to be evaluated.

For *instance*, is I the set of all varieties of wheat, or even all major crop species in existence at time *j*, or *I* is the set of varieties obtained by plant breeders from suppler countries in *j*. If it is the former, V_i will be enormous. Or is *I measured* at the gene level? With regards to the time *dimension*, in quantifying total benefits for the purpose of determining contributions to the benefit-sharing pool, is the present value of past benefits to be included in *addition* to current benefits? If so, how far back? While scope and time *dimensions* have not been set in the multilateral negotiations, some definitions have been discussed.

For example, the African Region states that the contribution to the fund should be some percentage of "the value of the commodity produced using intellectual *property* rights [IPR] material...", which is similar to a definition set out in a proposal by Malaysia.

Since Intellectual Property Rights (IPR) for PGRFAs are a fairly recent concept, this definition addresses the time dimension as well as the scope of PGRFAs to be considered in *determining* benefits. Even if *I* is *precisely* defined, a complicating factor in estimating *(W\I)* is that the substitutability of other inputs for set *I* must be considered, given that *(Wj\I)* in the case where some substitutability of inputs is possible will be higher than *(Wj\I)* in which no other inputs can substitute for set *I*.

Substitutability may be in the form of *technological* innovations. If advances in *biotechnology* make the availability of PGRFAs for plant breeding less *necessary*, then $V_i(I)$ may decrease. Because of substitution possibilities between PGRFAs, one would expect *diminishing* marginal benefits to be associated with adding another randomly chosen PGRFA to *I*. The more varieties are contained in *I*, the smaller the difference in *V(I)* with and without any randomly chosen variety.

For example, if *I* is a set of PGRFAs currently in publicly owned gene banks, given that most of these accessions are never examined, then the market value portion of V(I) changes little when a randomly chosen PGRFA is added to the *collection* (though, as *discussed* in the next section, the value added to society may be greater). Also, if one wants t o consider the value associated with *I* supplied from any one country *i*, or $V_i(I_i)$, then the extent of *geographic substitutability* must be considered in estimating this value. The value V(I,) will be higher the more PGRFAs in *I;* are also grown in other countries other than *i*, and higher the less easily varieties from other countries can substitute for varieties in *I*.

COMPONENTS OF VALUE OF PLANT GENETIC RESOURCES FOR FOOD AND AGRICULTURE

The previous section discussed the general concept of benefit-sharing. In this section, we discuss the *composition* of V{I). From an economic standpoint, several value categories can be ascribed to PGRFAs. The most concrete one is its *use value, i.e.*, the value associated with the direct and indirect benefits resulting from the use of PGRFAs by plant breeders, farmers, food processors, and consumers. For plant breeders, PGRFAs are inputs to producing more productive or disease-resistant varieties.

To a large extent, this use value is a function of the breeding technology and of the income achievable from productive use of the improved seed. Improvements in breeding technology, through *biotechnology* for example, may increase breeders' demand for germplasm and thus raise its value and market price.

A share of the economic benefits of improved varieties goes ultimately to consumers in the form of lower food prices and another to farmers in the form of greater revenues due to higher yields or higher quality products. The second-value component, the *option value*, is the value to society (their *willingness* to pay) of avoiding *irreversible* decisions on the conservation of PGRFAs. In particular, the loss of native landraces (or "traditional" varieties) is irreversible.

As first noted by Hanemann and Fisher (1986) in the context of option value, other varieties may be close, but not perfect, substitutes, so once a *particular* landrace is extinct, its value for future plant breeding will remain unknown. In this context, option value suggests that a premium exists that is associated with actions that preserve flexibility. In other words, it is an option to be able to consume the product in the future.

However, this *premium* is held by society, and not necessarily by private industry, which has little incentive to maintain a *in situ conservation* program outside of the firm's own private lines. The *reasons* are that the likelihood that any particular PGRFA currently *in situ* will yield a useful input to the breeding of a new variety is very low and, given its characteristics of nonrivalness and nonexcludability as a breeding input, one firm's conservation of a PGRFA will not *necessarily* exclude any other firm from conserving the same PGRFA.

If the PGRFA-utilizing firm could be assured that the same PGRFA it has bought rights to will not be sold by the supplier to another firm as well, the price it is willing to pay could have a positive option value component to it. Depending on the structure of the market for PGRFAs, even in the case where *suppliers* of PGRFAs could control access, the social value of conserving PGRFAs may be higher than the private value.

In other words, even in the case of a hypothetical single-market supplier for PGRFAs, the option value may still not be fully revealed in market prices for PGRFAs. If society has a value for conservation actions that assure the potential for future use of the PGRFAs, then this value is part of the total benefits from PGRFAs. However, this option value makes sense as a part of the contributions to the fund only if distributions from the fund require conservation activities on the part of the recipients.

Otherwise, it would not be equitable to make option value part of a country's contribution to the fund. The *possibility* of an option value component demonstrates that, at least in principle, how the money is to be distributed affects the level of contributions. At any rate, the discussion of the exclusion or inclusion of option value in *contributions* is an academic one at this point, given that no data exist on this value. A third component of the value of PGRFAs is its *existence value, i.e.,* the value one holds for a variety or set of varieties just for its own sake or for some moral or cultural reason.

With respect to loss of PGRFAs, the main threat facing agriculture is the loss of intra-species

Table 1.1: Potential country level indicators for benefit sharing.

Indicator	*Observability*	*Equity*	*Correlation*
Value of agricultural output	+	—	—
Gross Domestic Product	+	+	—
Agricultural Gross Domestic Product	+	—	—
Seed industry profits and/or revenues	—	+	+
The value of agricultural commodities produced using intellectual property rights (IPR) material	—	+	+
Value of commodities produced using improved (e.g., "green revolution" varieties)	—	+	+
Royalties earned on agricultural patents	—	+	+
Agricultural research and development expenditures by country	—	+	—
Plant protection titles issued	—	+	—
Number of landraces used in agriculture	—	—	—
Domestic-origin patents used in the agricultural and food sectors Matrix of varietal exchange and matrix of parental exchange	—	+	—
Diversity measures	—	—	—

Notes: The scale "-" / "+" are used to construct relative ranking within the group of the 13 potentially feasible indicators.

diversity, and not inter-species diversity-species that may be lost are not major players in *agriculture*. Hence, pure *existence* values for *conserving* PGRFAs are likely to be only a small portion of their total economic value. Almost certainly, people hold a larger existence value for knowing a certain species exists than they hold for individual varieties within that species.

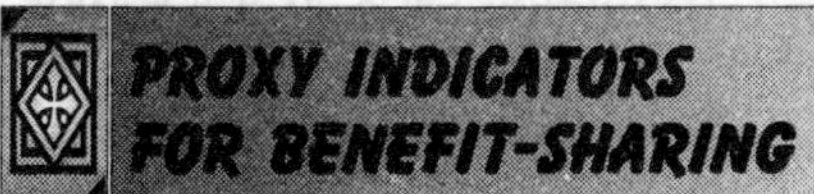

PROXY INDICATORS FOR BENEFIT-SHARING

In a world with full *information*, contributions towards a benefitsharing *arrangements* would be based on $V_{jl}(I)$, $I = 1,\ldots,L$ countries, where $\sum_{l=1}^{L} V_{jl}(I) = V_j(I)$, given that the benefit-sharing fund requires contributions from each of the L countries endorsing the IU. Of course, for the reasons discussed earlier, $(W_{jl} \setminus I)$ is unknown for almost any definition of I, and hence, $V_{jl}(I)$ is unknown. Hence, it appears that the best one can do is to base *contributions* on just the W_l portion of $V_{jl}(I)$.

While existing indicators cannot be explicitly based on benefits, given that a benefit-sharing fund will be created, the best one can do is identify existing *indicators* that are politically acceptable and that appear to have some connection to the benefits associated with PGRFAs. Table elsewhere in this chapter presents a list of potential country level *indicators* for benefit sharing that are at least somewhat within the realm of feasibility.

No claim is made that this list is all-inclusive, but it does attempt to cover a broad range of classes. Indicators 1-3 are easily *obtainable* for almost all countries. While the collection of indicators 4-13 is technically feasible, they cannot be found for more than a few countries and/or crops, and hence, may be of limited policy usefulness, regardless of their *correlation* with the value of PGRFAs.

Given that the *negotiations* on setting the total size of the Fund will essentially be independent from the negotiations that determine each country's *contribution*, a country's share of contributions to the Fund can be set to its share of the world total value of the indicator. *Analytically*, country *I*s contribution to the Fund can simply be C_I = *(Total Fund Value),* $\left(\text{Indicator}_I / \sum_I^N \text{indicator}_I\right)$ where *Indicator*, is the value of the indicator for that country, and *I* = 1, ... , *N* countries. Due to the continuous cycle of imports and exports of varieties between regions, in some sense every country is a net user of PRFAs (Wright). Under this formula, every country makes a contribution to the Fund.

The first indicator in the table is the value of agricultural output (VAO). This *indicator* has several desirable *characteristics*. First, this indicator is readily available for each country and, among the *indicators* listed in Table elsewhere in this chapter, is most tied to the value of agricultural activities as it is the value of agriculture production at the farm gate.

Second, it is equitable in the sense that countries with greater agricultural value pay more, which is analogous to a progressive income tax on agriculture. For instance, if the U.S. payment to the Fund were based on its share of total average annual worldwide value of *agricultural* production, its *contribution* would represent 12.7% of the total value of the fund.

Since most developing countries represent only small shares of the total value of agricultural production, their contribution would be small. The justification for the use of Gross Domestic Product (GDP), Agricultural Gross Domestic Product (AGDP), and Seed Industry Profits and/or Revenues indicators are similar to those for VAP. GDP is the *value-added* at each stage of *production* of all goods and services during one year.

A potential benefit of VAP over GDP from a negotiation *standpoint* is that an indicator based on the former is at least directly tied to agricultural production, if not to the value of PGRFAs. AGDP, which is the value-added only of agricultural goods and *services*, is more closely tied to agriculture than is GDP but includes the value of services that are not directly related to agricultural production.

Its use as an indicator would explicitly acknowledge that consumers and food processors as well as farmers and seed companies benefit from PGRFAs. *Therefore*, when GDP or AGDP is used as the basis for a funding mechanism, countries who benefit on the *consumer* side but are without an agricultural production base will contribute to the Fund.

Similar to VAP, the large size of AGDP relative to any realistic prediction for the total size of the Fund suggests that a benefit-sharing payment at the retail level is unlikely t o produce *significant* market effects. In practice, the indicators VAP, GDP, and AGDP are all highly correlated with one another and a country's share of the *contributions* to the Fund would be similar using any of these

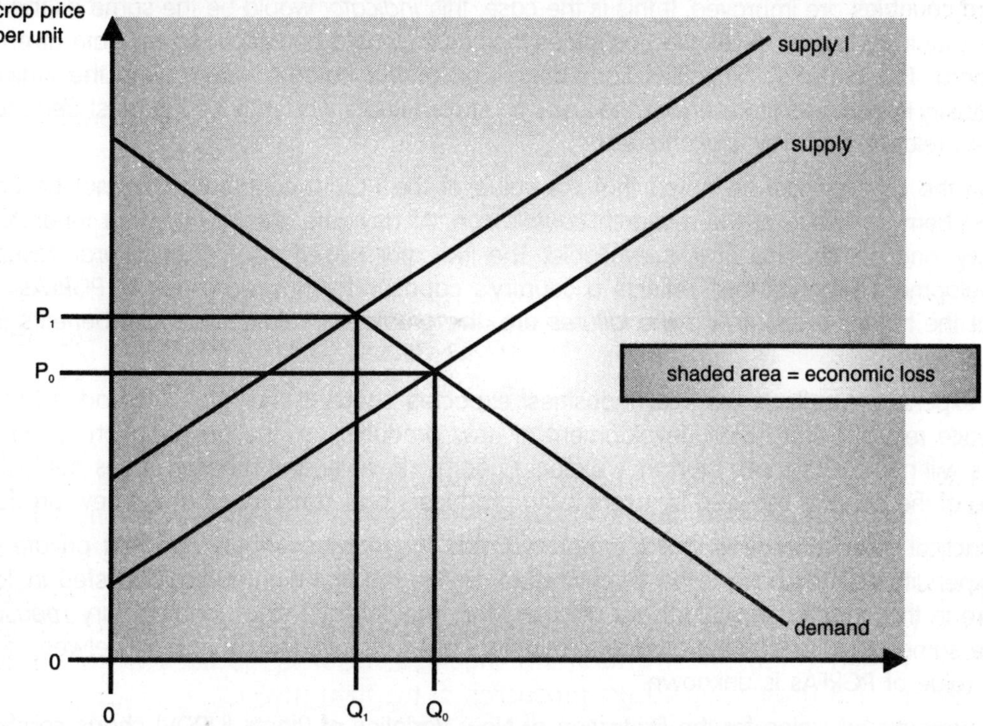

Figure 1.1: Economic losses stemming from loss ofPGRFAs set I.

as a basis. As a proportion of total revenues, commercial seed producers are probably the stakeholders most dependent on access to PGRFAs, and therefore, seed industry profits and/or *revenues* by country could be, among the class of measures that may be obtainable, the one most highly correlated with the value of PGRFAs.

However, this measure is not currently available for all countries, but with data available for 39 countries, it has better coverage than some other indicators. Given the *International* Association of Plant Breeder's (ASSINEL) definition of "commercial seed producer," the value at the world level of the *commercial* seed industry is $30 billion per year. If industry profits are around 10%, then a $50 million benefit-sharing tax levied on the industry represents 1.7% of profits, and a $300 million tax represents 10% of profits.

While the extent to which this increase in cost would be passed on to downstream *stakeholders* depends on supply and demand conditions in each stakeholder's market, the relatively large ratio of the tax to total profits (or revenues) suggests that a tax aimed only at commercial seed producers would produce more notable price impacts, at least in the market for seed, than a tax at the farm gate or retail level. The indicator "Value of Agricultural *Commodities* Produced Using *Intellectual* Property *Rights* (IPR) Material" is more narrow in scope than the previously mentioned indicators.

This indicator is available for only a few countries, making it difficult to develop a benefit-sharing mechanism. The indicator "Value of Agricultural Commodities Produced Using Improved Varieties" has a slightly broader scope than the IPR indicator, but the term "improved" must be clearly defined. Under many possible *definitions*, one could expect that all varieties grown in developed countries are improved. If this is the case, this *indicator* would be the same as the VAP for these countries, but for developing countries the values would be difficult to estimate given the lack of data. The indicator "Royalties Earned on Agricultural Patents," along with the indicator "Seed Industry Profits and/or Revenues," focuses on stakeholders who may be the most *dependent* on PGRFAs relative t o other *stakeholders*.

Again the caveat must be noted that the value of the PGRFA contribution cannot be easily separated from the value of the research *contribution*. At any rate, the royalty data tends to be proprietary, and hence, little of it is available. The indicator based upon "Agricultural Research and Development Expenditures" reflects a country's capacity for upstream use of PGRFAs. It is likely that the higher a country's *expenditures* are, the greater its relative financial benefits from the use of PGRFAs.

This expenditure reflects the R&D industries' expected share of the total value-added in the marketplace resulting from their development of new products, i.e, the presumption is that the industries will not spend more than they expect to earn. However, the measure does not include the share of the benefits received by users (both producers and consumers) of the new products.

A practical disadvantage is that it is not available for many countries and that private and public expenditures may have to be disentangled. Unlike the first eight indicators listed in Table elsewhere in this chapter, indicators are *nonmonetary* measures. These nonmonetary *measures* share the same drawback as the available monetary ones, that is, the connection between them and the value of PGRFAs is unknown.

The International Union for the Protection of New Varieties of Plants (UPOV) charts certificate applications and titles issued for plant variety protection, and has made the data available for 28 member countries (UPOV, 1995). The indicator "Number of Landraces used in Agriculture" is available for rice for select countries from the International Rice Research Institute, along with a tabulation by country of own and borrowed rice landraces. Data on "Domestic-Origin Patents Used in the Agricultural and Food Sector" are available only for some countries.

The indicator "Matrix of Varietal Exchange and Matrix of *Parental* Exchange" tracks international exchanges of varieties and is available for rice for selected countries. With nonmonetary measures such as these, one essentially has the problem of comparing apples and oranges. For example, one country could hold 20 plant protection titles, but the sum of their values could be less than one plant protection title held by another country.

Most likely, a large proportion of the total value of improved varieties is ascribed to a small percentage of improved varieties. This *asymmetry* is apparently the case for U.S. university-held patents, in which only a few patents (and universities) account for most of U.S. *university* royalties on patents (AUTM, 1993). Given the small number of *commercially* successful varieties relative to the total number of available varieties in genebanks (FAO, 1998), there is reason to believe that this asymmetry is also true for the economic value of PGRFAs.

That said, monetary measures as a basis for contributions have the advantage that contributions tend to be equitable (contributors with higher incomes often pay more) and may be rationalized from a development aid standpoint. The final category in the table (Diversity Measures) covers indicators of PGRFA diversity. Many different measures are available, and no single measure can satisfy all aspects of diversity.

These diversity measures are most useful for tracking changes in diversity over time and for assessing the impact of *conservation* measures. As with the other indicators, the relationship between this indicator class and the value of the PGRFAs is unclear. Unless the diversity itself is of value (which may be the case for wild species and ecosytems), the degree of crop diversity in a country is somewhat irrelevant as a *proxy* for the value of PGRFAs.

Instead, it is their current or future contribution to agricultural production and to protecting agricultural production (e.g., from disease) that are of value, and not the resources themselves. For instance, while two countries may have the same genetic diversity index, there is little reason to assume that their respective contributions to the value of agriculture are equal. After weighing the advantages and *disadvantages* of each of the listed indicators, the VAP and AGDP appear to be "superior" in the case of PGRFA conservation.

If each countries' contribution to a benefit-sharing fund is the ratio of its indicator value to the total world value of the indicator, then *application* of either of these indicators will be progressive, primarily with respect to equity considerations, but also with respect to efficiency considerations. That is to say that although all countries contribute, those countries with higher VAP and AGDP benefit from PGRFA more than those with lower values and also pay more.

CONCLUSION

If a multilateral *environmental* agreement (MEA) requires member countries to meet specific environmental targets or to fund the provision of a global public good, then environmental indicators become necessary. An environmental indicator measures environmental quality, whether as a measure of the physical quantity itself or of the monetary impact of that quantity.

For agriculture, in particular, indicators generally include measures of land-use changes between agriculture and other land uses, both on-farm and off-farm impacts of soil erosion, total agricultural water use, nutrient balances, pesticide use, and water quality (*Organization* for Economic Cooperation and Development [OECD], 2000).

Obviously, some decision on the choice of indicators is necessary if one is to make inter-regional comparisons of the *environmental* benefits from the MEA. The OECD has developed a "core set . . . of commonly agreed indicators for OECD countries and for international use, . . ." (OECD, 2001). According to the OECD, the purpose for developing this set of core indicators is to:

- Allow countries to track environmental progress.
- Ensure integration of environmental concerns into sectoral policies (e.g., agriculture).
- Ensure integration of environmental concerns into economic policies.
- Measure environmental *performance*.
- Determine whether countries are on track towards sustainable development.

Among the criteria for selecting a core set of indicators are that the indicators are policy relevant, analytically sound, measurable, and easily interpreted (OECD, 2000). This set of OECD indicators could be used to assess whether or not countries are meeting environmental commitments specified under an MEA. Environmental indicators can also be used to determine a country's funding obligations to an MEA, as well as *disbursements* from an MEA to member countries, say, for conservation efforts. With respect to the agricultural sector, the OECD is seeking to develop indicators to better integrate environmental and economic policies with the goal of sustainable agriculture in mind. While developing such *indicators* may be a relatively easy task for an MEA seeking to address some fairly concrete externality, such as loss of forest cover, as pointed out earlier, it is not as easy in the case of conserving agricultural genetic *resources*.

From an economic standpoint, it seems reasonable to tie a country's contribution to the benefit-sharing fund to the benefits it receives from its use of PGRFAs. *Unfortunately*, as discussed in this chapter, these benefits cannot be quantified, except perhaps in limited case studies.

Hence, an alternative can be to appeal to indicators that take equity and development considerations into account in determining contributions. Furthermore, to ease the process of multilateral negotiations, these indicators must be available for all (or almost all) the countries participating in the negotiations.

Among *indicators* satisfying these conditions, the VAO and AGDP appear to be the most applicable. Both are progressive in the sense countries with higher values pay more. A potential advantage of VAO over AGDP in multilateral negotiations is that the former is explicitly focused on agricultural interests.

Use of the latter in the negotiations is appropriate if the benefits of PGRFAs t o consumers are to be accounted for in setting country contributions to the Fund. Furthermore, the latter is somewhat more *equitable* (in terms of income *distribution*) than VAO.

For all practical purposes, given that these two indicators are highly correlated, the choice between the two will not have much impact on the relative size of each country's contributions. Indicators such as the value of commercial seed production focus better on stakeholders most dependent on access to PGRFAs, but in addition to data limitations, their downside is that they ignore *downstream* benefits to farmers, food processors, and consumers.

It is worth mentioning in closing that no indicator exists that will be more than an imprecise guide for how to distribute the cost of any benefit sharing fund in an economically efficient fashion. Obviously, the ideal indicator relies on fully observable environmental and economic benefits, and thus increases the *possibilities* for efficient and equitable distribution of any and all benefit sharing funds.

However, at least having some reasonable indicators available as a guide may reduce the potential for the funds being distributed through an opaque process. Finally, as is discussed in the "Components of value" section, how the funds are to be *distributed* has some bearing on what is included in the valuation of the PGRFAs. However, some proponents of benefit-sharing assert that since PGRFA suppliers are due compensation in return for utilization of their genetic resources by others, they should not be restricted to what use they put the compensation to.

In fact, it is the political considerations regarding this view that are the primary reason why multilateral negotiations have not *simultaneously* discussed contributions to, and distributions from, the fund. However, data limitations aside, the often-blurred distinction between users and suppliers of PGRFAs, as well as other complications, make explicit compensation impossible. Hence, unrestricted distribution from a benefit-sharing fund would appear to be a simple income transfer. To help ensure that a benefit-sharing fund is maintainable over time, the major contributors must perceive that the funds are being *redistributed* in a constructive fashion. *Earmarking* the fund for conservation and sustainable use of PGRFAs or related food security activities will help the fund distinguish itself from other forms of development aid.

Chapter 2

CROP GENETIC DIVERSITY

Agricultural *productivity* enhancements have often been based upon *development* and dissemination of a small number of highly competitive plant varieties or animal species and, thus, have been associated with a decrease in crop genetic diversity (CGD) (FAO, 1998). *Mechanical* innovations such as tractors and combines, supported by breakthroughs in chemical fertilizers and pesticides, have facilitated the *mono-cropping* of vast areas of land, so it is not surprising there is concern that the advent of agricultural biotechnology may exacerbate these trends.

In this chapter, we argue that agricultural *biotechnology* may instead offer unique opportunities to preserve CGD, but the speed and extent to which this potential is realized depend upon *institutional* factors, including the distribution and level of protection afforded to intellectual property rights (IPRs), transaction costs associated with licensing, technology transfer, and biosafety regulations. As is *common* in the literature (FAO, 1998), CGD is used to describe the genetic diversity of agricultural crops.

The number of different varieties or landraces being used by farmers is an important indicator of the *in situ* CGD of a particular crop species. Biotechnology introduces a fundamental change in the way that seeds and other genetic materials can be produced. With *traditional* breeding techniques, *existing* varieties are selectively combined to develop new varieties or hybrids.

This is a lengthy process and involves a significant degree of randomness. The outcome is usually a novel variety that has a number of new traits and characteristics, not all of which are desirable. Biotechnology, however, allows the *targeted* introduction of selected genetic materials into existing crop varieties. Once the genetic sequence coding for a desirable trait such as insect resistance has been identified, a "transformation event" is created by transferring this genetic sequence to a *particular* receptor variety.

Additional genetically modified varieties (GMVs) are then developed by crossing existing conventional varieties with this transgenic receptor variety. Although several backcrossing *generations* are necessary to eliminate unwanted *characteristics*, this process is far quicker, easier, and cheaper than developing a new conventional variety through cross-breeding.

It usually results in GMVs that are virtually identical to their conventional *counterparts except* for the new desirable trait. In other words, biotechnology permits a separation between the act of developing a specific *agronomic* trait and the breeding of a *particular*, locally adjusted variety. As such, it has important institutional and economic implications that will affect the diversity of crop plants in agricultural production. In this chapter, we evaluate the potential for adoption of seeds and genetic materials developed using *biotechnology* and assess the impact on the diversity of crop varieties produced under alternative industrial and policy *structures*.

We examine the roles of IPRs and the research capacity and efforts of private and public sectors in determining the *utilization* of agricultural biotechnology to obtain policy *implications* for agricultural research efforts. The next section provides an overview of our framework and main findings. It is followed by a detailed mathematical derivation of the main results.

The final section presents a summary of our results, and it demonstrates their validity using available *information* on a number of GMVs of various crops grown in different countries.

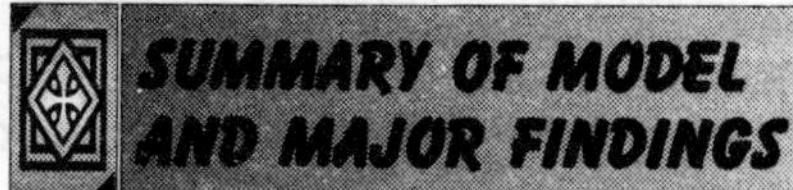

SUMMARY OF MODEL AND MAJOR FINDINGS

The legal framework in the United States, Europe, and other developed countries has gradually evolved so that those who decipher genetic structures, discover the functions of genes, or identify mechanisms to alter genes can register patents and own the IPRs for the *utilization* of these discoveries. Private parties have an incentive to conduct *research* leading to new discoveries because they expect to gain financially from selling the rights to utilize the IPRs, or to utilize them directly in their own *commercialization* efforts.

However, the extent to which IPRs are protected and traded varies among nations, and these *variations* may affect the way the products and *processes* of biotechnology are managed and utilized. In our *conceptual* analysis, we consider the case of an agricultural industry sector that produces a single crop where producers are characterized by high degrees of heterogeneity. Variation in land quality, *topography*, and climatic conditions, even within a region, may result in growers adopting different varieties of the same crop.

We assume that, technically speaking, all the existing varieties can be modified using biotechnology. This *modification*, for example, may reduce susceptibility to pests and diseases or increase the efficiency of nutrient uptake. For simplicity, we assume that the effect at the farm level is an increase in yields.

It is assumed that utilization of a biotechnology-based innovation, such as *a Bacillus thuringiensis* (Bt) gene that codes for the expression of an *insecticidal* protein in plant tissue, requires a large fixed cost in infrastructure for technology development, a modest fixed cost to obtain the capacity to incorporate the technology into each specific variety, and a *relatively* small variable cost of seed production. We argue that biodiversity and the impact of a new biotechnology-based innovation on *grower* and consumer welfare depend on the manner in which its *introduction* and pricing strategies relate to the existing *cropping* system.

In particular, we distinguish between situations where private sector companies introduce the technology and situations where it is introduced by the public sector. We also distinguish between situations where traditional local varieties are replaced by "generic" GMVs and situations where specific genes are introduced to local varieties, which continue to be planted in the new, modified form.

The generic GMV may be an imported variety that, on average, performs well, but since it may not be highly suited to each location's conditions, will likely not perform as well as GMV versions of local varieties. These features will determine the outcomes of farmer and *consumer* welfare and biodiversity preservation. When a private company introduces GMVs, we assume that it charges a monopoly price. This price is set at a level where marginal revenues are equal to marginal costs of producing the *modified* seeds.

When farmers are heterogeneous in their conditions, the impacts of the technology may vary. GMVs will be adopted only where the gain from adoption is sufficient to cover the technology fee. The technology fee will increase with the variable cost of modification (and when the cost of modification is assumed to be fixed, only the cost of seed production is variable).

Therefore, low variable costs increase the adoption of GMVs. When a local variety is genetically modified (GM), the effect of environmental heterogeneity is likely to be an increase in the yield effect and cost savings compared to the case in which the local variety is replaced by a generic GMV. However, the modification of each variety entails a fixed cost.

In deciding whether to genetically modify a specific variety or to offer growers a generic GMV, the private company will compare the extra cost of *modification* with the extra revenues earned from a modified variety relative to a generic variety. Two factors that will affect this decision are the cost of modification and the size of the market.

When modification costs are high or when the market is small, it will be more profitable to sell a generic GMV. Thus, regions where the crop-breeding industry has low *capacity* and the cost of modification is relatively high are more likely to adopt generic GMVs, and their introduction may lead to a reduction of crop *biodiversity*. Thus far, private sector *companies* have developed and introduced most of the GMVs globally.

The history of the Green Revolution suggests that eventually public sector institutions will also develop and introduce GMVs, and these varieties will be distributed through small seed companies. We analyze outcomes where the public sector makes the choice of whether or not to introduce a GMV to a region, and whether there will be a genetic modification of the local variety or a generic GMV imported from elsewhere.

The public sector *organization* is assumed to *maximize* the domestic economic surplus, including seed producers, farmers, and consumers. Under these scenarios, the seed companies will charge competitive prices that are less than the price charged by a *monopoly* seller. As a result, adoption levels are likely to be higher than under the monopolistic scenario if all the other parameters are the same. Furthermore, since the total domestic surplus is larger than the profit of monopolistic seed companies, the public sector is more likely to invest in development of GMVs than a private sector monopolist. The public sector is more likely to develop local GMVs rather than import generic GMVs, since the development of local GMVs is likely to have a larger yield effect that benefits *growers* and consumers rather than seed producers.

The decision whether to introduce a local or an imported generic GMV, in any context, depends on the difference between the gain and the cost of development. Countries with more *advanced* crop-breeding capabilities and sufficient public sector resources for developing GMVs are more likely to modify existing varieties, which will lead to preservation of biodiversity.

On the other hand, in situations where public sector resources are limited and where investment in GMVs is not profitable for the private sector, the public sector may develop or import a small number of generic GMVs and, in spite of *limited* adoption, it may lead to a loss of biodiversity.

A MODEL

An agricultural sector is producing a single crop in M locations, with varying climatic and agroecological conditions. Let j be a location indicator. Then j takes values from 1 to M. Before GMVs are introduced, farmers at each location use the best traditional variety given their conditions, and we *assume* that each location has its own distinct traditional variety.

For simplicity, we assume that the gross benefits of production are measured in monetary terms. Output price denoted by P is assumed to be constant. In the analysis, we consider several scenarios with respect to market and properties of genetic materials.

Let s be a scenario indicator; $s = 0$ is the initial situation where only traditional varieties are used; $s = m$ is the case in which locally optimized varieties are modified *(GMV),* and $s = g$ is the situation in which only one generic modified variety *(GMV) is* adopted. Let X_j^s denote output produced at location j under *scenario* s, which is assumed to be a function of land and genetic *materials.*

The production function f_j^s (A) denotes the output produced on A acres of land in location j and scenario s. The marginal productivity of land for scenarios is $Mp_j^s\ (A)=\delta f_j^s/\delta A$. It is assumed that $Mp^s_j\ (A)$ and $\delta Mp^s_j/\delta A < 0$ (decreasing marginal productivity of land in a location), which may reflect heterogeneity of land *quality* within a location.

Prior to the introduction of GMVs, when $s = 0$, the demand for land of *variety j* at location *j is* denoted by $PMp_j^0(A)$. The marginal cost of land grown with variety j is assumed to be constant and is denoted by Mc_j and the seed cost *prior* to biotechnology is *assumed* to be 0.[3] Before the introduction of biotechnology, the acreage allocated to variety j, A^0_j was at the point when marginal benefits *equal* marginal cost of acreage, or $(PMp^0_j(A^0_j) = Mc^0_j)$.

Assumptions about the GM Technology and Seed Industry Structure

Before analyzing several outcomes from GMVs, let us specify assumptions about their benefits and cost structure. We assume that each traditional variety can be modified, but it may also be replaced by a "generic" GMV. The exact outcome depends on costs, constraints, and *decision* making about seed supply.

For simplicity, we assume that GMVs improve seed productivity. This corresponds to yield-increasing GMVs (e.g., Bt cotton in India or South Africa). Let Mp_i^m *(A)* denote the marginal productivity *(Mp)* of land *planted* with *GMVj*. We will assume that (a) GMVs have higher marginal productivity than the traditional variety, $Mp_i^m(A) > Mp_i^0(A)$ and (b) the marginal productivity gain declines with *A*,

$$\delta(Mp_i^m(A) - Mp_i^0(A))/\delta A < 0$$

This assumption that better-quality lands gain more from genetic modification is done for convenience, and most results hold for broader circumstances. It corresponds to situations, say, where a certain percentage of the crop is lost to pests. GMVs reduce pest damage and thus provide higher gains to locations with higher potential output.

If traditional varieties are replaced by a generic GMV, then $Mp^g_i(A) = \delta f^g_i(A)/\delta A$ *is* the *marginal* productivity of land in location *A*. We assume that the *Mp* of the unmodified version of the generic variety is less than that of variety *j*.

Because modification increases the marginal *productivity* of the target variety, it is assumed that the marginal benefit of the generic GMV, while less than that of the GMV_i is greater than the traditional variety. Hence,

$$Mp_i^m(A) > Mp_i^g(A) > Mp_i^0(A)$$

and also that $\delta(Mp_i^g(A) - Mp_i^0(A))/\delta A < 0.$

The introduction of the GM technology is associated with several cost *categories*. The first is the fixed cost to introduce the GMVs at the crop level. It consists of research, testing, *registration*, and regulatory compliance costs to introduce, say, Bt cotton in South Africa or Bt sweet potato in Kenya. We assume that this cost has been incurred and consider two other costs.

(a) F_i = fixed cost to modify variety *j*. This cost includes both the technical cost to insert genetic materials into a specific variety and the cost of IPR transactions and biosafety regulation. Countries with a more advanced breeding sector are likely to have lower modification costs. The IPR *component* depends on specific circumstances. If a private company already controls a local variety, it will not face any IPR cost. On the other hand, it may have significant cost if a competing company controls a specific *variety*.

(*b*) V_j = per acre costs of GMVs of variety *j*. We assume that these variable costs are constant. They include the physical variable cost of producing GMV seeds and the IPR and *marketing* costs that a manufacturer has to pay in order to be able to sell the seeds. Let V_g be the per acre cost of the generic seed variety when it is used. It is assumed to be smaller than V_i. It is imported from a low-cost *production* center and does not require payment to owners of the rights for local seeds. The difference decreases as the seed sector producing GMV_j becomes more advanced and when the seed producer does not face IPR costs for the use of the local variety.

The outcomes with GMVs depend on the cost to introduce them as well as the structure of the seed industry and the constraints it faces. We consider two patterns. The first pattern applies to countries where public research *institutions* develop GMVs, which are sold to farmers by a competitive seed sector. In the second pattern, GMVs are *introduced* and marketed by a monopolist.

This scenario is appropriate for the developed world, where major companies such as Monsanto control the GMVs available in the market. In the case of a monopoly, the seed industry may face IPRs and other constraints that impede its capacity to modify traditional local varieties. *Recognizing* these constraints, we derive outcomes for four *stylized* scenarios presented below.

Competitive markets for seeds of GM local varieties

Assuming that the fixed cost of introducing the technologies are covered by the public sector, *competitive* seed sellers will charge farmers V_i per land unit of GM seeds. The GMVs may be either fully or partially adopted.

In the case of partial adoption, some land will continue to be grown with traditional variety *j*. Let A^1_i be total acreage of variety *j* (traditional and GM), and let A^1_i be the GM acreage. The acreage of the *traditional* variety *j* will be equal to $A^1_i - A^{m1}_i$. There are three possible outcomes under competition:

- $C_i^m 1$: *No adoption when* $PMp_i^m(0) - PMp_i^0(0) < V_i$. In this case $A_i^{m1} = 0, A_i^1 = A_i^0$.
- $C_i^m 2$: *Partial adoption when* $PMp_i^m(0) - PMp_i^1(0) > V_i > PMp_i^m(A_i^0) - PMp_i^m(A_i^0)$. Here $A_i^{m1} < A_i^0$

  where $PMp_i^m(A_i^{m1}) = PMp_i^0(A_i^{m1}) + V_i$ and $A_i^1 = A_i^0$.
- $C_i^m 3$: *Full adoption when* $PMp_i^m(A_i^0) - PMp_i^0(A_i^0) > V_i$.

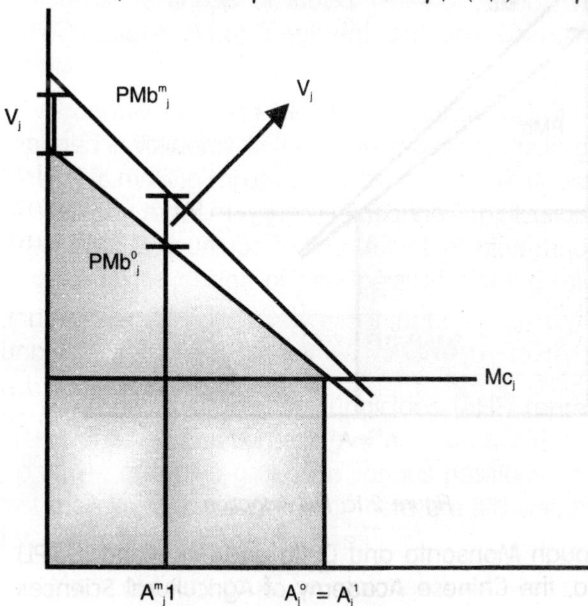

Figure 2.1 a: No adoption.

$$A_i^1 = A_i^{m1} \text{ when } PMp_i^m(A_i^{m1}) = Mc_i + V_i.$$

The assumption that the *Mp* gap between the GMV and traditional variety declines with acreage is the key to the above results. If gains from adoption cannot cover the variable cost even for the first acre, adoption will not occur. If *marginal* gains from adoption are greater than the variable cost, but only for a subset of the acreage, there will be partial adoption; and if the marginal gains from adoption are greater than the variable cost for all acres, there will be full adoption and the acreage under the crop might even increase (recall that we have assumed the total output market is *sufficiently* large that this change has no impact on output prices).

Figure elsewhere in this chapter depicts case C^m_i1 with no adoption when $PMp^0_i(0)+V_j > PMpC^{m1}_i(0)$. Figure elsewhere in this chapter depicts case C^m_i2 of partial adoption, and Figure elsewhere in this chapter depicts case C^m_i3 with full adoption.

These figures demonstrate that high variable costs may discourage adoption entirely, while reduction in these costs may result in partial or, eventually, full adoption. In the figures, the same notation should be used as in the text, that is, replace *Mb* by *PMp*.

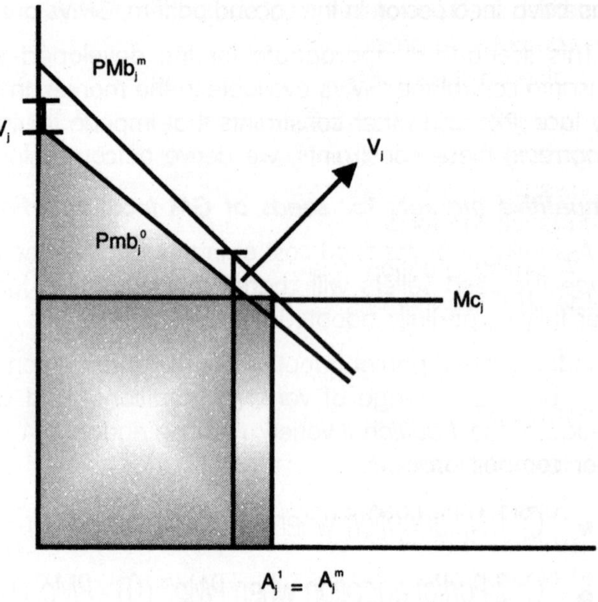

Figure 2.1b: Partial adoption

The introduction of GMVs affects output and net benefits of land *utilized* with variety *j*. The output when GMV_i is available becomes

$$X_i^{m1}\begin{cases} f(A_i^0) & \text{if } C_i^m1 \\ f_i^m(A_i^{m1})+f_i^0(A_i^0-A_i^{m1}) & \text{if } C_i^m2 \\ f_i^m(A_i^{m1}) & \text{if } C_i^m3 \end{cases}$$

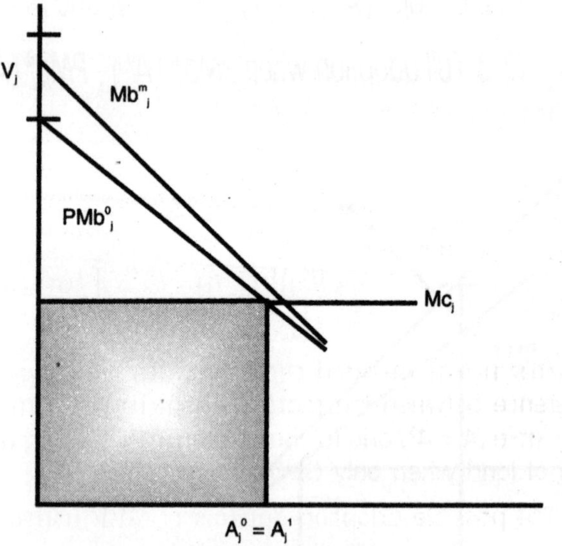

Figure 2.1c: Full Adoption.

The net change in *social* benefit is

$$Nsb_i^{cm}=P(X_i^{m1}-X_i^0)-V_iA_i^{m1}-F_i \qquad (1)$$

and it represents the difference between the production gain and the variable and fixed cost of the new variety. Since up until now almost all crop biotechnologies have been developed and *commercialized* by private companies in monopoly *situations*, introduc-tion of GMVs by the public sector has rarely occurred in reality.

However, the case of Bt cotton in China closely *corresponds* to this scenario. Although Monsanto and Delta and Pine Land (D&PL) have introduced U.S. Bt cotton varieties in China, the Chinese Academy of Agricultural Sciences has developed and *commercialized* its own Bt cotton *technology*, which can be freely used by public and private sector breeders. Due to weak IPR protection, the Monsa-nto technology also has been incorpor-ated into Chinese cotton varieties by local organiza-tions, without payment to the company. At present, there are 22 officially registered local Bt varieties and five imported ones available on the market.

Adoption of Bt varieties has occurred in *approximately 35%* of the Chinese cotton sector and is increasing rapidly. There is no indication that GM technology has a negative effect on cotton biodiversity. On the contrary, the Bt varieties imported from the United States appear to have broadened the local *germplasm* base.

A similar situation could occur in other countries that do not protect IPRs but have a strong breeding capacity. If foreign GMVs are introduced in these countries, breeders can freely use these varieties to cross-breed the transgenic traits into their own germplasm.

The trade-off, however, is that without IPR enforcement, private seed industry development is hampered and *technology* transfer from abroad is *discouraged.*

Monopolistic markets for seeds of GM local varieties

Consider the case when GMVs are produced and marketed by a monopolist. The monopolist is assumed to have access to the traditional local varieties and to modify them. Let A^{m2}_i denote the area of GMV_i and let A^2_i denote total area (traditional and GM) of variety *j*. In this case the inverse demand function, denoting the *maximum* price (W^m_i) farmers are willing to pay per acre for GMV seeds, as a function of acreage, is

$$W_i^m(A) = D^{-1}(A) = \begin{matrix} P[Mp_i^m(A) - Mp_i^0(A)] \text{ if } A < A_i^0 \\ PMp_i^m(A) - Mc_i \quad \text{if } A > A_i^0 \end{matrix} \tag{2}$$

The marginal revenue from the sale of seeds for *A* acres is

$$MR_i^m(A) = \begin{matrix} P\left[Mp_i^m(A) - MP_i^0(A) + A\frac{\partial}{\partial A}[Mp_i^m(A) - Mp_i^0(A)]\right] \text{ if } A < A_i^0 \\ P\left[Mp_i^m(A) + A\frac{\partial}{\partial A}\right][Mp_i^m(A)] - Mc_i \quad \text{if } A \geq A_i^0 \end{matrix} \tag{3}$$

This inverse demand curve indicates that buyers will not be willing to pay more than (a) the difference between the marginal benefits per acre of GMVs and traditional varieties when both are viable ($A < A^0_i$) and (b) the difference between marginal benefits of land with GMV and marginal cost of land when only GMVs are economical.

The possible adoption patterns of GMV_i under *monopoly* includes

- $M_i^m 1$: No adoption if $PMp_i^m(0) - PMp_i^0(0) < V_i^m$.
- $M_i^m 2$: Partial adoption if $PMp_i^m(0) - PMp_i^0(0) > Vj > MR_i^m(A_2^0) - MR_i^0(A_i^0)$.

  In this case, $A_i^{m2} < A_i^0$, at A_i^{m2}; $MR_i^m(A_i^{m2}) - MR_i^0(A_i^{m2}) = V_i, A_i^2 = A_i^0$.

- $M_i^m 3$: *Full adoption if* $Mc_i > V_i$. At A_i^{m2}, $MR_i^m(A_i^{m2}) = Mc_i + V_i$.

The monopolist will sell the amount of seeds where its marginal revenue is equal to the variable per unit cost. When the marginal revenues intersect V_i at a quantity smaller than A^0_i (case $M^m_i 2$), there will be partial adoption and when the marginal revenues intersect V_i at a *quantity* greater than A^0_i, there will be full *adoption*.

It can be verified that (a) higher gains in marginal productivity (high $PMp^m_i(A) - PMp^0_i(A)$) result in an increase in adoption of the GMV, and (b) adoption rates under monopoly are smaller than under competition. This is so because the monopoly price $PMp^m_i(A_i^{m2})$ for GMV_j will be greater than the competitive price, V_i.

The profit of the monopolist, presented in Eq. (4), is smaller than the net social benefits considered by public sector decision makers when *determining* whether or not to assume the fixed cost of introducing a new variety. Thus, a monopoly outcome will provide a less-than-optimal *introduction* and adoption of *GMV.*

There may be cases when profit does not cover the fixed cost of *modification*. Then, a monopolist will not introduce GMV_j, even though the net social benefits might be positive. Figure elsewhere in this chapter denotes the *monopoly* outcome for the case of partial adoption. Curve ABC denotes demand for GMV seeds and has two segments—AB is $PMp^m_i - PMp^0_i$ and *BC* is $PMp^m_i - Mc_i$.

The marginal revenues AE, *associated* with AB, intersect V_i to establish $A_i^{m2} < A^0_i$. Figure *anywhere* else in this chapter denotes the monopoly outcome for the case of full adoption. With the marginal benefits of using GMV, PMp^m_i*'s* are much higher than those of the traditional variety. Demand for GMV seeds is represented by ABC, and the relevant marginal revenues are EF, which intersects with V_i at $A_i^{m2} > A^0_i$.

The output of the industry in a monopoly situation is

$$X_i^{m2} = \begin{cases} f_i^0(A_i^0) & \text{if } M_i^m 1 \\ f_i^0(A_i^0 - A_i^{m2}) + f_i^m(A_i^{m2}) & \text{if } M_i^m 2 \\ f_i^m(A_i^{m2}) & \text{if } M_i^m 3 \end{cases} \tag{4}$$

The price of seeds is equal to

$P\left[MP_i^m(A_i^{m2}) - MP_i^0(A_i^{m2})\right]$ if $M^m_i 2$ and $PM_i^m(A_i^{m2}) - MC_i$ if $M_i^m 3$. Thus, taking into account the fixed cost *i*, net profit of the monopolist is

$$\pi_i^m = \begin{cases} 0 \text{ if } M_i^m 1 \\ \left[P\left[Mp_i^m(A_i^{m2}) - Mp_i^0(A_i^{m2})\right] - V_i\right] A_i^{m2} - F_i \text{ if } M_i^m 2 \\ \left[PMp_i^m(A_i^{m2}) - Mc_i - V_i\right] A_i^{m2} - F_i \text{ if } M_i^m 3 \end{cases} \tag{5}$$

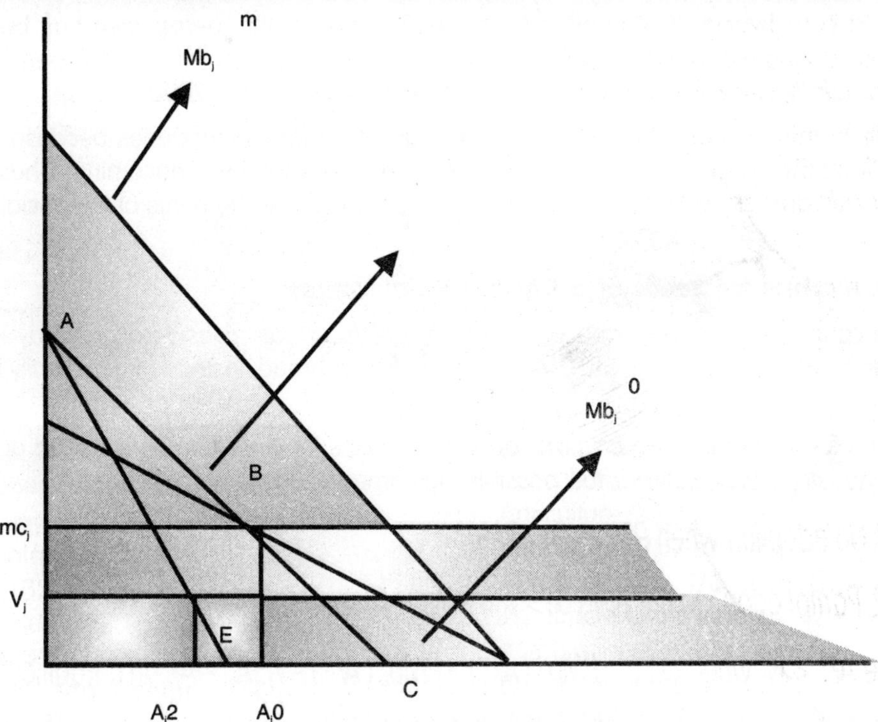

Figure 2.2a: Partial adoption.

Whether the monopolist has access to all local varieties and markets GMVs itself or the *technology* is licensed to other seed producers does not matter for the scenario outcome. Technology licensing under strong IPRs is rather typical for GMVs in the United States. For Roundup Ready (RR) soybean and RR and Bt corn, biotechnology firms have issued nonexclusive licenses to all breeders and seed *companies* interested in *endowing* their own breeding lines with the transgenic traits.

Due to high demand and a relatively low fixed cost to modify local varieties, numerous GMVs are available on the U.S. market. The number of RR soybean varieties increased along with increasing techn-ology adoption rates. In 2002, around 200 different seed companies marketed over 1,100 RR soybean varieties, which were adapted to diverse local conditions. On average, this implies an area of less than 20,000 hectares per variety.

Likewise, several hundred RR and Bt corn hybrids are available from seed companies of all sizes, with an average area of less than 10,000 hectares per hybrid. Although the patent owners capture a *significant* share of the rent through monopoly pricing, U. S. farmers and consumers also benefit significantly, as the licensors have not set prices to take *advantage* of variations in marginal product across regions or varieties. For RR soybean and Bt corn in Argentina, the scenario is similar.

Although IPR protection is weaker than in the United States, the technologies have been licensed to various seed companies that incorporated them into their own breeding lines. Today, there are seven different *companies* providing *56* RR soybean varieties and four companies providing over

20 different Bt corn hybrids in Argentina (ASA, 2002). Most of this *germplasm* has been locally bred or adjusted, and the total number of soybean varieties and corn hybrids in Argentina did not change *significantly* since the *introduction* of GM technology (INASE, 2000).

Especially in the case of RR soybeans, farmers are the main beneficiaries because *Argentine* legislation allows the on-farm reproduction of seeds (Qaim and Traxler, forthcoming). Thus, average royalty payments are relatively low, so that the situation has some elements of the social optimum scenario.

Competitive Markets for Seeds of a GM "Generic" Variety

In some countries, a limited capacity to modify GMVs or cost *considerations* may lead even the public sector to introduce a generic GMV, imported from abroad, instead of genetically modifying local varieties.

The results for the competitive and *monopolistic* markets for GMV_i can be modified accordingly. If the seed industry is competitive, the possible outcome includes

- C_i^g1: *No adoption when* $PMp_i^g(0) - PMp_i^0 < V_i$.
- C_i^g2: *Partial adoption if* $PMp_i^g(0) > V_i > PMp_i^g(A_i^0) - PMp_i^0(A_i^0)$.

  Here $A_i^{g3} < A_i^0$ when at A_i^{g3} $PMp_i^g(A_i^{g3}) = PMp_i^0(A_i^{g3}) - V_i$, $A_i^0 - A_i^{g3}$ in a traditional variety.
- C_i^g3: *Full adoption when* $PMp_i^g(A_i^0) - PMp_i^0(A_i^0) > V_i$. At A_i^{g3}, $Mp_i^g(A_i^{g3}) = Mc_i + V_i$. Output in this case is

$$X_i^{g3} = \begin{cases} f_i^0(A_i^0) & \text{if } C_i^g1 \\ f_i^g(A_i^{g3}) + f_i^0(A_i^0 - A_i^{g3}) & \text{if } C_i^g2 \\ f_i^g(A_i^{g3}) & \text{if } C_i^g3 \end{cases}$$

and net social benefit is

$$NSb_i^{Cg} = p[X_i^{g3} - X_i^0] - V_i A_i^{g3} - F_g \tag{6}$$

The generic GMV has a lower marginal *productivity* than GNV_i, but its variable cost is lower. Thus, comparing $C_i^m1 - C_i^m3$ with $C_i^g1 - C_i^g3$ suggests that more (less) acreage will be utilized with the GMV_i than the generic *GMV* if the marginal benefit gain $P\left[Mp_i^m(A_i^{m1}) - Mp_i^g(A_i^{m1})\right]$ is greater (smaller) than $Mc_i - (V_i - V_g)$. When the variable cost of GMV_i is very high due to an undeveloped crop-breeding sector (and the yield disadvantage of generic variety is not overwhelming), adoption rates of a generic GMV will be higher.

In these *situations*, it may be that, despite lower yield per acre, the actual output of the generic GMV will be greater than those of the GMV_i. In most situations, however, we do not expect the extra cost of GMV_i to be dominant and expect both adoption and output to be higher with a GMV_i. At this stage, there are no empirical examples of a generic GMV being introduced by the public sector. As indicated before, the *introduction* of most GMVs worldwide has been done by private monopolists, and outcomes with generic GMVs in such *situations* are discussed below.

Monopolistic Markets for Seeds of a GM "Generic" Variety

If a monopolistic firm that controls the GM technology is precluded from access to local varieties, or if the fixed cost of modifying local *varieties* is too high relative to *expected* profits, a generic GMV will be introduced to the area grown with variety *j*.

In this case total acreage is A_i^4. and acreage of the generic GMV is A_i^{g4}. The inverse demand $W_i^g(A)$ and marginal revenue functions $Mr_i^g(A)$ for this case are defined *similarly* to the ones in Eqs. *(2)* and (3), and only the indicator g replaces m. The possible outcome includes

- $M_i^g 1$: No adoption when $PMp_i^g(0) - PMp_i^0(0) > V_i$.
- $M_i^g 2$: Partial adoption occurs when $PMp_i^g(0) - PMp_i^0(0) > V_i > MR_i^g(A_i^0) - MR_i^0(A_i^0)$.

  At $A_i^{g4} < A_i^0$, $PMp_i^g(A_i^4) - PMp_i^g(A_i^4) = V_i$ and $A_i^0 - A_i^{g4}$ is the non-GMV area.
- $M_i^g 3$: Full option when $MR_i^g(A_i^0) - MC_i > V_i$. In this case adoption of A_i^{g4} is when $MR_i^g(A_i^{g4}) - Mc_i - y = 0$.

With this notation, the output and the profits of the monopolist are given by

$$X_i^{g4} = \begin{cases} f_i^0(A_i^0) & \text{if } M_i^g 1 \\ f_i^g(A_i^{g4} + f_i^0) + (A_i^0 - A_i^{g4}) & \text{if } M_i^g 2 \\ f_i^g(A_i^{g4}) & \text{if } M_i^g 3 \end{cases} \tag{7}$$

$$\pi_i^g \begin{cases} 0 & \text{if } M_i^g 1 \\ [Mp_i^g(A_i^{g4}) - Mp_i^0(A_i^{g4}) - V_i]A_i^{g4} - F_i & \text{if } M_i^g 2 \\ [Mp_i^g(A_i^{g4}) - Mc_i - V_i^g]A_i^{g4} - F_i & \text{if } M_i^g 3 \end{cases}$$

Comparison of $M_i^g 1$ to $M_i^g 3$ with $M_i^g 1$ to $M_i^g 3$ suggests that, under a monopoly, adoption of GMV_i is greater (smaller) than *adoption* of the generic GMV if the gain in marginal revenues of the GMV_i, $[PMR_i^m(A_i^{m2}) - MR_i^g(A_i^{mg})]$ is greater than the extra variable cost. Thus, there may be a situation when adoption of GMV_i will be less than that of the generic variety.

There can be different reasons why rights cannot be freely traded. Lack of IPR *protection* may prevent the innovator from trading because it is difficult to enforce compliance of licensing *agreements*.

A similar situation might occur when IPRs are strong but high *transaction* costs hinder a smooth transfer. Due to diverging objective functions, licensing agreements can be difficult to negotiate for the private innovator, especially when germplasm is owned by the public sector.

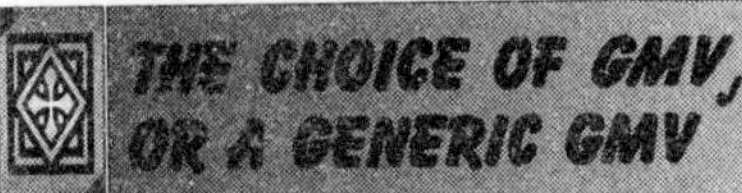

THE CHOICE OF GMV, OR A GENERIC GMV

Thus far, we have analyzed the area to be planted with GMVs in the four scenarios considered. However, a more fundamental question is whether to introduce a generic GMV at a given location or to modify the local varieties. The decision rule for a monopolist is different from that of a public sector entity, which *introduces* seeds to be distributed by *competitive* seed companies.

We will solve the public sector case first and then consider the monopolist problem. Let δ^m_i be an indicator variable that assumes the value 1 if option GMV_i is selected and 0 otherwise. The indicator variable δ^g_i assumes the value 1 if option *GMV* is selected and 0 otherwise.

The public sector can either (a) modify variety *j*, (b) introduce a generic GMV, or (c) neither. The public sector aims to *maximize* net social welfare so its decision problem for region *j* will be

$$NSb = \max_{\delta^m_i \delta^g_i} NSb^m_i \delta^m_i + NSb^g_i \delta^g_i$$

subject to $0 \leq \delta^m_i + \delta^g_i \leq 1, \delta^m_i, \delta^g_i.$

The public sector will select GMV_i if net social benefits with this technology are positive and greater than Nsb^g_i. The generic GMV is chosen when it generates positive social surplus greater than Nsb^m_i.

From conditions (1) and (6), the public sector decision whether to introduce *GMT., or a* generic GMV hinges on three factors-revenue differential $P[X^{m1}_i - X^{g3}_i]$, variable cost differential $[V_i A^{m1}_i - V_g A^{g3}_i]$, and fixed cost *differential*, F_i. GMV_i is selected when its extra revenues are greater than the extra variable and fixed cost that its *introduction* entails.

When the production advantage of the local variety is not substantial or the breeding sector is not well developed (so F_i is high), the public sector will prefer to introduce a generic GMV to location *j*.

In cases where the monopolist controls the introduction of GMVs, let δ^g_i and δ^m_i be defined similarly. The *optimization* problem becomes $\max_{\delta^m_i, \delta^g_i} \delta^m_i \pi^m_i + \delta^g_i \pi^g_i$ subject to the constraints $0 \leq \delta^m_i + \delta^g_i \leq 1.$

The monopolist will elect to introduce GMV_i if *Ire > g5* and *Tce > 0.* The generic variety will be introduced if $\pi^g_i > \pi^m_i$ and $\pi^g_i > 0$.

A comparison of Eqs. (5) and (7) suggests that in *determining* whether to genetically modify the local variety or to introduce a generic GMV, the monopolist compares the likely extra revenues of GMV_i with the extra fixed and variable cost its introduction entails. The generic GMV is likely to be introduced (a) in locations with small acreage, where volume of seed sales will not cover the extra cost of *Gk1k*, (b) in cases when the yield differences between the generic and local varieties are not substantial, and (c) when the variable and fixed costs of modification are substantial. The high fixed cost may reflect cost of access to local varieties or undeveloped local breeding sectors that make it worthwhile to import a generic GMV.

CASE OF INELASTIC DEMAND FOR AGRICULTURAL COMMODITIES

The price of internationally traded *agricultural* commodities, such as corn and cotton, is determined according to international demand and supply. Neglecting *transportation* costs and quality differences, the assumption of price-taking behaviour is appropriate for most regions in the world. Yet, there are likely to be situations where commodities, especially food staples, are primarily produced for local consumption within a region.

This may occur in regions with high transportation cost to major markets or low degrees of integration to the global economy for other reasons. These regions are not likely to be served by *monopolistic* seed companies but, rather, by small companies or direct *provision* of seeds by public extension programs.

When GMVs are introduced in these situations, then the output price effect may be significant. Suppose the demand for output at location *j* is $X_i = D_i(P_i)$, where P_i is output price in region *j*. The initial *equilibrium* with traditional variety *j* consists of acreage A^0_r, output X^0_r, and price P^0_r.

These values are determined solving simultaneously the equilibrium condition in the output market, $P^0 = D^{-1}(X^0_i)$ ($D^{-1}(X)$ in inverse demand); the equilibrium condition in the land market $P^0_i Mp^0_i(A^0_i) = Mc_i$; and the production function $X^0_i = f_i(A^0_i)$.

Suppose GMVs are sold by small seed companies (or distributed by extension programs) at price V_r. Let P^{m1}_i be output price under competition when GMV_i is introduced. The equilibrium conditions in this case determine output price P^{m1}_r, total output X^{m1}_r, total acreage A^1_i and acreage with GMV_r, A^{m1}_r.

A procedure that can yield the equilibrium value consists of using the formulas we developed to obtain A^{m1}_r, A^1_r, and X^{m1}_i for cases of fixed output prices for a range of plausible values of P^{m1}_r. P^{m1}_i will clear the output market, so that $P^{m1}_r = D^{-1}(X^{m1}_i)$ is the equilibrium output price.

The equilibrium price P^{m1}_i also establishes acreage level $\hat{A}^0_i$ solved from $P^{m1}_i MP^0_i(\hat{A}^0_i) = Mc_i$, which is the amount of land that would have been utilized if the initial price was P^{m1}_r. Since the GMVs are assumed to increase yield, it can be shown that their adoption (even partially) will increase supply and, with *negatively* sloped demand, $P^{mi}_i < P^0_r$, which implies that $\hat{A}^0_i < A^0_i$.

Thus, in the case of partial adoption, when the total acreage $A^1_i = \hat{A}^0_i$, adoption of *GMV* will reduce acreage. There may be cases of reduced acreage even in cases of full adoption. The reduction in output price associated with the *introduction* of GMV_j will lead to increases in consumer surplus, denoted by ΔCS^{cm}_i,

$$\Delta CS_i^{cm} = \int_{X_i^0}^{X_i^{m1}} D^{-1}(X)\,dx \tag{8}$$

The change in net social benefit becomes

$$NCb_i^{CM} = \Delta CS_i^{CM} + P_i^{m1} X_i^{m1} - P_i^0 X_i^0 - A_i^{m1} V_i - [A_i^1 - A_i^0] M¢ - F_i$$

While consumers will gain from the introduction of *GMV,* the impact on farmers is mixed. They produce more, yet receive a lower price. They may plant fewer acres but have to pay a technology fee.

The case with inelastic demand suggests that the introduction of GMVs to a location by the public sector may raise social welfare, reduce farmed acreage (and thus may improve environmental conditions), and improve *consumer* well-being but not *necessarily* help farmers.

The analysis of the impact of introducing a generic GMV is similar to that of introducing a modified local variety to location *j*. The generic variety, with its lower yield effect, will have less of an impact on output prices and may lead to a smaller reduction in acreage. Its introduction may benefit farmers more than the introduction of GMV_i and, thus, the introduction of a *GMV* will benefit consumers more.

The likelihood of *introducing GMV* increases, the smaller the fixed costs to modify variety *j* and the variable cost to produce *GMV* seeds. The analysis of the public sector-led scenario is useful to provide some intuition about the private *monopoly* case, when demand for output is negatively sloped. Private companies that introduce new GMVs globally are aware of the output price effect of the new innovation and its impact on their sales and revenue.

We do not analyze the choices formally here but, rather, compare the outcomes when technology is provided by an idealized public sector that maximizes global welfare, versus when it is provided by a monopolist. Under the monopoly, we expect low rates of adoption and, thus, lower aggregate output, and a higher output price. The profit of the monopolist is likely to be less than the aggregate social welfare that also includes consumer and farmer surplus.

Thus, the *monopolist* will not introduce GMVs in some cases where they would have been introduced by the public sector. Furthermore, since generic varieties have a lower yield effect and lower fixed costs, the monopolist is more likely to introduce the generic varieties than modify local ones. The decision of whether or not to introduce a GMV at a location depends both on expected productivity gains as well as the fixed costs of introduction.

Reduction in the costs of *introduction,* due to improved efficiency of the breeding sector or a decrease in the regulatory costs, is likely to increase introduction of GMVs in general and modified local varieties in particular. In some cases, there may be substantial costs to the public sector to introduce the *technologies* because of domestic capacity *limitations.*

Then, the private sector may introduce a GMV, even though the private benefits are smaller than the public benefits. In other situations, low private benefit and high costs of introduction by the public sector may prevent the introduction of a GMV to a region. In these situations, if the public benefits of the GMV are sufficiently high, it may be introduced eventually as a result of policies that will enhance the capacity of the public sector and reduce its technology introduction costs.

IMPACT CROP BIODIVERSITY

The effect of the introduction of GMVs on crop biodiversity depends on the extent that traditional local varieties are replaced by a small number of GMVs. Continued planting of local varieties, even in GM form, can be a mechanism for preserving crop biodiversity. Our analysis has identified a wide array of circumstances in which local varieties may be *preserved* after the introduction of agricultural biotechnology. In situations where the revenue gains from genetic modification of local varieties relative to a generic variety are substantial, and fixed and variable costs of modification and production are low, local varieties will be modified and biodiversity will be preserved.

Even in situations where the introduction of *agricultural* biotechnology will lead to replacement of areas of local varieties with a generic GMV, adoption of the modified varieties need not be complete. In *particular,* when a monopolistic private company controls the technology, it will charge a technology fee that may not warrant *adoption* of the technology on much of the land.

Hence, a significant portion of the acreage will continue to be planted with traditional varieties. Full adoption of GMVs rarely occurs, and with partial adoption local varieties may be preserved.

DISCUSSION AND CONCLUSIONS

Biotechnology may preserve CGD more than conventional breeding. The reason is that biotechnology allows for separation between the act of developing novel crop traits and the process of breeding plant varieties. As a result, a given biotechnology innovation may be incorporated into a large number of plant varieties.

This chapter has shown some of the conditions under which this might happen. Modern biotechnology is still a fairly recent phenomenon, so that empirical evidence about the actual impact on biodiversity is limited. Table elsewhere in this chapter shows adoption levels and the number of GMVs available in different countries for selected *innovations.*

So far, widespread adoption occurred only for RR soybeans and Bt corn in the United States and Argentina, and Bt cotton in the United States and China. In all these cases, the technology has been incorporated into a large number of varieties, which supports our general hypothesis that biotechnology can preserve CGD.

In other empirical cases, technology diffusion is still at an early stage so that conclusive statements are difficult to make. Biosafety regulations for GM crops can play an important role in this respect. The cost of regulatory *compliance* has become a major component in the overall budget to develop new biotechnologies. In most countries, only the transformation event is regulated, so that the regulatory cost for each technology occurs only once, regardless of the number of varieties into which it is incorporated later on. However, in countries such as India, each GMV is regulated separately. Such varietyspecific approval procedures may foster loss of CGD and can be challenged on this basis.

Table 2.1: Number of available varieties for different GM technologies in selected countries.

Country	Technology	Area under technology (ha.)	Number of local varieties/ hybrids	Number of imported varieties/hybrids
USA	RR soybean	22 million	> 1,100	0
	Bt corn	7 million	> 00	0
	Bt cotton	2 million	19	0
Argentina	RR soybean	10 million	45	11
	Bt corn	0.7 million	15	6
	Bt cotton	22,000	0	2
China	Bt cotton	1.5 million	22	5
India	Bt cotton	40,000	3	0
Mexico	Bt cotton	28,000	0	2
South Africa	Bt cotton	20,000	1	2

Based on the conceptual analysis and empirical observations, we suggest a fourfold classification of situations according to the expected biodiversity outcome.

1. *Strong IPRs, a strong breeding sector, and low transaction costs:* Most situations within the private sector in developed countries, and some advanced developing countries such as Argentina, Brazil, and South Africa, belong in this category. The private technology owner will license the innovation to different seed companies that incorporate it into many or all local varieties, so that CGD is preserved. Adoption will be fairly widespread, and the innovator captures a rent through royalty payments. This outcome is equivalent to case II of the above analysis.

2. *Strong IPRs and a strong breeding sector, but high transaction cost to trade rights.* High transaction costs can occur, *particularly* when licensing contracts between a private technology owner and a public breeding organization have to be negotiated. Examples are biotech companies trying to reach agreements with international agricultural research centers, or public breeding stations that serve certain regional niche markets. If an agreement cannot be reached, the most likely outcome is that the biotech company will directly introduce GMVs that are not locally adapted. A widespread adoption of these varieties would lead to a loss of CGD, which is the outcome described under case IV of our analysis.

3. *Weak IPRs and a strong breeding sector.* Countries such as China and India belong into this category. Because IPRs are weak, every breeder or seed company can use commercialized GMVs, in order to cross-breed the technology into their own germplasm. Thus, many different GMVs will be available on the market. Due to the competition, the

innovator's ability to capture rents are limited, so that farmers and consumers are the main beneficiaries. This outcome almost corresponds to case I, the social optimum.

4. *Weak IPRs and a weak breeding sector.* This situation is typical for most of the least-developed countries in Africa, Asia, and Latin America. In none of these countries have GM crops been *commercialized* so far. If biotechnology developed by the private sector abroad should reach these countries, the most likely outcome is that foreign GMVs are directly introduced without adaptation. A widespread adoption of these varieties would lead to a loss of CGD, which is the outcome described in case IV of the analysis.

There are different policy implications for each category. In category (1), strong IPRs and smooth *transactions* ensure that biodiversity is preserved. In category (2), widespread realization of the benefits of biotechnology will likely require *international* efforts to create an effective mechanism that reduces the transaction costs of trading IPRs.

An intellectual property clearinghouse is one model that may be effective, if it is *designed* in a manner that addresses the needs of the poor as well as environmental concerns, while recognizing that much of the technology will be developed by profit-driven firms in the developed countries. In category (3), the outcome is socially optimal in the short run, but the situation might look differently from a dynamic perspective.

Lack of IPRs deters international technology transfer and innovation in the private sector. If biotechnology is entirely acquired and provided by the public sector, this might be less problematic. However, this may not be feasible or even advisable in more advanced developing countries. It may require significant amounts of funds and retard the evolution of private seed companies.

Our analysis suggests that, in such situations, introduction of IPR protection and enabling sale of rights will be desirable. The alternative will lead to takeovers and concentration in the seed industry, which would be associated with a loss of CGD and underutilization of the economic potential of biotechnology. For situations in category (4), the implications are somewhat different. *Realizing* the biotechnology *opportunities* in least-developed countries will require that there is a minimum absorptive R&D capacity. Since a local private seed industry is hardly existent, funding for the incorporation of biotechnology into local varieties will have to come from noncommercial sources that recognize the total gain in social welfare rather than strictly private financial returns.

In practice, this means that the existing system of agricultural *experiment* stations may increasingly become centers of biotechnology adaptation and application. Also, the international agricultural research centers could play a bigger role in this respect. Lumpsum royalties to private innovators will have to be paid if such technologies are being used.

In some cases such royalties might be waived for humanitarian purposes. An *international* clearinghouse mechanism would also be very beneficial in these situations. In summary, biotechnology-based innovations in agriculture have the potential to preserve CGD, yet the actual impact will depend on the specific institutional conditions, R&D capacities, IPR policies, and biosafety *regulation* schemes in the individual countries.

PLANT GENETIC RESOURCES

Agricultural biological *diversity*, or more specifically, genetic resources for food and agriculture, is the storehouse that provides humanity with food, clothes and medicines. Its management is essential in the development of sustainable agriculture and food security.

The conservation and sustainable use of genetic resources and the management of related biotechnologies may appear to be technical issues, but they have strong socioeconomic, political, cultural, legal, and ethical implications, in that poor management of these matters could put the future of humanity at risk. *According* to present estimates, food *production* in developing countries will have to increase more than 60% in the next 25 years just to keep pace with population growth.

The possibilities for expanding the areas used for terrestrial and aquatic *farming* are relatively limited, and most of the world's wild fish stocks are already overexploited. Production must, therefore, be intensified, productivity increased. and productive natural systems must be optimally managed, all in a *sustainable* manner.

This will require the combined application of new and old *biotechnologies,* including *innovative* approaches to plant and animal breeding and to farming practices. Success in this endeavor will depend on the sustainable utilization of a broader range of species, and of the genetic material within each species, including genes from the wild relatives of domesticated species. In spite of its vital importance for human survival, agricultural biodiversity is being lost at an *alarmingly* increased rate.

It is estimated that some 10 thousand species have been used for human food and agriculture. Currently no more than 120 cultivated species provide 90% of human food supplied by plants, and 12 plant species and five animal species alone provide more than 70% of all human food. A mere four plant species (potatoes, rice, maize, and wheat) and three animal species (cattle, swine, and chickens) provide more than half.

Within the so-called "main food species," a tremendous loss of genetic diversity has occurred in the present century. Hundreds of thousands of farmers' plant varieties and landraces that existed until the beginning of the 20th century in farmers' fields, have been *substituted* by a small number of modern and highly uniform commercial varieties.

In the United States alone, more than 90% of the fruit tree and *vegetable* species that were grown in farmers' fields at the beginning of the century can no longer be found, and only a few of them are maintained in gene banks. Similarly, alarming figures can be given for the genetic erosion of domestic animal breeds and varieties. The picture is much the same throughout the world. For example, in the 1970s a virus was decimating rice harvests across Asia.

Scientists turned to crop diversity collections and screened 7,000 rice lines for resistance to the grassy stunt virus. They found genetic resistance in just one species of wild rice collected in India, and bred this into new rice varieties. This species *(Oryza nivara)* can no longer be found in the wild, and if not for its *conservation* in a crop gene bank, grassy stunt virus would have likely gone unchecked, to the great harm of poor farmers across Asia.

Loss of agricultural biological diversity has drastically reduced the capability of present and future generations to face unpredictable *environmental* changes and human needs. The rapid process of *globalization* and economic integration is creating an increasing interdependence between nations and regions. This can raise important ethical questions. One of the oldest forms of interdependence, starting in the Neolithic period and continuing today, is in relation to agrobiodiversity, involving the spread of crops from their centers of origin to destinations throughout the world.

Essentially no country on the planet is today self-sufficient with respect to the genetic resources for food and *agriculture* they are using, and the average degree of interdependence among countries with regard to the most important crops is 70%. Paradoxically, the countries which are poorest from the economic point of view, and which are in general located in tropical or *subtropical* zones, are also richest in terms of genetic diversity needed to ensure human survival.

International cooperation is needed to develop a more fair and equitable sharing of the benefits derived from the use of genetic resources, and provide incentives to ensure that countries continue developing, conserving, and making available to *humanity* their genetic diversity. There is also interdependence between generations.

Agricultural biodiversity is a precious inheritance from previous *generations,* which we have the moral obligation to pass on intact to coming generations, allowing them to face unforeseen needs and problems. Until now, how-ever, the interests of future generations, who are neither voters nor consumers, have not been adequately taken into account by our political and economic systems. *Interdependence* also exists between genetic *resources* and biotechnology.

In general, genetic resources provide the raw material for biotechnologies. New and more powerful biotechnologies drastically increase the potential of using genetic resources, but, in some instances, they can also raise new risks for the environment, and socioeconomic concerns, if the only purpose of the users is immediate profit without ethical *considerations*. Regulatory *mechanisms* need to be developed to maximize the potentials and minimize the risks.

It is important to ensure that both the new and traditional biotechnologies contribute together to the sustainable and efficient utilization of biological diversity, and to the fair and equitable distribution of the benefits derived from their use. The industrialized world has developed 1 egal-economic mechanisms, such as *intellectual* property rights (e.g., patents and plant breeders' rights) to provide incentives for the development of new biotechnologies and to *compensate* their inventors.

However, there are no economic or legal mechanisms to compensate or provide incentives for the developers of the raw material, the genetic resources themselves. An important step in this direction has been the *unanimous* recognition by Food and Agriculture *Organization* of the United Nations (FAO) Member Countries of Farmers' Rights, whereby farmers are *recognized* as donors of genetic resources, as a *counterweight* to plant breeders' rights (rights of the donors of technology).

Unlike plant breeders' rights, however, Farmers' Rights are not yet operative. It is the inescapable responsibility of our generation to develop ethical solutions to the problems and issues raised above, within a political *framework* that allows an equitable sharing of benefits for all countries, and ensures food and agriculture for future generations.

The United Nations, as a universal intergovernmental forum, has a fundamental role to play in the *facilitation* of the necessary inter-governmental *negotiations* to accomplish this task. In the 1970s, action to address the global management of plant genetic resources began within the FAO resulting in the establishment, in 1983, of the first permanent intergovernmental forum on this subject: the *Commission* on Genetic Resources for Food and Agriculture (CGRFA), which is currently composed of 164 Member Countries and the European Community.

In the 1980s this forum made possible the negotiation and development of an *International* Undertaking on Plant Genetic Resources, which recognized Farmers' Rights as being complementary to plant breeders' rights. Farmers' Rights were adopted by the FAO Commission in 1989 as "rights arising from the past, present and future *contributions* of farmers in conserving, improving, and making available plant genetic resources, *particularly* those in the centers of origin/diversity", with the aim of allowing "farmers, their *communities*, and countries in all regions, to participate fully in the benefits derived, at present and in the future, from the improved use of plant genetic resources."

The members of the Commission then *negotiated* a revision of the International Undertaking in harmony with the Convention on Biological Diversity (CBD), which allows the regulation of access to genetic resources for food and *agriculture*, the fair and equitable sharing of the benefits derived from their use and the realization of Farmers' Rights.

This *international* agreement, which is binding, was adopted by consensus by the FAO Member Countries on November 3, 2001, with the name of International Treaty on Plant Genetic Resources for Food and Agriculture (ITPGRFA). It entered into force on June 29, 2004. The FAO/CGRFA also negotiates other international agreements related to the *conservation* and sustainable *utilization* of plant genetic resources for food and agriculture (PGRFA).

In 1993, it adopted the International Code of Conduct for Plant Germplasm Collecting and Transfer; and it is currently discussing a Code of Conduct on Biotechnology as it relates to genetic resources for food and agriculture. This chapter provides a description of some of the key economic, technical, and legal issues that arose in the negotiations and *ultimately* the design of the ITPGRFA, together with a description of the main *provisions* of the ITPGRFA.

This is followed by a brief description of the motivation and key provisions of the two other international *agreements* negotiated in the FAO/CGRFA.

THE INTERNATIONAL TREATY ON PLANT GENETIC RESOURCES FOR FOOD AND AGRICULTURE

Background: The uniqueness of PGRFA

In the long negotiating process to develop and agree on the ITPGRFA, a key consideration was the fact that PGRFA differs substantially from other plant genetic resources and, therefore, specific solutions were needed for their conservation and development and the fair and equitable sharing of the benefits derived from their use, which would not *necessarily* be similar to those required for other kinds of biodiversity.

Unique features of PGRFA include:

(*i*) They are essentially *man-made,* that is, biological diversity devel-oped and consciously selected by farmers since the origins of agriculture, who have guided the evolution and development of these plants for over 10,000 years.

In recent times, scientific plant breeders have built upon this rich inheritance. Much of the genetic diversity of cultivated plants can only survive through *continued* human conservation and maintenance.

(*ii*) They are not randomly distributed over the world, but rather concentrated in the so-called "centers of origin and diversity" of cultivated plants and their wild relatives, which are largely located in the tropical and sub-tropical areas.

(*iii*) Because of the *diffusion* of agriculture all over the world, over the last 10,000 years, and because of the association of major crops with the spread of *civilizations,* many crop genes, genotypes, and populations have spread, and continue to develop, all over the planet. Moreover, PGRFA have been systematically and freely collected and exchanged for over two hundred years, and a large *proportion* have been incorporated in *ex situ* collections.

(*iv*) There is much greater inter-dependence among countries for PGRFA than for any other kind of biodiversity.

Continued agricultural progress implies the need for continued access to the global stock of PGRFA. No region can afford to be isolated, or isolate itself, from the germplasm of other parts of the world.

For such reasons, the second session of the Conference of the Parties to the CBD, in 1995, adopted decision 11/15, "*recognizing* the special nature of agricultural biodiversity, its *distinctive* features and problems needing distinctive solutions." The Conference of the Parties also supported the negotiations for the ITPGRFA, in order to provide such solutions.

Economic, Technical, and Legal Issues Involved in a Multilateral System

During the negotiations for the ITPGRFA, a number of complex economic, technical, and legal issues needed to be examined and understood in order to develop, negotiate, and reach consensus on innovative concepts and provisions, based on an *interdisciplinary* approach.

To facilitate this negotiation, the secretariat of the negotiating body, the Commission on Genetic Resources for Food and Agriculture *commissioned* a number of technical papers, as part of its series "Background Study Papers" (available on the internet). Many of the concepts discussed in this section were presented and developed in such papers.

Economic Issues

Wild and weedy crop-relatives and landraces provide the *foundationbreeding* materials for crop *improvement* and sustainable agriculture. They allow value to be added or provide a "value of use" in breeding and farming activities. This value is realized through the use of germplasm from in *situ* conditions, as well as material in *ex situ* collections. Besides the current use-value of plant genetic resources, there are a variety of other values that can be derived from plant genetic resources.

The *portfolio value* is the value of *retaining* a relatively wide range of assets within biological production systems, to smooth yield fluctuations. The *option value* is the value of retaining a wide range of known agrobiodiversity across time, as a source of currently unknown potential usefulness. The *exploration value* is the value retaining unexplored biodiversity, for the same reasons.

Another way of grouping those values is to see them as *insurance values* (diversity acts as an insurance against crop yield fluctuations) and *information values*, (specific information coded in the germplasm may later prove to be of concrete value). It is clear that the conservation of PGRFA generates a use-value.

A further question is how an *exchange-value* may arise, that is, how is it possible to set a price, or determine an appropriate level of economic remuneration, for the exchange of these resources? An understanding of this matter is necessary in *attempting* to identify effective incentives for the conservation and sustainable use of PGRFA. Traditional farmers, their communities, and countries maintain agrodiversity *in situ*, and thereby conserve and further develop the diversity contained in their *landraces* and related materials.

A problem arises, however, in that they often have an economic incentive to replace their heterogeneous landraces by homogeneous modern varieties, as these *frequently* offer higher yields and productivity, and thus, higher incomes. While this process of conversion (the replacement of landraces by modern varieties) may be a rational decision on the part of an individual farmer, increasing conversion means a continuous and irreversible loss of diversity, which is not in the global interests.

The question of the realization of an exchange-value for PGFRA is complex, because the farmers and communities developing and cultivating landraces, and other related genetic resources in their farming systems are, in fact, creating a global economic value, much of which they are unable to appropriate. In other words, they have no mechanism for obtaining a price, or other form of compensation, for the valuable germplasm they *generate* and conserve.

It is the germplasm, which they have developed within their farming systems, that is the world's main source of PGRFA (whether it is still *maintained* in the fields, or in ex *situ* collections). This germplasm is, however, mostly available at no cost. Traditional farmers thereby generate externalities, as providers of a "public good" (that is, a good that cannot be appropriated by its producers, and which may be used by many without exhausting it, and without adding cost).

To the extent that traditional farmers, and their communities and countries, are not able to *appropriate* the values that they generate, they lack economic incentives to continue developing and conserving the diverse PGRFA, on which *agricultural* development will continue to depend.

That is, they lack economic incentives to maintain this biodiversity, rather than *converting* to improved varieties. In more general terms, where public goods are created, the investments for producing or *preserving* them necessarily tend to be suboptimal, because their producers are unable to fully benefit from the rents such goods may generate.

This is a typical market failure, and is also often found in areas such as the funding of basic science. The public nature of the goods generated by *traditional* farming does not mean, however, that other agents do not appropriate and benefit from these values, at a later point of the development and production process. Plant *breeders* and seed companies do, for example, capture at least part of the rents generated by the farmers' germplasm which they have *incorporated* in their varieties, especially when these are protected by plant breeders' rights, or other forms of intellectual property right.

But this value is not *appropriated* at the correct point in the *production* cycle to provide the necessary incentives to promote ex and *in situ* conservation. If it is in the global interest to maintain landraces and other diverse PGRFA, it is necessary that farmers and communities, who develop and conserve diversity, and their countries, either appropriate the value of *maintaining* diversity directly, or are compensated for the costs of *conserving* diversity, including the potential benefits that they forego by not converting to modern varieties.

A major difficulty arises with *agrobiodiversity* in the design of incentive mechanisms, as values are both difficult to estimate and to appropriate. In fact, an essential part of these values, specifically those of global nature, cannot be *appropriated*. Economic analysis suggests that, for an agreement to be economically effective, it should be forward-looking and include *structural* incentives to favour and reward conservation in a clear, transparent manner.

These incentives must be greater than the benefits foregone by renouncing conversion to specialized agriculture. If necessary, they could be linked to conservation for precise periods of time. The implementation of such incentives would require *international arrangements*, within the *framework* of an overarching *multilateral* agreement.

Such a system might, in principle, be based on market mechanisms (for example using intellectual property rights or contracts), on nonmarket mechanisms (such as an international fund), or on a mixture or combination of mechanisms (such as a system of payments from countries on the basis of the *commercial* benefits derived from the use of foreign PGRFA to an international fund, and utilized to pay countries and farming communities maintaining diverse PGRFA, for making specific commitments). These three possible mechanisms- especially the latter two-provide ways in which Farmers' Rights and benefit-sharing could be implemented. The design of such mechanisms, in turn, raises a number of technical and legal questions, which condition their feasibility and enforceability; and these are discussed in the following sections.

A further approach, which avoids the burden of many separate bilateral agreements and the need to trace material in use with all the technical difficulties involved, is the development of a *multilateral* system shaped to the needs of agriculture, and which would not compensate individual farmers but facilitate access generally and, as benefits, mobilize finance for projects, programs, and activities for *internationally* agreed priorities, which promote conservation and sustainable use of PGRFA and, in particular, support small farmers holding biological diversity in their farming systems.

Technical Issues

For the design and *implementation* of mechanisms for the appropriation of, or compensation for, values generated by PGRFA, the identity and origin of material must be identifiable, at least when bilateral instruments are concerned. A major question is how far this is possible, that is, whether it is possible to identify and track the *geographic* origin and *distribution* of plant genetic resources in use, over time.

A document commissioned by FAO reviews the capabilities and limitations of genetic fingerprinting, and related modern techniques, in identifying PGRFA, and establishing their geographical origin. In this analysis, a *distinction* is made between an original accession, the population from which that accession was sampled, a single genotype from that accession, and a particular gene from an accession.

While any individual organism appears as a phenotype,' genetic *fingerprinting* and related techniques help to analyze the genotype and the particular combination of genes and gene variants (that is, alleles) it contains, independently of the environment in which it may be *expressed.* Diverse populations can be described in terms of genotype and allele frequencies. It must also be noted that there are important differences in the genetic structure, as well as the genetic variability contained in landraces, when compared with the modern varieties that are the subject of plant breeders' rights.

Current plant breeders' rights legislation applies only to *propagating* materials that are distinct, uniform, and stable, and can thus easily be identified, that is, to modern varieties. These contain much less variation than is usually present in a landrace. A landrace is the product, at a particular moment of time, of continuous, changing *evolutionary* processes that result in great *variability* in the gene pool, but which also provide the capacity to adapt to changing human needs (expressed through selection by farmers) and *environmental* conditions (expressed through evolutionary pressure).

It is these characteristics that give landraces their high value as sources of plant germplasm. However, these same dynamics mean that the identification of a landrace is much more difficult than the identification of a modern variety. Genetically inherited traits, such as flower colour, growth habits, and disease resistance, can be used to identify PGRFA. More precise identification can also be obtained at the level of biochemical and molecular *composition*, especially through proteins and DNA-sequences. The examples given in the Hardon, Vosman, and Van Hintum (1994) review for FAO show that, in specific instances, a number of techniques have been used to *distinguish* varieties and accessions. However, it is unlikely that such techniques can be routinely used to prove the identity of specific genotypes, or gene *sequences*, and even less the origin of unknown genetic material. There are several reasons for this:

(*i*) The high costs of some of the techniques, *particularly* sequencing and restriction fragment length *polymorphisms* (RFLPs).

(*ii*) The same, or similar, genetic material may exist, and be detected, in more than one place, especially in neighboring countries.

(*iii*) Different methods of analysis may give different genetic estimates for the same accessions, which may lead to disputes.

(*iv*) The complex pedigrees of most improved varieties resulting from a plant-breeding program complicate *attempts* to trace specific genes, and to infer their possible relative values.

In addition, it must be borne in mind that, on the rare occasions when the ultimate geographical origin can be identified, it may not necessarily benefit the country or region of origin, since this might not be the provider of the accession, which, in line with the CBD, will usually be the subject of any rights.

All these were strong arguments in favour of multilateralism as the preferred option to deal with PGRFA.

Legal Issues

There is a need to establish a clear distinction between *sovereign* rights and property rights, as well as between physical and intangible property. The recognition of sovereign rights over PGRFA is not equivalent to the attribution, or existence, of property rights over such resources: *sovereignty* only means that the State may, within the limits imposed by the nature of such resources, determine what type and modalities of property rights, if any, are recognized.

The values of PGRFA are derived from the genetic information contained in their *germplasm.* It is from this point of view that intellectual property rights become relevant. Intellectual property rights cover the intangible content of processes or goods: In the case of living forms, for instance, they may govern knowledge of the information contained in genes, or other subcellular components, in cells *propagating* materials or plants.

However, the existence of intellectual property rights over such information is not equivalent to property rights over the individual organism that carries such information, but is the right to exclude third parties from producing or selling such *organisms* without prior agreement. Intellectual property rights (in particular, patents and breeders' rights) cannot currently apply to crop landraces and farmers' varieties.

An important question is whether it is technically sound, and legally feasible, to extend such rights, possibly in a modified, *sui generis,* form to cover such heterogeneous populations, and whether this would create adequate *incentives* for the conservation of landraces. The issue raises a number of complex legal problems.

These include the definition of the subject of such rights, requirements for protection, who may become titleholders, the territorial validity and administration of the system, and the actual enforceability of rights. A *proposal* to extend *intellectual* property rights to landraces, if feasible, would also have to consider the transaction costs involved in the establishment and operation of the system. In certain cases, the value of plant genetic resources may also be *appropriated* by contractual *arrangements*, whereby the suppliers of germplasm are remunerated, or otherwise ensured an equitable sharing in the benefits of their exploitation.

Most contracts concluded until now relate to genetic resources of specific pharmaceutical or industrial value under a bilateral arrangement, rather than PGRFA, where multilateral approaches are likely to be more efficient. Under either a *multilateral* or a bilateral approach, "material transfer agreements" (a form of contract) may be useful in regulating the transfer of material.

Material transfer agreements typically regulate the use of the *materials* by the receiver, issues relating to intellectual property rights, and economic compensation to the *supplying* source.

Figure below shows the relationship between access to genetic resources and related biotechnologies for food and agriculture, and between plant breeder's rights and Farmers' Rights, as well as the possible role of *sui generis* systems in ensuring *harmony*, coordination, and synergy among the various related international agreements in all relevant sectors, especially agriculture, environment, and trade.

Intellectual property rights allow individuals and plant breeders to appropriate the benefit of applying biotechnologies to make commercial products. These rights are individual and have already been established for many types of biotechnologies.

In contrast, farmers' rights and other forms of *communal* rights are the means by which benefits from the genetic resource inputs to the development of *commercialized* products can be obtained. These rights are collective and yet to be operationalized. The *institutions* governing intellectual property rights include the World Intellectual Property Organization (WIPO), the World Trade *Organization*/Trade Related Intellectual Property Agreement: (WTO/TRIPS) and the Union for the Protection of New Varieties of Plants (UPOV).

Institutions *governing* farmers' rights and other types of communal rights are the FAO *ITPGRFA* and the CBD Article 8j. National legislation for countries that are members of these international binding agreements needs to comply with international legal provisions contained in the agreements noted on both sides of Figure elsewhere in this chapter.

This can be done through the establishment of *sui generis* systems of rights as referred to in Article of the TRIPs agreement and the recognition of rights through national *legislation* as ways in which to balance the appropriation of values from genetic resources and the *biotechnologies* that use them. One example is recently enacted Indian legislation, "The Protection of Plant Varieties and Farmers' Rights Act".

Implications for the Design of the ITPGR

Many of these economic, technical, and legal issues have been overcome during the negotiation process of the ITPGRFA by innovative *provisions*. A key outcome was the development of a multilateral system shaped to the needs of agriculture, and which would not compensate individual farmers but facilitate access generally and share benefits, by mobilizing finance for projects, programs, and activities for *internationally* agreed priorities, which promote conservation and sustainable use of PGRFA, and, in particular, support small farmers holding biological diversity in their farming systems.

This approach was developed to overcome the difficulties that arise from the frequent lack of knowledge of the origin of specific germplasm contributions; the difficulty of attributing value; the fact that the same diversity may be found in *in situ* conditions in a number of countries; and the onerous transaction costs of bilateralism. Under the ITPGRFA, Farmers' Rights are the responsibility of national *governments*, and separate from the concept of the sharing of the monetary and other benefits of *commercialization*.

The latter are expected to contribute to the implementation of a rolling Global Plan of Action periodically agreed by the governing body of the ITPGRFA. In order to identify needs and priorities, national and regional plans of actions were developed in a *country-driven* process that culminated with the *negotiation* and adoption by 150 countries of the Global Plan of Action for the Conservation and Sustainable Utilization of Plant Genetic Resources for Food and Agriculture, at the Fourth FAO International Technical Conference on Plant Genetic Resources for Food and Agriculture in Leipzig (Germany) in 1996. This first Global Plan of Action identified 20 priority activities on PGRFA and estimated the funds necessary for their *implementation*.

International Treaty on Plant Genetic Resources

On November 3, 2001, the 31st Session of the Conference of the FAO adopted, by consensus and as a binding international agreement, the ITPGRFA, which entered into force on June 29, 2004. The Treaty's *objectives as stated in Article 1,* are "the conservation and sustainable use of PGRFA and the fair and equitable sharing of the benefits arising out of their use, in harmony with the CBD, for sustainable agriculture and food security." Article 3 establishes that its *scope* "relates to plant genetic resources for food and agriculture."

For the first time in a binding international agreement, the *Treaty* makes provision for *Farmers' Rights,* in recognition of the collective innovation on which agriculture is based. Under Article 9, contracting parties to the Treaty are called upon to take measures to protect and promote Farmers' Rights. Specific measures called for include the protection of traditional knowledge, the right to equitably participate in sharing benefits arising from the utilization of PGRFA; and the right to participate in making *decisions,* at the national level, on matters related to the *conservation* and sustainable use of PGRFA. A further important and innovative part of the Treaty is contained in Articles 10 to 13 (Part IV), which establish a *Multilateral System of Access and Benefit-Sharing,* which applies to a list of 64 crops, selected according to criteria of food security and interdepe-ndence.

The crops in question cover about 80% of the world's food calorie intake from plants. Under this part of the Treaty, contracting parties agreed to include the plant genetic resources of these crops that are under their management and control and in the public domain into the multilateral system. In addition, they will encourage natural and legal persons within their jurisdiction to include the plant genetic resources they hold into the multilateral system.

The *ex situ* collections of the International *Agricultural* Research Centres of the Consultative Group on International Agricultural Research (CGIAR) will also be brought into the Multilateral

System, through agreements with the Governing Body. The multilateral system is designed to provide access to PGRFA solely for the purpose of utilization and conservation for research, breeding and training for food and agriculture, provided that "such purpose does not include chemical, *pharmaceutical* and/or other nonfood/feed industrial uses".

Recipients of materials from the multilateral system cannot claim any intellectual property or other rights that would limit access to the resource or their genetic parts and components, in the form that it is received from the system. Access to the materials is to be provided under a standard material transfer agreement, which has yet to be developed by the Governing Body of the Treaty.

Benefit-sharing takes several forms under the multilateral system. One benefit which is shared among all contracting parties is facilitated access to the resources. Other benefits provided are information sharing, as well as access and transfer of technology. The latter are to be shared through measures such as:

> "the establishment and maintenance of, and participation in, crop-based thematic groups on *utilization* of plant genetic resources for food and agriculture, all types of partnership in research and development and in commercial joint *ventures* relating to the material received, human resource development, and effective access to research facilities".

Monetary benefits from the commercialization of resources accessed under the *multilateral* system are also to be shared through payments of an equitable share of the overall benefit to an international funding mechanism which will support conservation and sustainable utilization of PGRFA. Agreements on how an equitable share of the benefits is to be calculated and how the funding mechanism is to be managed are yet to be developed by the *Governing* Body of the Treaty.

Following its adoption by the FAO Conference, other universal fora have expressed unanimous support for the ITPGRFA:

The Sixth Meeting of the Conference of the Parties to the CBD (April 719, 2002, The Hague), in its Ministerial Declaration, which was agreed by the delegations of 176 countries, including some 130 Ministers, "urged all States to ratify and fully implement [...] the International Treaty on Plant Genetic Resources for Food and Agriculture."

The Declaration adopted by the World Food Summit five years later (10-13 June 2002, Rome) recognizes "the importance of the International Treaty on Plant Genetic Resources for Food and Agriculture, in support of food security objective," and calls "on all countries that have not yet done so to consider signing and ratifying the International Treaty on Plant Genetic Resources for Food and Agriculture, in order that it enter into force as soon as possible."

In the Johannesburg Declaration on Sustainable Development, countries which were represented at the highest level in the World Summit on *Sustainable* Development (August 26-September 4, 2002) stated that they committed themselves to the *Johannesburg* Plan of Implementation, in which they "invite countries that have not done so to *ratify* the international Treaty on Plant Genetic Resources for Food and Agriculture." The adoption of the Treaty marks a milestone in international cooperation.

It entered into force after 40 governments ratified it. Governments that have ratified it make up its Governing Body. At its first meeting, likely to be held in 2005, this Governing Body will address important questions such as the level, form, and manner of monetary payments on *commercialization*, a standard Material Transfer Agreement for plant genetic resources, mechanisms to promote compliance with the Treaty, and the funding strategy.

An up-to-date listing of governments that have signed and ratified the Treaty is available on the Internet.

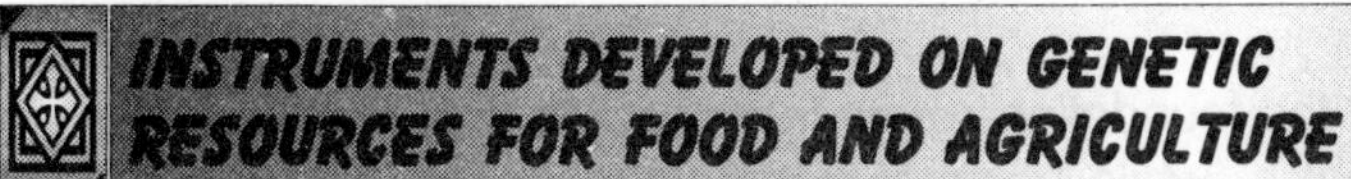

INSTRUMENTS DEVELOPED ON GENETIC RESOURCES FOR FOOD AND AGRICULTURE

The International Code of Conduct for Plant Germplasm Collecting and Transfer was negotiated by the Commission on Genetic Resources for Food and Agriculture and adopted by the FAO Conference at its 27th session, in November, 1993. A Code of Conduct on Biotechnology, as it relates to Genetic Resources for Food and Agriculture, is currently under development.

The International Code of Conduct for Plant Germplasm Collecting and Transfer

This Code of Conduct aims to promote the rational collection and sustainable use of genetic resources, to prevent genetic erosion, and to protect the interests of both donors and collectors of germplasm. The Code, a *voluntary* one, has been developed by FAO and negotiated by its Member Nations through the FAO Commission on Genetic Resources for Food and Agriculture. The Code proposes procedures to request and to issue licenses for collecting missions, provides *guidelines* for collectors themselves, and extends responsibilities and obligations to the sponsors of missions, the curators of gene banks, and the users of genetic material.

It calls for the participation of farmers and local institutions in collecting missions and proposes that users of germplasm share the benefits derived from the use of plant genetic resources with the host country and its farmers. The primary function of the Code is to serve as a point of reference until such time as individual countries establish their own codes of regulations for germplasm exploration and collection, conservation, exchange, and utilization.

Towards a Code of Conduct on Biotechnology as It Relates to Genetic Resources for Food and Agriculture

Following the proposal of the Commission on Genetic Resources for Food and Agriculture. in the 1990s, a survey of more than 400 international experts in plant genetics identified four major areas that should be covered in the Code of Conduct: biosafety, intellectual property rights, the substitution of traditional agricultural products and the development of biotechnologies appropriate for developing countries.

Biosafety: Public interest groups are concerned over the possible environmental and health risks resulting from biotechnology, especially from the field testing and release of genetically engineered organisms and plants in the area of food and *agriculture*.

They argue that there is a lack of scientific data to evaluate such risks and only the rudiments of a valid risk assessment procedure are currently possible. In the absence of an international agreement, countries that neglect to adopt adequate regulatory policies may become attractive as test sites for *genetically* modified *organisms* and plants in ways forbidden in other countries.

Once released, however, organisms modified by biotechnology will not be limited by political boundaries. It is critical that means of regulation be developed at the international level. It was *therefore* suggested that the Code could set *international standards* for testing and release of such organisms.

Intellectual property rights: Some developed countries have extended legislation on intellectual property rights to cover biotechnology processes and products, in order to stimulate and protect research. But often such measures tend to restrict the exchange of *germplasm*, scientific *information*, and technologies.

It was suggested that the Code lay the foundations of an international system of cooperation, through a set of agreed principles which, while being in harmony with existing international agreements, would promote research and transfer of technologies, and prevent the appropriation of existing genetic resources, traditional knowledge, and local technologies.

Substitution of traditional agricultural products: Biotechnology offers future possibilities for developing substitutes for existing crops, such as laboratory-produced vanilline flavor substituting for vanilla, which provides the livelihood of 70,000 farmers in Madagascar alone. Cocoa and sugar are two other crops threatened by substitutes.

Current international economic *equilibrium* could dramatically shift if biotechnologies *displace* workers and markets in developing countries. It was proposed that the Code of Conduct offer options to minimize such effects, resulting in less drastic economic change.

Development of appropriate biotechnologies to fit the needs of developing countries: Biotechnology research is expensive, and thus tends to concentrate on cash crops and commodities of major economic interest. Unless suitable provisions are taken, crops of local and social importance in developing countries could be neglected. It was proposed that the Code promote economic incentives and suitable institutional arrangements needed to stimulate research for *developing* biotechnologies more *appropriate* for the needs of developing countries.

A first draft Code of Conduct, with four modules corresponding to the above-mentioned issues, was discussed by the FAO intergovernmental Commission on Genetic Resources for Food and Agriculture in 1993. The draft Code is aimed at biotechnologies insofar as they "affect the conservation and utilization of plant genetic resources."

It recognizes that the new *biotechnologies* have tremendous possibilities both for improving the conservation of plant genetic resources and for stimulating throughout the world the creation of improvement programs. It further recognizes the risks inherent in these *technologies,* as well as how their application could have a negative effect, in *particular* in developing countries.

The purpose of the Code is to enhance the positive effects of these new biotechnologies, and mitigate the negative effects foreseen. *Subsequently,* and taking into account that the CBD was developing a Biosafety protocol, the *Commission* recommended that the module on biosafety in relation to PGRFA be sent to the Executive Secretary of the Convention as FAO's contribution to the development of the protocol.

In 1995, the Commission examined a report on recent international developments of interest to the draft Code of Conduct, and recommended that its further development and the negotiation be postponed until the negotiations for the ITPGRFA had been completed. Following the adoption of the ITPGRFA by the FAO Conference in 2001, the *Secretariat* of the Commission carried out a new survey among FAO Member Nations and a large number of *stakeholders* to revise the possible components for the Code on the line of recent biotechnology developments.

Subjects suggested by countries and stakeholders, and currently being considered as possible components of this Code, include access to and transfer of biotechnology, capacity-building, biosafety and *environmental* concerns, public awareness, development of appropriate biotechnologies for poor farmers and developing countries, ethical questions regarding new biotechnologies, genetic use restriction technologies ("terminator" technology), GMOs, gene flow and the question of liability, voluntary certification schemes, and possible FAO universal declarations on plant and animal genomes.

The Ninth Regular *Session* of the Commission (October, 2002) discussed a working paper on the subject, and requested the Secretariat to prepare a study covering all the issues raised in the survey which would identify what is being done in other forums and what remains to be done on this issue so it would help the Commission to identify the issues on which it should concentrate in the future, with respect to a Code, guidelines, or other courses of action.

FINAL REMARKS

The Treaty is the outcome of many years of intense negotiations in FAO's intergovernmental Commission on Genetic Resources for Food and Agriculture, to revise the voluntary International Undertaking on Plant Genetic Resources for Food and Agriculture. As the 30th Session of the Conference of the FAO noted, these negotiations were at the meeting point between agriculture, the environment and commerce.

The Conference agreed that there should be consistency and synergy in the agreements being developed in these different The innovative provisions of the Treaty provide for facilitated access to PGFRA and an agreed way of benefit-sharing, without deriving these benefits from individual *negotiations*, on a case-bycase basis, between the provider and the user of these resources.

They provide for both access and benefit-sharing to be through multilateral *arrangements*. This avoids the high transaction costs that such bilateral contracts involve, which are hard to justify in the context of plant breeding, which has for thousands of years been characterized by repetitive exchange, crossing, selection, and local *adaptation* of the intraspecific genetic resources of crops, within and between countries and regions.

This ensures that plant breeders, in both the public and private sectors, can have access to the widest possible range of the resources crucial for world food security. This will benefit consumers, by providing a stream of improved and varied agricultural products, and it will benefit the seed and biotechnology industries, by providing an agreed international framework, within which to plan their investments.

It also provides a firm international framework for IARCs of the CGIAR and other international organizations, whereby they hold PGRFA in trust, under the IT. The IT and other relevant international agreements need to be fully enforced at the national level.

The development of national legislation for *implementation* of their provisions will be essential in deterring genetic erosion, protecting indigenous germplasm and Farmers' Rights, facilitating access to genetic resources for food and agriculture, and ensuring benefit sharing. *Political* and economic support for implementation of these agreements can be stimulated, if the public is informed about the importance of genetic diversity and the dangers of its depletion, and encouraged to act to stop genetic erosion.

It should not be forgotten, however, that genetic erosion is but one consequence of man's abusive exploitation of the planet's natural resources, which has broken the balance of many ecosystems and brought about an increasing degradation of the biosphere. Safeguarding genetic resources by protecting them *ex situ* or *in situ* is crucial, if the process that has been *unleashed* is to be reversed, or controlled, at all.

The fundamental problem remains man's lack of respect for the rest of nature, and any lasting solution will have to involve establishing a new relationship with our small planet, in full understanding and recognition of its limitations and fragility. If *humanity* is to have a future, it is *imperative* that children learn this in primary schools, and that adults make it part of their life.

AGRICULTURAL BIODIVERSITY

In this volume we have brought *together* a unique set of analyses on *managing* agricultural biodiversity and biotechnology in the context of development. One of the key features of the book is using the common thread of plant genetic resource *management* as a point of departure in analyzing the potential for, and barriers to, jointly managing agrobiodiversity and biotechnology to achieve a range of development-related goals.

These include increasing agricultural productivity and *sustainability*, reducing poverty, and improving the conservation of genetic diversity. Policies governing the management of both genetic diversity and biotechnologies have the potential to affect the ability of countries to achieve any of these objectives, depending on prevailing *socioeconomic* and *environmental* circumstances.

One of the strengths of the approach taken in this volume is that it allows for an assessment of where there are overlaps, synergies, and *contradictions* in policy approaches to managing biodiversity and *biotechnology*. Just as importantly, the approach helps to identify where there is not likely to be any interaction between biodiversity and biotechnology management, and if not, the types of policy intervention needed in each separate arena to achieve desired outcomes.

Increasing the productivity of agricultural production systems through the development and *dissemination* of improved genetic resources is a primary means of accelerating economic growth and addressing the problems of food insecurity and poverty in developing countries. Biotechnologies provide one important vehicle to achieve this *improvement*, albeit with several caveats.

Both *institutional* and technological obstacles need to be overcome to realize its potential. Since molecular biotechnology is a major innovation, the full ramifications of its impacts are still unknown, and safeguards for preventing *undesirable consequences* are necessary, but difficult to formulate in the presence of uncertainty.

Advances in biotechnology have also generated radical institutional changes in plant breeding with a major shift of funds and control from the public to private sector. Harnessing the benefits of biotechnology for developing countries and poor farmers thus requires new institutions and new ways of managing existing institutions. There are other means of improving genetic resource *productivity* in agriculture besides biotechnology, which may be more effective, *particularly* in marginal production areas.

These include approaches such as production ecology, participatory plant breeding, or reducing the costs of accessing a diverse set of genetic resources by increasing the supply of diversity. However, these strategies face their own set of constraints, including cost effectiveness. Biotechnology-based approaches to improving genetic resource productivity are not mutually exclusive with alternative approaches: In fact, enhancing the productivity of *conventional* approaches is likely to be one of the most valuable *contributions* of the technology to development.

Regardless of the approach taken to improve the access to and performance of genetic resources, greater focus on improvements in the development and delivery of genetic resources that meet the specific *production* and consumption constraints of the poor is necessary in order to achieve poverty-reduction goals. The *conservation* and *sustainable* utilization of agricultural biodiversity is another important policy objective in developing countries, which is separate but linked to that of increasing the productivity of genetic resources for food and agriculture.

Agricultural biodiversity is a broader concept than genetic diversity, encompassing human knowledge and ecosystem functions as well as the genetic variability of plant, animal and microorganisms. Its *conservation* generates benefits that are realized globally as well as nationally and locally. As signatories to the International Treaty on Plant *Genetic Resources* and the Convention on *Biological Diversity*, many developing countries have assumed obligations to promote the conservation and sustainable utilization of agricultural biodiversity.

The relevant decisions that face developing country policymakers are how to optimize the benefits from agricultural biodiversity management, including the potential for increasing productivity and sustainability of agricultural production, as well as receiving compensation for providing public *environmental* goods and services. Much of the world's valuable plant genetic diversity is located in developing countries and the *conservation* of this resource generates both private and public goods.

To the extent that *maintaining* genetic diversity results in private benefits to the farmers who provide it through their planting decisions, incentives to conserve exist, although oftentimes rapidly eroding under processes of social and economic change. Maintenance of the public good aspects of diversity conservation (e.g., *reduced vulnerability* to pest and diseases, and options for future *genetic* inputs to *plant-breeding* efforts) *requires* some type of policy intervention.

One way to generate a socially desirable level of conservation is by setting up mechanisms to allow for flows of payments from the *beneficiaries* to the providers. However, two major problems arise with such mechanisms. First, there are concerns that *establishing* property rights and rights to *compensation* for diversity conservation will lead to a reduction in the free exchange in genetic materials among crop breeders and farmers that has prevailed thus far, thus reducing their capacity to generate new varieties, and ultimately reducing farmer access to genetic *resources* in developing countries.

Second, even if payments are desirable, payment mechanisms are difficult to design due to difficulties in valuing the benefits associated with conservation. This debate over compensation and benefit sharing is part of a bigger discussion about the ownership of genetic materials and the benefits that farmers, breeders, and other groups obtain from conserving agrobiodiversity, which have been the focus of the International Treaty on Plant *Genetic Resources* and for Food and Agriculture whose *implementation* mechanisms have yet to be designed.

The challenges outlined in the paragraphs above are being faced by developing country policymakers in a rapidly changing and high-stakes environment, where current policy choices may have *significant* consequences for current and future generations.

This book has been designed to provide insight into the key problems of managing agrobiodiversity and biotechnology efficiently and equitably in the context of economic development. It is structured in an incremental fashion, first looking at the key forces and factors which are shaping the overall "rules of the game" under which biotechnology and biodiversity can be managed, and the impact of these on the ability to achieve efficient and equitable management regimes.

Next, specific considerations of *biodiversity* conservation, biotechnology development and dissemination, and sharing benefits from genetic resource management are *addressed.* The last part includes a series of policy-oriented chapters drawing upon the analyses in earlier sections. In the first part of the book, the chapters describe a series of major changes in global economic and environmental settings.

On the environmental side, there is a decline in the global natural capital asset base of plant genetic diversity, together with a rising appreciation of their value and institutions designed to promote them. On the economic side, increased integration of global agricultural markets gives rise to changes in the structure of production and marketing, resulting in the *expansion* of markets and the economic opportunities associated with them.

However these same changes may also result in increased barriers to market *participation,* particularly among small-scale and low-income producers. Three key lessons can be summarized from the chapters in the first part. The first is that markets are increasingly important as mechanisms for transmitting incentives for production and consumption decisions, as they are expanding in terms of participants on both the supply and demand side for a wide range of agricultural input and output products as well as for environmental goods and services.

Secondly, demand and supply are *increasingly* determined at supra-national levels-e.g., consumers and suppliers beyond national borders have increasing impact on market signals at a national and subnational level. One example is the rise of environmental concerns in developed countries leading to increased willingness to pay for agricultural products grown under specific environmental conditions, e.g., organic, no genetically modified organisms (GMOs), etc., affecting the production decisions of farmers in developing *countries* who supply international markets.

Another is the potential impact of the privatization and commercialization of genetic resources in developed countries on the cost of accessing these resources in developing countries. The third, and perhaps most key point, is that we cannot rely on market forces alone to *generate* socially desirable levels of *poverty* alleviation, agricultural biodiversity conservation and biotechnology development.

In some cases this is because markets are non-existent for the socially desirable goods and services, as is the case with agricultural biodiversity conservation. The value and sources of diversity are difficult to identify and quantify and, thus, so is the *establishment* of *marketbased* payment mechanisms. In other cases the problem arises from distortions and poorly functioning markets.

Examples here include the inability of poor farmers to express their demand for improved genetic resources in commercial seed markets due to their limited purchasing power and poorly functioning seed, credit, and other markets. Increased reliance on a market-based research and development system in this era of increased privatization would bypass the needs of the poor farmers in technology development. Another example is concentration and vertical *integration* in agricultural input and output markets that lead to noncompetitive and inefficient markets.

Finally, markets are a means of achieving efficient, but not *necessarily* equitable, allocations of resources and thus interventions may be required to achieve socially desirable levels of equity. The tension between the increasing importance of markets as a means of improving genetic resource management for both *conservation* and development on the one hand, contrasted with the increasing recognition of a need for policy interventions to either improve, supplement, or substitute for markets on the other hand, is a theme that recurs *throughout* the analyses presented in this volume.

STRATEGIES FOR CONSERVING AGRICULTURAL BIODIVERSITY

Agricultural *biodiversity* is a major and valuable form of natural and human capital, comprised of several components, including plant genetic diversity, which has been the focus in this book. The gains from the conservation and enhancement of agricultural biodiversity spread far beyond the *location* such activities take place.

Agricultural biodiversity has strong public good properties, but its benefits are uncertain and vary across locations and over time. Thus, market forces by themselves will lead to a socially undesirable rate of loss, and global *collective* action and cooperation are required to efficiently manage agricultural biodiversity.

Several means of *attaining* conservation have been identified in this volume, associated with varying costs and benefits. The socially desirable levels of activities in conservation and enhancement of crop biodiversities may be most *efficiently* achieved through *compensation* in exchange for these activities. Ideally, conservation funds should be allocated across the portfolio of potential activities, in order to maximize the expected benefits of agricultural biodiversity conservation, and coordination between the various forms of conservation promoted in order to enhance cost effectiveness.

In reality this is difficult to achieve, due to several issues raised in the chapters of this volume. These include the following: the valuation of conservation benefits, the identification of criteria for establishing *efficient* conservation programs, the design of mechanisms to provide incentives to developing countries for conservation and developing means of *incorporating* diversity conservation into overall agricultural and economic development concerns and strategies. A key question which arises in this design of *effective* policies for conservation, is just how exactly should diversity be defined-what is it that we are trying to conserve?

The answer is complex, depending on the type of value focused upon, as well as assumptions about how best to generate or maintain it. Elsewhere in this book discusses the controversies over defining genetic narrowing in crop genetic diversity, noting several relevant dimensions, including spatial vs. temporal diversity, variation within vs. among varieties, and variation within landrace vs. modern varieties. In the other chapter of this book states that the biological diversity of crops encompasses phenotypic as well as genotypic *variations*, resulting in differences in the perception of crop genetic *diversity* between farmers and plant breeders.

The author also notes the importance of conserving rare alleles in centers of crop origin, which requires a different type of conservation strategy than one targeted at maintaining high levels of varietal heterogeneity. In the other chapter describes several forms of agricultural diversity, including species and varieties, ecosystems and human knowledge, all of which, the authors argue, are important to consider in conservation programs. Clearly agricultural biodiversity conservation generates several types of goods and services and conservation programs will vary depending on which are of key concern.

However, there is considerable uncertainty about the most effective means of generating goods and services from conservation, as well as uncertainty about the relative values of these services. Thus one of the biggest problems in designing effective *conservation* programs is defining what should be conserved and where. Several chapters in the book provide insight into where and how the conservation of agricultural biodiversity in general, and plant genetic diversity specifically, can be most effective. A variety of conservation methods exist, ranging from *ex situ* gene collections to *in situ* farm-based diversity management.

However the high degree of *uncertainty* associated with both the private and public values of diversity, as well as a lack of information about the actual and opportunity costs involved, means that conservation efforts are often not efficient. The private benefits associated with plant genetic diversity conservation are realized by farmers whose maintenance of diverse cropping systems can be thought of as the outcome of a *constrained* utility maximization problem.

These values are described and analyzed in some detail in other chapters. Elsewhere in this book summarizes the results of several studies where risk management, responsiveness to highly heterogeneous production conditions, labor management, and preferred consumption characteristics have all been found to be important determinants of on-farm diversity.

These private values of diversity are determined by agroecology, population density, and the level of *commercial* market development. In other chapter adds another important determinant of the private values of crop genetic diversity: the seed system, which affects the availability and accessibility of genetic resources and *information* at the farm level.

The incidence of natural disaster and political strife that can disrupt supply systems can also be important determinants the private value of maintaining crop genetic diversity. The market failure in diversity conservation arises from the fact that conservation generates several types of public goods. One is in the form of reduced vulnerability to pests and disease incidence, which occurs mostly as a local public good, but also with *potentially* wider benefits.

Much of the use benefits associated with diversity conservation have not yet been realized and, as such, remain as potential. In such cases, the benefits of biodiversity conservation are primarily in the form of an option value. In other chapter anywhere else in this book presents one approach to the measurement of this value, using an empirical *example* from teak breeding.

This chapter concludes that the value of increasing the number of potential parents for breeding is actually quite low at the margin, measured in terms of changes in the consumer and producer surplus. This chapter raises the *important* question of how much effort and cost should be made in maintaining crop genetic diversity as a source of input to future breeding efforts.

The question is still open to *considerable* debate, and is likely to vary considerably among crops. Rausser and Small (2000) argue that the option values of *agricultural* genetic resources are likely to be sufficiently high to support market-based bioprospecting activities, since researchers have prior knowledge about where the most promising leads are likely to be found.

In other part of this book discusses various criteria for assessing option values, and discuss the disincentives to plant breeders in broadening the genetic base of their breeding lines. Public sector interventions to promote genetically diverse "pre-breeding" activities could lead to higher option values for crop genetic resources. Moving towards consideration of criteria for designing conservation programs, one approach identified is minimizing associated costs.

Elsewhere in this book examines this issue in detail in the context of *ex situ* conservation, which is the term applied to all conservation methods in which the species or varieties are taken out of their traditional ecosystems and are kept in an *environment* managed by humans. An estimated 6.2 million accessions of 80 different crops are stored in 1,320 gene banks and related facilities in 131 countries at local, national, and international levels.

Elsewhere in this book also discusses the inefficient *management* of these facilities, finding *significant* differences in the degree of national commitment and expenditures on PGR conservation, which are not necessarily tied to the level and value of domestic genetic diversity. The chapter concludes that better collaborative relationships are the primary vehicle for reducing costs and improving the management of *ex situ* sites at a regional level, between public and private entities, and within the *multilateral* system.

The primary costs associated with *in situ* conservation are opportunity costs, which are addressed in other chapters. In other chapter provides a conceptual framework for assessing public/private tradeoffs in maintaining *in situ* conservation, differentiating between situations where the private and public values of diversity maintenance coincide, versus come into conflict.

Ostensibly, situations where they coincide require no intervention to maintain desired levels of conservation. This implies that the least-cost means of *in situ* conservation is to focus on areas where private values of diversity are high and, thus, opportunity costs of *conservation* are low. However, in a dynamic setting, problems arise as high private values of diversity conservation are often negatively associated with processes of economic development, particularly increasing *integration* of farmers into markets.

Assessing the future *opportunity* costs farmers may face in maintaining diversity given efforts to promote economic development thus becomes a critical issue. Reducing the future opportunity costs farmers may face in maintaining on-farm diversity, and thus providing incentives for its maintenance, can be achieved by either addressing the change in *conditioning* factors that reduce the value of diversity or through compensation programs.

A key strategy for reducing future opportunity costs of conservation is to increase the supply of diversity and reduce access costs. This strategy includes increasing the supply of a diverse range of improved crop varieties (that is, more *diversity* in modern, i.e., genetically uniform, varieties) as well as enhancements to existing varieties and *populations* that encompass a high range of diversity (e.g., enhance the performance from varieties that encompass genetically diverse populations, landraces, and seed lots).

Increasing diversity supply is an issue which is addressed throughout the book, with various pathways identified. The chapters in Part II analyze the potential for changes to traditional and conventional breeding systems that may lead to higher levels of diversity supply.

Participatory plant breeding, broadening the genetic base of conventional breeding programs, and the establishment of community seed banks and *registers* are all examples of programs that fit here. A major problem identified with these programs is their cost effectiveness. The inability of such programs to cover costs does not mean they are undesirable.

However, some level of public support will be required to achieve the desired objective of increasing the supply of genetic diversity and thus increase the provision of both private and public goods associated with conservation. In other chapter of this book argues that plant genetic diversity *conservation* requires consideration of the costs and benefits of all available options.

The chapter analyzes a variety of mechanisms which may be appropriate for *promoting* efficient conservation in both *in situ* and *ex situ* situations, ranging from direct approaches such as payments to farmers for growing diverse crop varieties and royalty payments on genetic resource inputs to commercialized products, to more indirect methods such as provision of access to biotechnologies and other forms of technology and institutional support.

Agricultural research and development and plant-breeding management, seed regulation, input and output market development, information transmission, and seed provision under disaster conditions all have implications for the costs and values of *in situ* conservation, and these have not been well researched to date.

LINKAGES BETWEEN BIOTECHNOLOGY AND PLANT GENETIC DIVERSITY

Advances in biotechnology will have a significant impact on both the demand for, and supply of, plant genetic diversity conservation, and several chapters in the book address this issue. In this discussion clear definitions are critical: Within both biodiversity and biotechnology, there is a range of meanings, and the relationship between the two depends on which specific aspect is being considered.

Improving *information* about the nature, source, and value of biodiversity is one critical function biotechnologies offer to the improvement of genetic diversity *conservation*. As raised in several points throughout the book, lack of information is a serious problem hampering effective agricultural biodiversity conservation efforts.

In other chapter, Virchow *suggests* that the value of genetic *collections* is reduced by the uncertainty regarding the properties and impacts of genetic materials stored in specific seed varieties. The existing and emerging tools of biotechnology will expand the capacity to utilize the information stored within *in situ* and *ex situ* collections, allowing analyses of the genetic content and potential of stored seeds.

Emerging *techniques* of molecular and cell biology, and in particular the tools of computational genomics, allow for rigorous classification and documentation of genetic materials and, thus, a reduction in the cost of accessing and utilizing genetic materials stored in various collections. This improves the ability of researchers to identify promising genetic materials for *incorporation* into breeding products, which is likely to increase their marginal value and demand for conservation.

The production and *dissemination* of GMOs are another aspect of biotechnology development likely to have significant impacts on agricultural biodiversity in general, and plant genetic diversity specifically.

The introduction of GMOs may affect the number as well as genetic content of new varieties available for adoption in *developing* countries.

Adoption patterns will affect both spatial and temporal patterns of diversity through two processes: The replacement of one type of germplasm for another, and the integration of new genetic materials into *existing* gene pools through gene flows. The first is a *human-driven* process, dependent on the supply of and demand for GMOs. The second is governed by the natural process of gene flow and integration.

Ultimately, the impact on genetic diversity depends on (1) a series of forces which drive supply and demand patterns, (2) the baseline situation with regard to crop genetic diversity, (3) vulnerability of the crop to geneflow (reproductive characteristics, presence of weedy relatives), and (4) the way in which diversity is defined.

Elsewhere in this book looks at factors that determine the supply of GMOs in developing countries. They argue that the capacity to adapt biotechnologies to local materials is a critical determinant of the potential benefits of GMOs in agricultural development as well as impacts on crop genetic diversity.

The strength of intellectual property rights (IPRs) over plant genetic *resources* and their *enforcement* within a country, together with the level of competence in the plant-breeding sector and the level of transactions costs associated with accessing biotechnologies, are identified as the key determinants of the numbers and genetic content of GMO varieties likely to be supplied in developing countries.

Countries with strong IPRs, advanced *breeding* capacity, and relatively low *transactions* costs are most likely to develop a wider range of GMOs for any one crop, as the marginal costs of adding a transgenic trait to an increasing number of varieties of a sexually *propagated* species is smaller than the marginal benefits. In addition, the degree of local materials *incorporated* and thus conserved into GMO varieties is likely to be higher under these conditions.

The authors argue that GMO development under these *conditions* can lead to an increase in crop genetic diversity, as incentives exist to modify local materials with improved traits and generate several varieties, resulting in a higher number of improved varieties, with a higher content of local materials preserved.

The impacts on diversity also depend on what the GMO varieties are replacing; the implications are quite different if they are replacing a few *conventionally* bred modern varieties versus landrace populations. The introduction of GMO varieties may also affect diversity through gene flows from transgenics to other planted varieties.

Managing undesired gene flows is an important aspect of biosafety regulation, but the degree to which gene flows pose a risk to biodiversity conservation and the degree to which *regulations* will be effective in managing such risks are still unknown.

Apart from the technology and products of biotechnology per se, several authors raised concerns about the impacts of the institutional changes accompanying biotechnology on diversity conservation. In other chapter argues that biotechnology-induced changes in IPR regimes increase the *privatization* of knowledge and could increase the costs of accessing breeding materials.

Therefore, stringent IPR regimes may well reduce the capacity of breeders in developing countries and the CGIAR centers to access new materials and technologies. They also note that the absence of transparent and *well-functioning* biosafety regulations are likely to restrict access, as suppliers of the technology may be unwilling to enter such markets.

Public sector access to genetic materials is a critical concern since it is this sector that will be focused on crops of most importance to the poor, which in many cases are not commercially attractive. IPRs have also been associated with increases in the number of new varieties developed. In other chapter argues that IPRs were a crucial stimulant to the development of private sector research and development in canola, leading to an explosion in the number of new varieties developed.

However, Graff and Zilberman describe the current situation with IPRs in agricultural biotechnology as an *anticommons climate,* restricting both public and private sector access to technologies and thus development of new varieties. Anywhere else in this book discusses the implications of changing IPRs under impetus from the TRIPS *agreement* of the World Trade Organization on agricultural biodiversity, finding the potential for both positive and negative impacts.

These chapters indicate that the numbers, genetic content, and accessibility of *improved* varieties are changing in response to institutional changes associated with biotechnology; however, assessing the impacts on plant genetic diversity conservation is again a function of how diversity is defined. *Overall,* the analyses in this book indicate that agricultural biodiversity and biotechnology are co-evolving, with a number of different points of intersection.

The adoption of *transgenic products* may harm or enhance crop biodiversity. The new tools of biotechnology improve our capacity to interpret and utilize agricultural biodiversity. Improvements in the conservation of plant genetic diversity are likely to increase the productivity and value of agricultural biotechnology. The analyses in this book suggest that recognition of the interdependency between biotechnology and biodiversity is critical to the achievement of sound policy design for the management of agricultural biotechnology and biodiversity in the context of economic development.

MANAGEMENT OF PLANT GENETIC RESOURCES

Sharing the benefits (and costs) of plant genetic diversity conservation and maintaining access to genetic resources and biotechnologies for lowincome groups are critical concerns addressed *throughout* this volume. Equity (and efficiency) criteria would suggest that since much of the natural capital embodied in agricultural biodiversity is in developing countries, companies and nations in the North, which are potential beneficiaries of this conservation, should contribute to crop biodiversity *conservation* funds.

Developed countries tend to be in regions whose original genetic endowment in the major agricultural crops was lower than in biodiversity hotspot areas. As Tables elsewhere in this chapter demonstrate, primary centers of agricultural genetic diversity are mostly in developing countries.

Thus, private breeders largely from developed countries develop and market *varieties* that rely on genetic materials that originated at some point (perhaps many generations ago) from the developing world. Many less-developed countries (LDCs) or associated interest groups claim that these breeders are benefiting from utilization of their native landraces without compensating the farmers responsible for their maintenance.

Furthermore, they assert that developed countries are benefiting more from the utilization of plant genetic resources for food and agriculture (PGRFA) from developing countries than do the LDCs themselves and that these LDCs are not being compensated in return for using these resources. The issue has become particularly acute with the development of biotechnology and privatization of agricultural research and development.

This perspective leads to active demand for compensation of farmers and others in LDCs for past conservation efforts. However, at least from the economics standpoint, there is some difference between biodiversity funds that aim to compensate for past *conservation* and funds that aim to *encourage* future conservation.

Paying LDC farmers for past conservation efforts is largely an equity issue given that insufficient data are available to establish compensation payments based on the economic value of conservation efforts in the past and, as such, one must appeal to equity, even though it is a weak mechanism for allocating funds. Paying for current and future conservation activities can have more potential to be made using notions of economic *efficiency* (i.e., making conservation payments such that the marginal benefit of conservation effort equals its marginal cost).

For example, demonstrates a proxy measure for economic value that can at least be used as a rough mechanism for distributing conservation funds to world regions with an eye on increasing the economic benefits to society of conservation efforts. The analyses suggest that on efficiency as well as equity grounds, direct beneficiaries from agricultural biodiversity conservation would be made to reward the providers of the benefits, based both on actual and expected gains.

However, there are also significant benefits to maintaining a free flow of genetic resources among breeders and other researchers, and this is a difficult issue to address in the design of compensation and incentive mechanisms. On the one hand, improved property rights over genetic resources and their embodied values would facilitate the establishment of exchange and compensation mechanisms. However, at the same time, economic efficiency and equity criteria suggest that the continued sharing of the benefits associated with these goods be promoted. While this book *suggests* some possibilities for cost-sharing mechanisms, their exact design still needs further research. In other chapter describes in detail how issues of equity and benefit sharing have been incorporated into the design of the *International* Treaty on Plant Genetic Resources for Food and Agriculture.

The chapter discusses the economic, technical, and legal reasons for the *establishment* of a multilateral system to facilitate access to and sharing of benefits from the utilization of plant genetic resources. The chapter also discusses the role of various forms of property rights, including intellectual property rights and farmers' rights and how the two systems can *complement* each other to ensure that *incentives* to innovate are maintained, while at the same time ensuring the capacity of rural communities to benefit from their conservation of plant genetic resources.

BIOTECHNOLOGY: MAXIMIZING THE BENEFITS AND MINIMIZING THE COSTS

One of the major points made about biotechnology in this volume is that it is much more than just a tool for genetically modifying crop varieties. Aside from the crop sector, livestock, fisheries, and forestry biotechnology products of relevance to the poor are currently under *development.* As noted above, one of the key benefits of biotechnology is through increasing information on genomics and, thus, values of *biodiversity,* which are necessary for developing effective conservation and compensation strategies and programs.

The main focus of the potential benefits of biotechnology in economic development has been on the increased potential to generate breeding materials that are *specifically* relevant to the production and consumption conditions in developing countries, and in a much more targeted fashion and shorter time frame than is possible with conventional breeding methods. *Several* chapters in the book describe the *experience* that has already been seen with biotechnology adoption in developing country agriculture.

Note the dominance of tissue culture technologies in developing countries, and their importance in generating disease-free plants. Other chapters focus on the experience with GMOs in both developed and developing countries. Transgenics are in the early stages of their development, yet GMOs that control pests have high adoption rates for major crops in Latin America and China.

Nevertheless, the adoption of transgenics in the majority of developing countries has been minimal, and no GMOs have been introduced for several major staples consumed by the poor (rice, wheat, cassava) in developing countries. At this point it is not possible to draw firm conclusions about the potential impacts of the adoption of transgenics in developing countries. However, the current evidence provides valuable insights, including:

- Adoption patterns and impacts of GMOs vary over different economic and agronomic circumstances. In other chapter cites evidence on how differences in pest incidence, land quality, and credit availability affect adoption rates. The availability, effectiveness, and prior use of *pesticides* determine the extent to which GMOs reduce chemical use and affect output levels, and GMOs may increase agricultural *production* where other approaches have not been effective in controlling pest damage at lower *environmental* costs.

- By reducing the *variability* of crop yields, GMOs can serve as an insurance *strategy* allowing the farmer to cope with the randomness of pest infestation within and between seasons. The benefits of GMOs consist both of their average yield effect and yield riskreducing effects.
- The yield gains from the adoption of GMOs are likely to be smaller if the modified varieties are generic, as opposed to those based on local materials adapted to the local conditions. Elsewhere in this book suggests that using GMOs not adapted to local conditions is likely to introduce new sources of yield risks.
- Whether transgenics increase yields or reduce pest-control costs, they tend to increase the overall supply of the crops. Elsewhere in this book suggests that this may lead to reduction in the prices of the modified *commodity*, which will benefit consumers including urban population, the rural landless poor, and net-consuming farm households. However, lower prices may harm the nonadopting farmers.
- Transgenics are a highly divisible technology with low fixed costs and low management requirements-e.g., they have limited *requirements* for human capital inputs. These characteristics make GMOs accessible and attractive to small- and low-income producers. Nonetheless, the traditional constraints to technology adoption among the poor-such as lack of credit, poorly developed input and output markets, and the presence of risk-are likely to impede adoption among smallholders.
- The adoption of GMOs may generate *environmental* and human health benefits through the reduction of pesticide use, and yield effects may lead to reduction of land *conversion* to *agricultural* use and thus reduce deforestation and land degradation. These benefits have to be weighed against the risks that may be introduced with GMOs, such as irreversible changes in genetic populations through geneflow.

The substantial rates of adoption of *agricultural* biotechnologies in some developed and developing countries and their realized net benefits suggest that these technologies are likely to play a significant role in global agriculture as they evolve. Elsewhere in the other chapter of this book highlight the applications of *agricultural* biotechnology currently available and in the development pipeline, which could be highly beneficial to lowincome farmers in *particular* and to developing countries in general.

However, the degree to which these potential benefits of biotechnology are realized by poor farmers and developing countries is likely to be determined more at a macro than at a microlevel. The benefits of biotechnology to farmers and the poor will depend on the degree to which biotechnology innovations address production and consumption constraints, and are affordable and accessible to farmers.

Farmer access to biotechnology is determined by the type, amount, and cost of technologies produced by plant breeders-either nationally or internationally. As argued in other chapter of this book, these factors are, in turn, driven by the combination of intellectual property (IP) regimes, local breeding capacity, commercial seed industry development, and biosafety regulation regimes.

Transactions costs (affected by IPRs and biosafety regulations) associated with obtaining breeding materials will determine the degree to which private sector materials would be available to local breeders, while local breeding capacity will *determine* the costs of adapting them to local conditions, and the development of the commercial seed sector drives the degree to which such innovations could be disseminated to farmers.

High transactions costs in obtaining breeding materials and local breeding capacity are the two most critical *determinants* of potential beneficial effects of biotechnology in developing countries, which can be addressed by *institutional* reforms at both the national and international levels. Chapter elsewhere in this book gives one example of such an institutional reform, arguing that the transaction costs can be significantly reduced by establishing clearinghouses for IP, which will provide crop breeders with information on the status of IP over various crops and technologies and assist them to obtain access to it.

The recently established Public Intellectual *Property Resources* for Agriculture (PIPRA) is one example of a clearinghouse that aims to reduce the transaction cost constraints of agricultural *biotechnology*. In other chapters also argue that reducing *registration* requirements for agricultural biotechnology will reduce transaction costs and lead to a more *diversified* portfolio of modified varieties.

A clear example of a policy that reduces transactions costs is the *requirement* of registration and safety testing only for new biotechnology events (such as development of a parent GMV, which through back crossing can lead to insertion of the modification from the parent into all the varieties of the crop) rather than for every modified variety.

In other chapters argue that the CGIAR centers have an important role to play in filling the gap created by a lack of local breeding capacity in many developing countries, as well as greater integration of NARS *research* work over agroecological regions In many developing countries, agricultural biotechnology may not be the least cost or most *efficient* means of improving agricultural productivity.

The national breeding capacity, type of farming systems present, and constraints to increases in agricultural productivity are key determinants of the degree to which developing countries will benefit from agricultural biotechnology. The future of agricultural biotechnology and its impacts on economic development will be affected by the management of the human health and environmental risks associated with it. Continuous research and monitoring of the potential risks involved are clearly critically important.

The efficiency of the regulation of biotechnology applications would increase significantly with greater quantification and definition of the risks associated with these applications. However, the high degree of uncertainty and the lack of information on the risks prevent precise estimation. Building this uncertainty into biosafety regulatory structures would provide more meaningful information than that associated with simply providing mean measures of risks.

One way to overcome the lack of information at the initial stage may be to quantify the potential risks under plausible pessimistic scenarios, and assess their costs relative to the expected economic and environmental benefits of the technology. It is important to recognize that, beyond a certain stage, estimates of outcomes and the technology itself will not improve significantly without field experience, which implies that the efficiency of assessment and regulation of technologies will be increased if they can incorporate adaptive learning and through taking advantage of findings in the laboratory as well as outcomes in the field.

Poorly designed biosafety regulations that lead to excessive delay in the introduction of biotechnologies may generate significant economic costs in terms of foregone opportunities for technological development, including learning by doing, and improvements in agricultural productivity.

An important reference point for the development of biosafety regulations in the context of agricultural biotechnology is the Draft Code of Conduct on Biotechnology as it relates to Genetic Resources for Food and Agriculture.

The objective of the code is to maximize the positive effects and minimize the possible negative effects, of biotechnology. The draft Code is based on the results of two major surveys of stakeholders in 1993 and 2001, which identified the key issues of concern.

Issues currently being considered as possible components of the Code include access to and transfer of biotechnology, capacity-building, biosafety and environmental concerns, public awareness, development of appropriate biotechnologies for poor farmers and developing countries, ethical questions regarding new biotechnologies, genetic use restriction technologies (*"terminator" technology*), GMOs, gene flow and the question of liability, voluntary certification schemes, and possible FAO universal declarations on plant and animal genomes.

A clear message that emerges from the analyses in this volume is that *appropriately designed* policies and institutions are essential for enabling agricultural biotechnology to fulfill its promise for developing countries. One policy implication arising from the analyses presented is the potential benefits to be reaped from strengthening of the capacity of developing country agricultural research and development and seed sectors to introduce desired traits into local varieties, rather than relying upon imports of generic transgenic varieties.

A second policy implication that emerges is the need for regulations to manage the risks associated with the new technology as well as the importance of including cost considerations-particularly the costs of foregoing opportunities to improve productivity-when designing such regulations. Thirdly, barriers to access the intellectual property needed for the development of transgenic crops for developing countries should be reduced through institutional arrangements for technology transfer and sharing of knowledge about IPR and technology management.

A clear message that emerges from the analyses in this volume is that designing appropriate policies and institutions is essential for enabling agricultural biotechnology to fulfill its promise for developing countries. First, benefits to developing countries will be greater if the capacity of the seed sector in these countries is enhanced to allow introduction of desired traits into local varieties than with simply importing generic transgenic varieties.

Second, economic efficiency suggests that the level of regulation of new varieties to allow control against risks has to be balanced against cost considerations-particularly the costs of foregoing *opportunities* to improve *productivity-when* designing such regulations. Third, the most efficient way to reduce barriers to access the IP needed for the development of *transgenic* crops for developing countries would be likely be through institutional arrangements for technology transfer and *sharing* of knowledge about IPR and technology management.

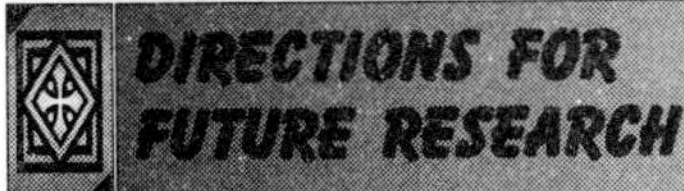

DIRECTIONS FOR FUTURE RESEARCH

Several areas where new research is needed on issues related to managing plant genetic diversity and agricultural biotechnology for economic development have been identified *throughout* this volume.

With regards to biotechnology, it is important to continue to assess the economic impacts of adoption of various types of agricultural biotechnologies as they evolve. To best assess these impacts, we need quantitative understanding on how the features of various technologies, the economic and environmental conditions in various locations, the institutional setup in general, and the policies associated with the new technologies affect their impacts in terms of pricing and welfare of *various* groups.

This research will allow *identification* of the countries and situations where investments in agricultural biotechnology are likely to generate significant returns in terms of agricultural productivity increases and poverty alleviation, relative to other potential strategies. The potential *environmental* side effects of agricultural biotechnology are a *continuous* source of controversy that will affect the future of this technology.

Identifying and assessing the risks associated with biotechnology adoption relative to potential benefits is a major *priority* and, more importantly, designing institutions for *monitoring* the environmental impacts of agricultural biotechnology and effectively regulating to control potential risks is a major policy challenge. It has also emerged as an area where more research is urgently needed.

We also need to identify features of biotechnology products that are especially desirable from the perspective of the developing world and identify mechanisms that will help developing countries gain access to them, especially if they will not be pursued as part of the agenda of the private sector.

For example, it is important to understand to what extent can biotechnology enhance the micronutrient content of food consumed in developing countries and to what extent the biotechnology innovations that serve this purpose will be pursued by the private sector and, if they are not pursued privately, whether and how to provide the incentives for their introduction.

We need a better understanding of the role and effects of regulatory regimes, including environmental, IPR, and market structure regulations on the evolution and adoption of new biotechnology products and their impact on the environment.

As new *institutions* for the *management* and regulation of biotechnology are *introduced,* we need research that assesses their performance and suggests design modification and reform. Specifically, more work on policy and institutional reforms necessary to facilitate the potential benefits of biotechnology to the poor is necessary-particularly in reducing the transactions costs associated with access under *increasingly* restrictive property rights for genetic materials and associated technologies.

On the topic of genetic diversity conservation, first we need to have a better handle on the contribution and value of various forms of genetic resources and the costs associated with their loss. One could use emerging information technologies to collect data on use of various genetic collections and analyze it statistically.

It is crucial to understand how improved capabilities affect the usage and productivity of biodiversity in order to better their storage and distribution, an understanding which requires interdisciplinary research cooperation.

Determining how to optimize the value of both *in situ* and *ex situ* conservation to developing countries requires better information on what these values are, as well as the costs associated with obtaining them, considered in the dynamic context of economic development.

Some of this valuation work must be inferred indirectly from greater understanding of the improved economic value derived by bioresources. Valuation work on plant genetic diversity has focused at the farm level in looking at household decision-making over a portfolio of crops and varieties. More work is needed on the local and global public good values of diversity in terms of reducing vulnerabilities to pests and diseases.

In addition, further work on the value of maintaining diversity as an input to *agricultural* breeding programs is needed, following up and expanding on the work of Simpson and others. *Combining* research on valuation and costs could be a highly useful guide to developing countries on targeting strategies for conservation.

Together with an assessment of the most efficient conservation opportunities, there is a need for, analysis of the most effective and equitable mechanisms for providing incentives for conservation.

Markets, due to their increasing importance as a mechanism for the allocation of resources, need to be analyzed in terms of their role in providing incentives and disincentives for conservation. Here markets are taken in the widest sense, ranging from local commodity exchanges up to global markets for environmental goods.

The efficiency of markets in allocating plant genetic resources and the implications this has for diversity at the farm and local level are areas where more research is needed. The efficiency and optimal design of market-based mechanisms for maximizing global public good values associated with diversity conservation are other areas where gaps in the economic literature exist.

However, market forces are not the only drivers of interest in assessing conservation incentives: The impact of nonmarket forces, particularly government regulations in the agricultural and seed sectors, is also a critical area for further research.

Regulations of interest range from biosafety, to seed certification and release procedures, to agricultural pricing interventions. Finally, an important area for further research is the equity implications of alternative management schemes for plant genetic diversity conservation and agricultural biotechnology.

Designing mechanisms to compensate farm *communities* for their past services in *conserving* and providing genetic resources to the formal breeding sector, which do not create new barriers to exchanges and thus reduces access, is a challenging area where more work is needed.

Designing incentives for *in situ* conservation, which address not only current but also future opportunity costs associated with conservation in the presence of economic development, is another important equity issue where the analysis in the book indicates the need for more economic research. Finally, further analyses of the distribution of benefits and costs to agricultural biotechnology investment and adoption and the impact, particularly on low productivity agricultural populations relative to other means of productivity increases, is a highly *important* area of research both from an equity and *efficiency* standpoint.

MICROPROPAGATION

In vitro technologies are continuously expanding in the field of biology. Plant tissue culture has become a general title for a very broad subject. While in the beginning it was possible to culture plant cells either as established organs, such as roots or as disorganized masses, it is now possible to culture plant cell in a variety of ways, individually (as single cells in microculture systems), collectively (as calluses or suspensions, on petri-dishes, in Erlenmeyer flasks or in large-scale fermentors), or as organized units as shoots, roots, ovules, flowers, fruits etc..

In case of *Arabidopsis,* which has been the subject of the most intensive research effort into technology development, it is even possible to culture complete plants for generations from seed germination to seed set without having to revert to an *in vivo* phase. In its most general definition plant cell culture covers all aspects of the cultivation and maintenance of plant material *in vitro.*

The cultures produced are being put to an ever-increasing variety of uses. At the early stages, *in vitro* cultured systems were developed as experimental tools for basic research and studies on plant cell division, growth, differentiation, physiology and biochemistry. Such systems were seen as ways to reduce the degree of complexity associated with whole plants, providing additional exogenous control over endogenous processes, to enable more reliable conclusions to be made through simpler experimental designs.

However, in the recent past tissue culture technology has been increasingly used in highly applied contexts. Successes in a number of areas have been achieved. There has been major change in both the number of people making use of these techniques and also in an enhancement of the degree of sophistication associated with *in vitro* technology. Techniques of micropropagation and production of diseasefree plant stocks have been defined and refined to such an extent that they have become standard practice for a range of (usually vegetatively propagating) crop plants. Thus creating what is now a multi-million plant/ multi-million dollar industry.

Moreover, the discipline within this technology in which advances have been most rapid and will eventually have the greatest impact on both fundamental and applied plant sciences is that of genetic modification of plant cell. Micropropagation deals with the propagation of plants, *in vitro, has* many advantages over conventional vegetative propagation. Its application in horticulture, agriculture and forestry is currently expanding world-wide.

The goal of micropropagation is to mass-produce genetically identical, physiologically uniform, developmentally normal and pathogen-free plantlets, which can be acclimatized in a reduced time period and at a lower cost. Development of both automated environmental control systems and improved *in vitro* culture systems are essential for a significant reduction in production cost.

However, commercial use of micropropagation is still limited, because of its relatively high production cost resulting mainly from high labor costs, low growth rate *in vitro,* and poor survival rate of the plantlets during acclimatization. Altman and Loberant have elegantly reviewed principles and practices of microprop-agation. Micropropagation of woody/tree/forest plant is feasible.

However, with some exceptions traditional *in vitro* methods are not as yet practical or commercially viable for most forest trees. Therefore, improvement in current procedures and their scaling-up is required. Although it has been argued that from the environmental perspective, the genetic diversity of forest should be maintained/conserved, hence the traditional use of mixed population of seedling for forestation be applied and clonal forestry may not be appropriate.

The case of cuttage propagation and micropropagation for all types of woody perennials is strongly affected by ontogenetic age. Cloning *in vitro and in vivo* of adult and/or mature plants is adversely affected by characteristics accompanying maturation such as reduced growth rate, reduced or total lack of rooting ability or sometimes the unpleasant phenomenon of plagiotropy. Maturation, a complex phenomenon, is the major problem preventing a wider application of tissue culture technology among woody plant species.

Micropropagation of woody plants of stressed environments which experience types of (annually recurring) abiotic-stresses, become more difficult as the seasonal and environmental factor influence the behaviour of explant(s) in culture to a great extent. In simple terms, plant tissue culture can be considered to involve three phases.

First, isolation of the plant (tissue) from its usual environment. Second, the use of aseptic techniques to obtain clean material free of bacterial, fungal, viral and even algal contaminations. Third, the culture and maintenance *in vitro* in a strictly controlled physical and chemical environment.

The components of this environment are then in the hands of the researcher who gains a considerable degree of external control over the subsequent rate of the plant material concerned. Hall suggested an extra fourth phase where recovery of whole plants for rooting and transferring to soil is the ultimate goal.

The success of this technology is to a great extent dependent upon abiding by a number of fundamental rules and following a number of basic protocols. During last four decades a number of plant tissue culture technologies have been developed for a number of plant species in India. Govil and Gupta have reviewed commercialization of plant tissue culture in India.

It has been suggested that plant tissue culture would play a very important role in conservation, propagation and genetic improvement of plants of our country and also in restoration ecology and restoration of degraded habitats. Since 1980, we have been working on development of tissue culture protocols for application in propagation and genetic improvements of woody plants of arid and semi-arid regions (namely the Indian Thar Desert and the Aravallis).

Some of the woody plant species (as important biomass producer) are keystone species of these regions of the country. We developed tissue culture processes for cloning and mass propagation, using nodal shoot explants of mature and selected woody plants namely *Aegle marmelos, Capparis decidua, Celastrus paniculatus, Maytenus emarginata, Zizyphus* spp..

We also cloned shoots of *Prosopis cineraria* and *Tecomella undulata.* Micropropagated shoots were rooted by pulse treatment with root-inducing auxins. Several species of *Anogeissus* (Combretaceae) were first micropropagated in our laboratory using cotyledonary nodes. Later, Saxena and Dhawan of Tata Energy Research Institute (TERI), New Delhi reported the micropropagation of *Anogeissus latifolia* and *A. pendula,* also using juvenile explants.

Now we describe the development of micropropagation protocols for cloning of *Balanites aegyptiaca* (Hingota), *Citrus limon* (Nimbu) *and Syzygium cuminii* (Jamun). These woody species are economically and ecologically important as they yield valued products. *Balanites aegyptiaca* (Balanitaceae) is a tree of arid regions. This has multiple uses particularly for the aboriginals and rural people. The stem-bark is used as a fish-poison, and the pulp of fruit as detergent/soap for washing cloths/hair. Hard and durable timber is utilized for making agricultural appliances and household articles.

The powder of mature fruits is taken orally by the women to prevent unwanted pregnancy. The roots and fruits of *B. aegyptiaca* yield 'Diosgenin'—a sapogenin widely used for production of pharmaceutical steroid and oral contraceptives. *Citrus* (Rutaceae) is considered as number one fruit of the world for its nutritional values, the magnitude of fruit production and an array of commercial products which are derived from it.

Citrus limon is an important horticultural species. Similarly, the Black-plum *S. cuminii* (Myrtaceae) is a tropical fruit tree which has multiple uses. Also this tree has very high water use efficiency and thus is effective biomass producers. We developed cloning processes using nodal segments of rejuvenated (fresh shoot sprouts) shoots of selected mature trees.

MATERIALS AND METHODS

Source Plants

Selected mature tree(s) of *Balanites aegyptiaca, Citrus limon and Syzygium cuminii* were pruned during December-January. Shoot sprouts were harvested during the months of February-MarchApril. Fresh shoot sprouts collected during the months of March/April were used as explants. The nodal explants were dressed and treated with 0.1–0.2% Bavistin and 0.1% Tetracylin for 10–15 min.

These were surface sterilized with 0.1% $HgCl_2$ (5–6 min), then with 90% ethanol (60 sec) and were thoroughly washed with sterile water. These were finally treated with chilled sterile antioxidant solution (0.1% ascorbic acid; 0.05% citric acid and 0.1% PVP) for 15–20 min. The explants were inoculated on MS medium supplemented with different concentrations of BAP or kinetin.

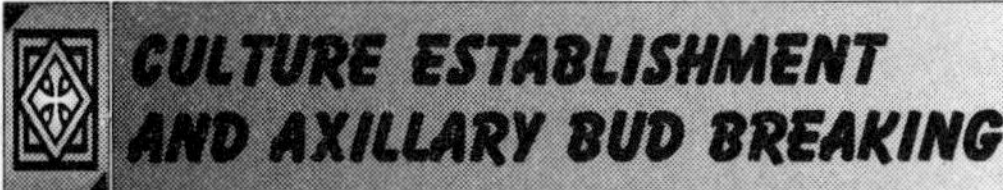

CULTURE ESTABLISHMENT AND AXILLARY BUD BREAKING

The nodal explants of the three species were inoculated in culture tubes on agar-gelled MS medium supplemented with different concentrations of benzylaminopurine (BAP) or kinetin. The cultures were incubated at 28 ± 2°C in a culture room with 10 h per day photoperiod. The responses of the explants were recorded regularly.

Amplification of Shoots in Culture

Shoot of *B. aegyptiaca* were multiplied by subculturing of nodal shoot segments of *in vitro* generated shoot on MS medium + 0.2 μ M BAP. Shoot amplification in *Citrus* limon was achieved when the mother explants were repeatedly transferred or nodal explants subcultured on amended MS (50% of NH_4NO_3, KNO_3) with 0.25 μ M BAP.

Multiplication of shoots of *Syzygium cuminii* was achieved by (a) repeated transfer of mother explants and (b) subculturing of *in vitro* produced shoots on above mentioned amended medium with K_2SO_4 (100 mgl^{-1}), KCl (70 mgl^{-1}) and ammonium citrate (50 mgl^{-1}). Subculturing was done after 20–25 days. The cultures were amplified in 250 ml flaks or bottles. These were kept under the controlled conditions of temperature (28 ± 2°C), light (40–50 μ mol m^{-2} s^{-1} SFP for 12 h/d photoperiod) and 60% RH.

Rooting of Cloned Shoots

Experiments were conducted to induce the roots *in vitro and ex vitro* from the micropropagated shoots. For *in vitro* rooting 4 to 5 cm long shoots were excised and cultured on agar-gelled full, half, one-third and one-fourth strengths of MS medium containing 0.1% of activated charcoal and different concentrations (1.23 to 16.1 μ M) of IBA or NAA or NOA.

These shoots were cultured at 30°C, under different regimes of light (8–10 h photoperiod per day). For *ex vitro* rooting, the individual shoots were pulse-treated with sub-lethal concentrations of root-inducing auxins and cultured on autoclaved soilrite in glass bottles (jam bottles). These bottles were kept in the green house at 30 ± 2°C.

Acclimatization of Micropropagated Plants

In vitro rooted plantlets were washed with sterile water to remove adhered nutrient agar and transferred to sterile soilrite in the culture bottles. In case of *ex vitro* rooted plantlets after roots were visible, the culture bottles were shifted from low temperature/high relative humidity (RH) regime of green house to the region which experienced relatively high temperatures and low RH.

Also the rooted plantlets were exposed gradually to external environment by loosening/ removing the caps of the culture bottles. Micropropagated and hardened plantlets were transferred to polybags containing mixture of organic manure, garden soil and sandy soil. These plants were watered regularly. The green-house-hardened plants were kept in nursery covered with agronet.

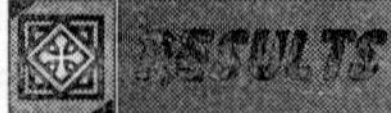

RESULTS

Selection of Explants

Nodal shoot segments harvested from pruned and non-pruned tree(s) were used as explants for establishment of cultures of three species. Explants prepared from fresh (rejuvenated) shoots regenerated from pruned plants during the months of March-April proved to be the most suitable for culture establishment.

The explants harvested from non-pruned tree(s) proved to be difficult to surface sterilize as these carried recalcitrant microbial contaminations; these rarely showed bud breaking, caused excessive browning of the culture medium and exhibited browning/darkening of cut ends/explants. Thus management and pruning of mother tree was found to be essential for harvesting shoots to be used as responsive explants.

Establishment of Shoot Cultures

The surface sterilized nodal explants could be cultured on MS media containing 0.45 μM BAP or higher concentrations of BAP or kinetin. The axillary meristems were activated and bud breaking was observed after 10-15 days of inoculation in 85-90% of the explants of three species.

Maximum number of shoots differentiated on MS medium supplemented with 0.45 μM BAR Shoot differentiated from each node were 2-3 in *Citrus Union,* 1-2 in *B. aegyptiaca* and 3-4 in *S. cuminii,* respectively. Kinetin proved to be less effective as compared to BAP in the activation of axillary buds. More than 0.45 μM of either of BAP or kinetin caused callusing from the explants and proved to be inhibitory.

Amplification of Shoots *in vitro*

After the activation of meristems, bud breaking and axillary bud differentiation, the shoots were further multiplied on suitable culture media. Shoots of *B. aegyptiaca* were multiplied by subculturing of segments of *in vitro* produced shoots.

Shoot amplification occurred on MS + 0.22 μM BAR Shoots of *C. Union* were multiplied by repeated transfer of mother explants on amended MS medium + 0.22 μM of BAR About 12-15 shoots differentiated from each mother explant.

Three-fold rate of shoot multiplication was achieved by repeated transfer of the mother explants. By repeated transfer of mother explants on amended MS medium, shoots of *S. cuminii* multiplied. Two- to three-fold rate of shoot multiplication was achieved. The cultures were transferred on to fresh media after 20-25 days.

The cultures were maintained at high light intensity (50-60 μ mol $m^{-2}s^{-1}$) at 28-30°C. The shoot cultures of all the three species are being multiplied and maintained for the last 3 years.

Rooting of Cloned Shoots

In Vitro Rooting

Isolated shoots of all the three species rooted on half-strength MS medium with 0.1% activated charcoal. Ninetyfive to 100% of the shoots of *C. Union* rooted on half-strength MS medium + 27.0

N M of NAA. From each shoot six to eight roots regenerated. Rooting was poor on media supplemented with IBA or NOA. Of the shoots of *B. aegyptiaca,* 80-90% rooted *in vitro* on halfstrength MS medium + 0.2 $_{N}$M IBA + 0.1 % activated charcoal. Ninety percent of the shoots of *S. cuminii* rooted on half-strength MS medium + 0.1% activated charcoal + 9.8 $_{N}$M of IBA.

Ex vitro Rooting

About 90-95% of the *in vitro* amplified shoots of *C. Union* rooted *ex vitro* if pulsed with 0.98-2.46 $_{N}$M IBA. The rooting percentage was 85-90% if the shoots were treated with equimolar NOA. The shoots treated with 1.07-2.68 μ M NAA showed maximum rooting. The *ex vitro* roots were visible after 10-12 days of pulse treatment.

The shoots of *S. cuminii* could also be rooted *ex vitro.* A pulse treatment with 2.46 μ M of IBA for 10–15 min was found to be sufficient to induce *ex vitro* roots from the shoots. Cent-per-cent of the shoots rooted on soilrite in the green house within 20–25 days.

If the shoots were pulsed with NAA, 65% of these rooted after 30–35 days. About 70% of the shoots of *B. aegyptiaca* rooted on soilrite after 12–15 days if treated with 1.0–2.5 mM IBA for 2–5 min.

Hardening of Micropropagated Treelets

In vitro rooted plants were hardened by transfer to soilrite containing bottles in the green house. These were kept near pad section for 8–10 days and gradually shifted towards fan section. After 10 days the caps of culture bottles were loosened and gradually removed.

Plantlets rooted *ex vitro* were acclimatized in the green house. After formation of roots the plants were exposed to low RH and high temperatures. *Ex vitro* rooted plantlets were found to be easy to harden and acclimatize than those rooted *in vitro.*

Hardened and acclimatized plants were transferred to black bags containing garden soil; sandy soil and organic manure. Several plants have been transferred to the field. These are growing normal. Flowering of these is yet to be recorded.

DISCUSSION

The research work presented in this article demonstrates that the mature woody plants can be cloned using appropriate *in vitro* methods. We have described the development protocols for cloning of *Balanites aegyptiaca, Citrus limon and Syzygium cuminii.* These are valuable woody species that yield products of economic value.

Selection of the individual plant with desired (superior) characters is possible only after certain age, when reproductive maturity is reached. Such selected and mature plants give high yield of quality product. Once the selection is done it is necessary to maintain genetic fidelity of the clone. This is done by vegetative propagation *in vivo* or *in vitro* (micropropagation).

Cloning of mature woody plants *in vitro* and *in vivo* is adversely affected by characteristics accompanying maturation such as reduced growth rate, reduced or total lack of rooting ability or sometimes the unpleasant phenomenon of plagiotropy.

Maturation, a complex phenomenon, is the major problem preventing a wide application of tissue culture technology among woody species. Nevertheless, a number of woody species/trees have been micropropagated. Success with several species have been achieved mainly by the use of special starting (explanting) material, by special pre-treatment to mother/source plant(s) *in vivo* or by *in vitro* culture. All of these tricks, which improve clonal propagation are often described by the general term rejuvenation.

It is clear that rejuvenation is a pre-requisite for possible cloning of adult trees and that the success in practice mainly depend on the ability to rejuvenate them. We found that in all the three species under investigation, pre-treatment (pruning during winter) of mother plant was desirable otherwise the explant did not respond in culture.

The shoot sprouts (flushes) from plants pruned during winter proved to be the only useful (suitable explants) for culture initiation. Rejuvenation (also known as phase reversal or return to the juvenile form) includes the complete reversal of maturation as a result of sexual reproduction or vegetative propagation via shoot formation (through activation of preexisting axillary- or apical-meristems) or somatic embryogenesis. Re-invigoration is defined as the reversal of ageing (which leads to reduced vigor and rooting ability).

Reinvigoration can be used when rooting-ability and vigor are increased as a result of, for example, pruning, hedging, repeated culturing, BAP-treatment and grafting. The nodal explants of *B. aegyptiaca, C. limon* and *S. cuminii* derived from fresh shoot sprouts, responded in culture and produced multiple shoots on BAP (0.45 μ M) supplemented medium.

The shoots could be further amplified *in vitro* by (i) repeated transfer of explants and (ii) subculturing, but on medium with comparatively lower concentration of BAP. In quite a number of plant species repeated subculture/transfer of adult shoots (mother explants) were reported to induce invigoration and complete rejuvenation, by which shoot multiplication and rooting ability are strongly improved.

It is suggested that once the cultures/explants were established these become conditioned and they required low cytokinin for further multiplication. In case of *Citrus* limon the cultures could be multiplied by lowering the concentrations of certain salts (NH_4NO_3 and KNO_3).

Chaturvedi et al. critically reviewed tissue culture employing vegetative explants in *Citrus* spp. It is stated that maximum tissue culture research has been done in *Citrus* during the last four decades however the results of practical value are meager. We have successfully established procedure for large-scale shoot multiplication of *Citrus* limon. This is important contribution in *Citrus* tissue culture with practical utility.

The micropropagated shoots were rooted *in vitro* on half-strength MS medium + 0.1% activated charcoal supplemented with IBA *(B. aegyptiaca* and *S. cuminii)* and NAA *(Citrus limon)*. Bonga and Von Aderkas suggested that roots from rejuvenated shoots of woody plants, is induced *in vitro* by IBA or NAA.

Probably the nature of auxin required and the concentration for *in vitro* root regeneration are species specific. In the present case the micropropagated shoots of all the three species could be rooted *ex vitro.* The main advantage of *ex vitro* over *in vitro* rooting is that root damage during transfer to soil is less likely.

Furthermore, rooting rates are often higher and root quality is better when the rooting takes place *ex vitro*. McClelland et al. studied the effect of *in vitro and ex vitro* root initiation on subsequent microcutting root quality in three woody plants. They suggested greater resistance of *ex vitro* rooted plants to stress. Arya et al. found that the *ex vitro* rooted plantlets of a woody climber, *Celastrus paniculatus* were easy to harden.

The duration of time and cost of plant production are also reduced by switching to *ex vitro* root generation. IBA proved to be more effective auxin for pulsing of the shoots for *ex vitro* root induction. The auxin, most commonly used for root formation is IBA. It is generally assumed that the greater ability of IBA as compared with other auxins to promote rooting is due to its relatively higher stability.

It has been possible to induce *ex vitro* induction in number of woody species of stressed environments. The rooted plantlets of all the species could be hardened in the green house and pot transferred with ease. The survival rates have been satisfactory. Development of protocol for micropropagation of *B. aegyptiaca* is important contribution as this could be applied for cloning of plants selected for higher yield of diosgenin.

Selected and tested plants of *Citrus* limon bearing desired attributes of horticultural importance can also be cloned using our protocol. Yadav et al., and Jain and Babbar reported *in vitro* micropropagation of *Syzygium cuminii.* They used explants from young seedlings.

This method of cloning is not preferred for fruit trees. Multiple shoot induction from 1- to 2-year-old seedlings of *S. travancoricum* was recorded by Anand et al.. Mathew and Hariharan reported *in vitro* shoot multiplication in *S. aromaticum*. Shah Valli Khan et al. reported *in vitro* micropropagation of mature *S. alternifolium*.

In this article we have described a process for cloning of mature tree of Black-plum (*S. cuminii*). This is the most desired level of cloning. The micropropagated plantlets of all the three woody species could be hardened and pot transferred. The processes defined are highly reproducible and efficient and these can be utilized for cloning of selected trees of these species.

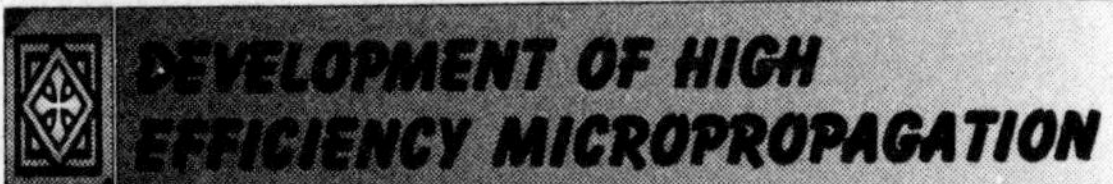

DEVELOPMENT OF HIGH EFFICIENCY MICROPROPAGATION

Protocol of an Adult Tree—Wrightia tomentosa

Wrightia tomentosa (Roxb.) Roem et Schult (*Apocynaceae*), once a common tree species of Aravallis in Rajasthan (India), has traditionally been exploited for its ivory-white wood in making toys and as fuel. High rate of seedling mortality, lack of suitable method for natural regeneration and overexploitation has reduced its population drastically and the plant has been listed as an endangered species.

There is, therefore, a strong need for an alternative method to produce large number of plants of superior types for conservation and regeneration. Micropropagation methods have been widely applied for clonal propagation of tree species for afforestation, woody biomass production and conservation of elite and rare germplasm.

These methods have been successfully integrated with modern forest tree management programs for rapid restoration of the degraded lands. A lab-scale protocol for micropropagation of *W. tomentosa* using adult material was reported by Purohit et al.. Only a limited number of plants could be produced by that method with major constraint being in hardening of *in vitro* developed plants.

The present article describes a highly efficient and reproducible micropropagation protocol that is being taken up for scaling-up production by this laboratory for large-scale plantation by the foresters.

MATERIALS AND METHODS

Trees of W. *tomentosa* selected and marked for quality of wood (age of tree more than 30 years) were used as a source of explants. Shoots were harvested from these plants round the year, divided into four distinct periods viz. April-June, July-September, October-December and JanuaryMarch. Management of donor tree was done by lopping one major fork and juvenile shoots produced near cut ends were collected for explantation.

Such newly flushed shoots were serially harvested fortnightly and successive flushes were termed as first (F_1), second (F_2), third (F_3), fourth (F_4) and fifth (F_5). Nodal shoot segments (1.5, 3.0 and 4.5 cm long and 0.2–0.5 cm in diameter) were prepared as explants. Two different orientations of explant on medium was tested.

Horizontal placement was done in two ways: (i) explant lying horizontally on medium (H_1) and (ii) one side of node, having axillary bud, inserted in medium while other side exposed to air (H_2). Vertical placement of explant was done in three different ways: (a) node completely immersed in medium (V_1), (b) node on the surface of medium (V_2) and (c) node one centimeter above the medium (V_3).

Explants were washed thoroughly with sterilized distilled water containing few drops of Tween-20 and then surface-sterilized with 0.1% (w/v) mercuric chloride for 5 min followed by thorough washing with sterile distilled water. Surface sterilized explants were inoculated on standard multiplication medium containing MS salts with 2 mg l–1 BAP. Explants were also inoculated on MS medium containing different concentrations of Kn (0.5–5.0 mg l^{-1}), TDZ (50–10,000 nM) and GA_3 (1.0–2.0 mg l^{-1}). Proliferated shoots from nodes of F_5 flush were further subcultured on MS medium with various concentrations of TDZ (0.1, 1.0 and 10.0 μM) or BAP (2 mgl^{-1}).

Phloroglucinol (50, 100 and 250 mg l^{-1}) was also added to standard shoot multiplication medium. Cultures in conical flasks (100, 150 ml) covered with non-absorbent cotton plugs were kept under controlled conditions of temperature (28 ± 2°C), light (45 μmol $_m$–2 s–1 for 16 h/day provided by fluorescent tubes) and 60–70% relative humidity. Once culture conditions for optimum shoot induction from explants were established, the shoots produced *in vitro* were subcultured on fresh medium every 3 weeks.

Shoots having passed through three, six and nine passages in multiplication medium were used for rooting. Shoots (2.0–3.0 cm) were excised and their cut ends were dipped in different concentrations of IBA solution (50, 100, 200 and 500 mg l^{-1}) for different duration (5–15 min) followed by their implantation on standard rooting medium containing quarter strength MS salts, sucrose (1%) and agar (0.6%).

Activated charcoal (50, 100, 200 and 400 mg l^{-1}) was also tested in standard rooting medium.

Initially the culture vessels were kept wrapped with black paper or in darkness for 5–7 days at 30 ± 2°C temperature at 60–70% relative humidity. Rooted shoots from 3-week-old cultures were hardened prior to *ex vitro* exposure. Hardening was attempted by three different methods. In first method, individual plantlets were planted carefully in Soilrite™ (Karnataka Explosives, Bangalore, India) filled netted pots (3 cm high) and placed in horizontally kept pickle bottles (30 cm long) which accommodated 25 such pots (W1).

In second method, individual plantlets, transferred to netted pots were placed in glass troughs (30 cm diameter) covered with polythene sheets (W_2). Thirdly, autoclaved 400 ml screw cap glass bottles one-fourth filled with soilrite™ irrigated with 40 ml inorganic salt solution (major salts of MS medium reduced to 1/4 strength, pH 5.0) were used (W_3).

Each bottle containing 4 plantlets were kept in culture room for 30 days. After 30 days, plantlets hardened by methods described as W1 and W_2 were shifted individually to plastic pots (10 cm high) and covered with polythene bags. Gradually, humidity was lowered by perforating polythene cover, then opening it for 1 h/day and finally completely removing it. Plantlets hardened *in vitro* (W_3) were kept in closed bottles till they touched the caps of bottles (nearly after 30 days).

The caps were loosened and finally opened in misthouse (with 70–85% RH). After one month, plants were transferred to pots and kept under greenhouse conditions where a gradient of humidity (80–40%) was maintained by a Fan-Pad evaporative cooling system.

Statistical Analyses

Standard analysis procedures were followed for CRD analysis. Abnormality, non-additivity and heterogeneity of variance in raw data of different experiments were minimized using square root $(\sqrt{x}$ and $\sqrt{x}+0.5)$ transformation. In ANOVA, test for significance (F test), standard error of mean and critical difference at 5 and 1 per cent probability was calculated on transformed data which were tabulated along with retransformed values in each experiment. In case of discrimination amongst two treatments '*t*' test was used.

RESULTS

Initiation of Shoot Cultures

Bud break frequency was strongly influenced by the nature and management of donor tree, season of explant collection and their orientation on the medium. April-June was found to be the best period for collection of explant to obtain maximum (98%) bud break response with minimum (5.0%) loss due to contamination.

The explants collected during the months of July-September developed fungal growth associated with shoot bud proliferation. Least bud break response was observed in the winter months of October-December. Explants prepared from oneyear-old mature branches responded poorly in cultures as compared to explants taken from freshly flushed branches.

Orientation of explant on medium was a significant factor in shoot proliferation from nodal segments. Between horizontal and vertical orientation of explants on medium, the latter was found significantly superior. Maximum number of shoots were produced in explants oriented in

Table 5.1: Effect of season of harvest on shoot initiation from mature node explants of W. tomentosa on standard multiplication medium

Period	*Per cent* contamination*	*Per cent* response*	*Callus* intensity
April–June	5 ± 1.08	98 ± 13.85	++
July–September	90 ± 8.78	78 ± 12.68	++++
October–December	81 ± 7.02	20 ± 4.12	+++++
January–March	72 ± 8.60	45 ± 8.02	+++

*Mean ± SE.

Explants of different size showed varied response in terms of bud break and amount of associated callus during shoot initiation. Maximum bud break was found in explant measuring

Table 5.2: Effect of source of explant on shoot initiation in W. tomentosa on standard multiplication (SM) medium.

Source of explant	*Per cent explant sprouted*	*Mean number of shoots per node*	*Mean shoot length (cm)*
Mature branches	52	2.17	0.85
Juvenile branches	98	3.83*	2.92*

*Significant at 1% level using *t* test.

3.0 and 4.5 cm in length. However, the size of explant did not make significant difference in shoot bud proliferation both in terms of their number and length. In very small explants (1.5 cm) the basal callus developed upto node, posing difficulty in further subculturing.

Table 5.3: Effect of length of explant on shoot initiation in W. tomentosa on standard multiplication medium.

Size of explant	*Per cent explant sprouted*	*Mean number of shoots per node*	*Mean shoot length (cm)*
1.5 cm	82	4.32 (2.08)	1.92
3.0 cm	98	4.64 (2.15)	2.33
4.5 cm	97	4.47 (2.11)	2.00
SEm±	0.07	0.18	
$CD_{.05}$		NS	NS'

*Figures in parentheses are $_J$ transformed values.

Table 5.4: Effect of node orientation on shoot initiation in W. tomentosa on standard multiplication medium.

Orientation of nodal explant	*Per cent explant sprouted*	*Mean number of shoots per node**	*Mean shoot length (cm)*
Horizontal placement			
Both sides of node on the medium (H_1)	98	2.45 (1.57) c	1.42 c
One side inserted into the medium (H_2)	94	3.29 (1.81) b	2.50 b
Mean		2.85 (1.69)	1.96
Vertical Placement			
Node immersed in the medium (V_1)	10	1.30 (1.14) d	1.00 d
Node on the surface of the medium (V_2)	95	3.98 (1.99) b	3.83 b
Node nearly 1 cm above the medium (V_3)	98	5.31 (2.30) a	3.33 a
Mean		3.28 (1.81)*	2.38*
SEm±		0.09	0.18
$CD_{.05}$		0.25	0.52
$CD_{.01}$		0.35	0.71

Table 5.5: Effect of serial harvesting on *in vitro* response by MN of W. tomentosa on standard multiplication medium

Flush number	*Per cent contamination*	*Per cent explant sprouted*	*Mean number of shoots per node**	*Mean shoot length (cm)*
I flush (F_1)	10	88	3.98 (1.99) d	2.54
II flush (F_2)	10	90	5.14 (2.27) c	2.18
III flush (F_3)	8	91	5.49 (2.34) bc	2.43
IV flush (F_4)	5	95	6.51 (2.48) b	2.17
V flush (F_5)	1	98	7.33 (2.71) a	2.31
SEm±	0.06	0.17		
$CD_{.05}$	0.17	NS		
$CD_{.01}$	0.23	NS		

vertical position V_3 followed by V_2 and horizontal position H_2, both statistically at par in terms of number and length of shoots.

When positioned vertically, callusing was associated with lower internodal region of explant only while it extended to whole surface in horizontally placed explants. Nodal explants placed 1.0 cm above the medium provided callus-free shoot proliferation. Management of donor tree from which the explants were collected was a very important step in accelerating the number of shoots per node during initial phases of cultures establishment.

Explants obtained from serially lopped branches producing different flushes of juvenile shoots exhibited graded increase in shoot bud proliferation. An increase in per cent bud break response and number of shoots per explant from first flush (F_1) to fifth flush (F_5) was noted with a concomitant decrease in per cent contamination.

Explants from F_5 flush exhibited initiation of ca 7.33 axillary shoots per node as compared to 3.98 shoots induced in explants from F_1 flush. However, effect of flushes on length of proliferated

Table 5.6: Effect of different PGRs on shoot initiation in W. tomentosa.

MS + PGR		*F_1 flush*		*F_5 flush*	
		Per cent explant sprouted	*Mean number of shoots/node**	*Per cent explant sprouted*	*Mean number of shoots/node*
BAP	2 mg l^{-1}	84	3.74 (1.93)a	97	6.52 (2.55)a
Kn	0.5 mg l^{-1}	48	2.91 (1.71)ab	89	3.73 (1.93)b
	2.0 mg l^{-1}	58	2.69 (1.64)ab	90	3.19 (1.79)bc
	2.5 mg l^{-1}	52	3.19 (1.79)ab	88	3.11 (1.79)bcd
	5.0 mg l^{-1}	65	2.44 (1.56)abc	95	2.91 (1.71)bcd
TDZ	50 nM	61	2.44 (1.56)abc	98	2.91 (1.71)bc
	250 nM	60	3.44 (1.85)a	92	3.48 (1.87)bc
	500 nM	52	2.69 (1.64)ab	98	2.91 (1.71)bcd
	1,000 nM	61	2.91 (1.76)ab	97	3.48 (1.87)bc
	10,000 nM	53	2.69 (1.64)ab	92	3.19 (1.79)bc
GA3	mg l^{-1}	35	1.72 (1.31)bc	95	2.23 (1.49)cd
	2 mg l^{-1}	24	1.45 (1.21)c	92	1.93 (1.39)d
SEm±		0.14		0.13	
CD.05		0.40		0.38	
$CD_{.01}$		NS		0.52	

Table 5.7: Effect of different concentrations of TDZ on shoot multiplication in W. tomentosa cultures.		
MS + TDZ (μM)	*Multiplication fold**	*Callus intensity*
Control (SM medium)	2.42b	+
0.1	1.92c	+
1.0	2.67ab	+++
10.0	2.92a	+++
SEm±	0.14	
$CD_{.05}$	0.42	
$CD_{.01}$	0.58	

**Means followed by different alphabets in the same column differ significantly.*

Table 5.8: Effect of different concentrations of phloroglucinol on shoot multiplication in W. tomentosa.		
MS + BAP (2 mgl⁻¹) +PG (mgl⁻¹)	*Multiplication fold**	*Callus intensity*
Control	2.34 b	+++
50	1.08 c	+++
100	3.33 a	+++
250	1.92 b	++++
SEm±	0.20	
$CD_{.05}$	0.61	
$CD_{.01}$	0.83	

**Means followed by different letters differ significantly.*

shoots was non-significant. Explants from F_5 flushes responded differently as compared to that of F_1 flush. About 90 per cent bud break response was observed when explants from F_5 flush were inoculated on the MS medium containing any of the four growth regulators tested.

In explants from F_1 flush, mean number of axillary shoots initiated were statistically insignificant on tested concentrations of any of the four growth regulators except GA_3 with least shoot formation. However, influence of PGRs on shoot initiation was marked in explants from F_5 flush, maximum being on BAP (2 mg l^{-1}).

None of the other three growth regulators (Kn, TDZ and GA_3) at any of the concentrations showed better response in terms of number of axillary shoots induced per node. Numerically, 2 mg l–1 BAP produced maximum number of shoots, followed by 250 nM TDZ.

Shoot Multiplication

Shoots after their initial proliferation from F_5 explants on medium containing 2.0 mgl^{-1} BAP along with the mother explant were further subcultured onto standard multiplication medium after every 3 weeks. Substitution of BAP with TDZ in subcultures increased shoot multiplication rate, highest being on medium containing 10 μ M TDZ (2.92 fold) which was significantly superior to the rate obtained on standard multiplication medium containing 2.0 mg l^{-1} BAP. Incorporation of phloroglucinol (PG) in standard multiplication medium increased the rate of shoot multiplication above 3-fold. PG also induced healthy cultures with dark green and lustrous leaves.

Rooting in Shoots

Those shoots having passed through six multiplication cycles, responded to rooting treatments. With the increasing number of shoot multiplication cycles, the rooting response was more favourable, showing early root initiation and reduced callusing.

Table 5.9: Effect of number of subcultures on rooting in IBA pulse treated shoots of W. tomentosa on standard rooting medium.

Shoots harvested after subculture	*Per cent rooting response*	*Mean number of days to rooting*	*Callus intensity*
III subculture	00.0	00.0	+++
VI subculture	15.4	35.6	+++
IX subculture	40.3	18.1	+

**IBA pulse treatment (100 mgl^{-1} for 10 min).*

Concentration of IBA and duration of treatment affected the root induction process considerably. Among various IBA concentrations tested for pulse treatment, rooting response was maximum in shoots treated with 100 mg l^{-1} IBA solution for 10 min. Such shoots exhibited ca 2.92 roots with 2 cm mean root length. Higher or lower concentrations of IBA did not improve rooting response. Rooting percentage was found to be positively related with concen-tration of activated charcoal (AC) added to rooting medium.

Maximum rooting response (69.7%) was observed when the IBA-treated shoots were placed in medium containing 400 mgl^{-1} AC. Addition of AC helped in early root initiation, increased root number and reduced callusing at the root-shoot junction.

Hardening and Acclimatization

Plantlets raised *in vitro* initially posed problems in hardening and acclimatization. Rooted plants when transferred directly to pots without prior hardening started wilting, no sooner they were transferred, and desiccated completely within 24 h. Seedlings were also used in experimentation to understand their requirements for hardening. Even the seedlings were prone to transplantation shock similar to *in vitro* developed plantlets.

Owing to fast desiccation of rooted plants on direct pot transfer, other methods of hardening were employed. Rooted plants with nearly 4 cm long shoot, 1 to 2 cm long root and 4 to 6 leaves in number were hardened by three different methods as described in materials and methods.

Apical growth was visible in more than 95 per cent of plantlets reared through any of the three methods. After 30 days, plantlets attained an average height of 5.4 cm with 2.1 cm long root

Table 5.10: Effect of IBA pulse treatment on rooting in shoots of W. tomentosa on standard rooting medium

Strength of IBA solution (mgl⁻¹)	*Duration (min)*	*Per cent rooting response*	*Mean number of roots**	*Mean root length (cm)*	*Callus intensity*
50	5	11	1.30 (1.14) d	0.78 d	–
	10	15	1.49 (1.22) cd	1.25 bc	+
	15	18	1.84 (1.36) bcd	1.50 b	+
100	5	40	2.00 (1.41) bc	1.50 b	++
	10	59	2.92 (1.71) a	2.00 a	++
	15	48	1.90 (1.38) bc	1.85 a	+++
200	5	18	2.37 (1.54) ab	1.85 a	+++
	10	21	1.96 (1.40) bc	1.00 cd	+++
	15	19	1.79 (1.34) bcd	1.00 cd	++++
500	5	0	–	–	
	10	0	–	–	
	15	0	–	–	
$SEm_{\pm}$	0.08	0.11			
$CD_{.05}$	0.23	0.31			
$CD_{.01}$	0.31	0.42			

**Means followed by different alphabet in same column differ significantly.*

(root shoot ratio being 0.37) and 6-8 broad leaves. Such plantlets were transferred to individual plastic pots and covered with polythene bags in case of W1 and W_2 while the caps of glass bottles were loosened in W_3 plantlets. Plantlets exhibited wide variation in growth during this period.

Most of the plantlets in all the three methods exhibited good shoot growth while root growth was better only in case of plantlets hardened through W_3 method. Generally, the plants were ca 9.08 cm long with 3.67 cm long roots and 8 to 10 leaves.

These plants on an average accumulated 20.22 mg dry matter in shoots (without leaves), 7.95 mg in roots and 13.54 mg in each leaf. During gradual exposure to *ex vitro* conditions in greenhouse, all plants remained green and healthy for initial 15 days. Plants hardened through W1 and W_2 modes exhibited yellowing of leaves and leaf fall in next 15 days. The rate of survival was 26.8% after 30 days of transfer to pots which declined to 2.4% after 60 days.

Plantlets hardened through W_3 mode grew vigorously having rigid and thick stem, highly branched root system and green, lustrous, healthy and broad leaves. Survival rate of such plants was more than 95 per cent. More than 5,000 plantlets have been successfully hardened and acclimatized and are ready for transplantation into field.

Table 5.11: Effect of activated charcoal (AC) on rooting in pulse treated shoots of W. tomentosa on standard rooting medium.

Medium + AC (mgl⁻¹)	*Per cent rooting response*	*Mean number of roots**	*Mean root length (cm)*	*Mean shoot length (cm)*	*Callusing intensity*
Control	59	2.89 (1.70) b	1.87 b	3.74	+
AC 50	45	2.93 (1.71) b	1.92 b	3.67	–
100	50	2.76 (1.66) b	2.11 ab	3.56	–
200	65	3.06 (1.75) b	2.41 a	3.89	–
400	69.7	3.78 (1.94) a	2.06 b	3.60	–
$SEm_{\pm}$		0.07	0.11	0.18	
$CD_{.05}$		0.21	0.34	NS	
$CD_{.01}$		0.30	0.46	NS	

**Values in parentheses are $\sqrt{}$ transformed values.*

Means followed by different letters in the same column differ significantly. IBA pulse treatment (100 mgl⁻¹ for 10 min).

DISCUSSION

An adult superior tree can be micropropagated for desired attributes by enhanced axillary proliferation. The buds residing in the axil of twigs are induced to proliferate and generate multiple shoot buds *in vitro*. Proliferation of these axillary buds may be difficult due to microorganism contamination, phenolic oxidation and tissue maturity.

Maturity of tissue is accompanied with reduced growth rate, reduced/lack of rooting ability and sometimes plagiotrophy. By reverting a part of tree to complete/partial juvenility by *in vivo and in vitro* methods problems associated with maturity can be minimized. In *W. tomentosa* the explants collected from previously lopped trees showed better proliferation when shoots were harvested serially.

Severe pruning has been found to be an efficient method for rejuvenation. In *Quercus robur* forced flushing method, related to severe pruning, was adopted. The season of explant collection greatly influenced establishment of *W. tomentosa* cultures *in vitro*. Effect of season on bud sprouting was also noted in many tree species viz. *Tectona grandis*, guava, *Tecomella undulata, Prosopis cineraria and W. tinctoria.*

Vertical orientation of *W. tomentosa* explants was found better than horizontal orientation in terms of number of proliferated shoots. On the contrary, horizontal orientation of explants was found better in *Fraxinus angustifolia* and *Quercus robur.* Cytokinins promote cell division in plant tissues under specific conditions and are found obligatory for shoot differentiation.

Hu and Wang reported superiority of BAP among cytokinins in differentiation of shoots from explants of trees. Buising et al. suggested that transient exposure of soybean embryonic axes to

BAP interrupted chromosomal DNA replication and reprogrammed the developmental fate of a large number of cells in shoot apex.

In our case, maximum number of shoots were obtained in medium containing BAP in comparison to other PGRs. Recently, TDZ has been found to be one of the most active cytokinin-like substances used for woody plant tissue culture. In present study, TDZ did not supercede the response obtained with BAP in shoot induction, but it did enhance multiplication of shoots in subcultures.

Incorporation of phloroglucinol in the multiplication medium improved shoot multiplication rate. Similar results have been reported in apple root stock M.7 by Jones. The effect could be related to hastening of rejuvenation process *in vitro* by phloroglucinol. Rooting by dip treatment of auxin has been recommended by Harry and Thorpe. It is supposed to eliminate the inhibitory effect on root growth when IBA is incorporated in the media.

Purohit et al. have recommended IBA pulse treatment for rooting in *W. tomentosa* shoots. This method of root induction has been successfully employed in the present studies also. Hardening is most critical factor for achieving success in pot transfer of regenerated plantlets. We have observed that a gradual shifting of plants from medium to culture bottles containing low salt concentration without sucrose allowed stress, compelling plants to become partially autotrophic.

This step proved useful in achieving more success in hardening. The results have demonstrated the feasibility of application of this protocol for raising large number of *W. tomentosa* which would greatly help in afforestation programmes in Aravallis in Rajasthan (India). Large numbers of plants are ready for field transfer that can be used for field evaluation studies.

Chapter 6

CONSERVATION OF MEDICINAL PLANTS

Traditional medicinal systems are part of a time-honoured and time-tested culture, that still intrigues people today. A culture that has successfully used plants to treat primary and complex ailments for over 3,000 years obviously has a contemporary relevance. In an age when toxic drugs are increasingly unwelcome and when people are using viable alternatives, this heritage of medicinal plants must be documented and conserved for effective use in future. During the past decade, a dramatic increase in exports of medicinal plants attests to worldwide interest in these products.

Nevertheless, most of these plants being taken from the wild, hundreds of species are now threatened with extinction because of overharvesting, destructive collection techniques, and conversion of habitats to crop-based agriculture. Preservation of these genetic resources is currently at the forefront of conservation activities and biotechnology has played an important role in international conservation programs.

Traditionally, plant genetic resource management involves conserving germplasm as seeds at low temperature, or as field plantings (field genebanks) for vegetatively propagated plant species. These approaches are now complemented by *in vitro* conservation methods that can be used in combination with traditional practices and offer added security for field genebank conservation.

The ideal genetic resource conservation program consists of active collections that are available for distribution or characterization and base collections held for the sole purpose of long-term preservation. Base collections of vegetatively propagated plants are more difficult to achieve and recently, cryopreservation has been identified as the best option for long-term conservation of germplasm of these species.

Table 6.1: Summary of different techniques used for cryopreservation.

Technique	Explants	Protocol
1. Vitrification	Shoot tips/embryogenic tissues/cell cultures	Explant is treated with LS (1 M glycerol) for 20 min at 25°C followed by dehydration with PVS_2 (30% glycerol, 15% EG, 15% DMSO) at 0°C for 90 min, rapid freezing in LN, rapid thawing at 40°C for 1–2 min, UL (1.2 M sucrose) and culture for recovery growth
2. Encapsulation dehydration	Shoot tips/embryogenic tissues	Explant is encapsulated in calcium alginate and precultured in high sucrose solution (0.5–0.75 M), followed by dehydration in laminar airflow for 4–5h, rapid freezing in liquid nitrogen, rapid thawing at 40°C for 1–2 min and culture for recovery growth
3. Encapsulation-vitrification	Shoot tips	Excised encapsulated meristems containing 2 M glycerol +0.4 M sucrose were dehydrated with PVS_2 for 2h at 0°C and subsequently plunged in LN
4. Pregrowth	Zygotic and somatic embryos	Pre-growth technique consists of cultivating samples in the presence of cryoprotectants, then freezing them rapidly by direct immersion in liquid nitrogen
5. Pregrowth desiccation	Stem segments	Pre-growth desiccation refers to the pre culture of the explant on a medium with high concentration of sucrose or ABA or Proline and desiccation/drying followed by freezing in liquid nitrogen
6. Desiccation	Large number of recalcitant and intermediate seeds	Desiccation is usually performed in the air current of a laminar flow cabinet, but more precise and reproducible dehydration conditions are achieved by using a flow of sterile compressed air or silica gel
7. Droplet freezing	Shoot tips	Apices are pretreated with liquid cryoprotectant in medium then placed on aluminum foil in minute droplets of cryoprotectant and frozen directly by rapid immersion in LN

Cryopreservation, i.e., non-lethal storage of plant tissues at ultra-low temperature usually that of liquid nitrogen (–196°C) is the only available method for the long-term conservation of germplasm of these problem species. Cryopreservation has manifold applications in conservation and biotechnology. A number of medicinal plant species have been subjected to cryopreservation.

The major advantage of storage of biological material at such a low temperatue is that both metabolic processes and biological deterioration are considerably slowed or even halted. Additionally, continued maintenance of plants in tissue culture can lead to loss of morphogenic, genetic and biosynthetic capacity, which may confound successful exploitation.

It scores advantages over other conservation strategies as it minimizes the risk of contamination, cost of maintenance and cost of labor. Through cryopreservation, it may be possible to establish a reserve of freshly initiated competent cultures which after thawing and recovery can be reintroduced into culture.

PROGRAMMES FOR CRYOPRESERVATION OF MEDICINAL PLANTS

Interests and concern of the international scientific community in this area has lead to formulations of several national and international level programmes, which are devoted to cryopreservation of medicinal plants. For example, G-15 Genebanks for medicinal and aromatic plants were initiated from the Summit Level Meeting of Group on South-South Consulting and Cooperation of the G15 countries held in Kuala Lumpur. Malaysia together with Indonesia and India represent the Asian region where India is a Regional Coordinator. India has also been given the overall responsibility for coordinating the activities of the G-15 nations for the establishment of gene banks for medicinal and aromatic plants.

Under the aegis of this programme, Department of Biotechnology, Government of India, has constituted a network of three national gene banks at Tropical Botanical Garden and Research Institute (TBGRI), Thiruvananthapuram; Central Institute of Medicinal and Aromatic Plants (CIMAP), Lucknow; and National Bureau of Plant Genetic Resources (NBPGR), New Delhi. One of the important mandates of this group is to develop cryopreservation protocols for long-term conservation of medicinal plants.

CRYOPRESERVATION TECHNIQUES

Some plant organs such as orthodox seeds and frost-hardy dormant buds contain very low amounts of water and can thus be cryopreserved directly, without any pretreatment. However, most of the experimental systems employed in cryopreservation (cell suspensions, calli, shoot tips, embryos) contain high amounts of cellular water and are thus extremely sensitive to freezing injury since most of them are not inherently freezing-tolerant. Cells have thus to be dehydrated artificially to protect them from the damages caused by the crystallization of intracellular water into ice.

The techniques employed and the physical mechanisms upon which they are based are different in classical and new cryopreservation techniques. Classical techniques involve freeze-induced dehydration, whereas new techniques are based on vitrification, i.e. the transition of water directly from the liquid phase into an amorphous phase or glass, whilst avoiding the formation of crystalline ice. Classical cryopreservation techniques involve slow cooling down to a defined prefreezing temperature followed by rapid immersion in liquid nitrogen.

They are generally operationally complex since they require the use of sophisticated and expensive programmable freezers. In some cases, their use can be avoided by performing the freezing step with a domestic or laboratory freezer. In the new vitrification-based procedures, cell dehydration is performed prior to freezing by exposure of samples to concentrated cryoprotective media and/or air desiccation.

This is followed by rapid cooling. As a result, all factors, which affect intracellular ice formation, are avoided. Glass transitions (changes in the structural conformation of the glass) during cooling and rewarming have been recorded with various materials using thermal analysis. Vitrification-based procedures offer practical advantages in comparison to classical freezing techniques.

Like ultrarapid freezing (above), they are more appropriate for complex organs (shoot tips, embryos) which contain a variety of cell types, each with unique requirements under conditions of freeze-induced dehydration. By precluding ice formation in the system, vitrification-based procedures are operationally less complex than classical ones (e.g., they do not require the use of controlled freezers) and have greater potential for broad applicability, requiring only minor modifications for different cell types.

A common feature to all these new protocols is that the critical step to achieve survival is the dehydration step, and not the freezing step, as in classical protocols. Seven different vitrification-based procedures can be identified:

a. encapsulation-dehydration;

b. a procedure actually termed vitrification;

c. encapsulation-vitrification;

d. desiccation;

e. pregrowth;

f. pregrowth-desiccation; and

g. droplet freezing.

SOME IMPORTANT CASE STUDIES

Cryopreservation of Shoot Tips of *Dioscorea spp.*

In vitro grown shoot tips of two medicinally important species of *Dioscorea, D. floribunda and D. deltoidea,* were successfully cryopreserved using vitrification and encapsulation-dehydration techniques. For vitrification the excised shoot tips were precultured for 16h on MS medium containing 0.3M sucrose followed by loading for 20 min at 25°C (2 M glycerol + 0.4 M sucrose), dehydration with plant vitrification solution (PVS_2) (30% glycerol, 15% Ethylene Glycol (EG), 15% DMSO and 0.4 M sucrose) for 90 minutes at 0°C prior to plunging in liquid nitrogen (LN).

After storage in LN for atleast 1 h the shoot tips were unloaded for 20 min at 25°C (1.2 M sucrose) and transferred to medium for recovery growth. During recovery growth, apices of *D.*

Table 6.2: Advantages and disadvantages of commonly used cryopreservation techniques.

Technique	*Advantages*	*Disadvantages*
Slow freezing	Stability from relatively nontoxic	Requires expensive equipment, slow recovery, low
cryoprotectants	applicability to tropical species	
Vitrification	No special equipment needed, fast procedure, fast recovery	Vitrification solutions are toxic to many plants, cracking is possible, requires careful timing of solution changes
Encapsulation-	No special equipment needed, non	Requires handling each bead several times, some
dehydration	toxic cryoprotectants, simple thawing procedures	plants do not tolerate high sucrose concentrations
Dormant bud	Easy, useful for many temperate	Requires freezing equipment, larger storage space,
desiccation	tree species	recovery requires grafting or budding, works best in cold temperate regions

floribunda and *D. deltoidea* regenerated directly with a frequency of 30 and 75%, respectively.

For encapsulation-dehydration the shoot tips pregrown in 0.3 M sucrose were encapsulated in calcium alginate followed by preculture in 0.75 M sucrose, dehydration for 51/2 and 5 h respectively, rapid freezing and rapid thawing.

The encapsulated shoot tips were recovered with high frequency direct regeneration in both the species. Interestingly, the diosgenin content in the plants recovered after cryopreservation was found to be stable using HPLC analysis.

Molecular studies using RAPD analysis proved that the plants were genetically stable.

Cryopreservation of Somatic Embryos of *Dioscorea bulbifera*

Somatic embryos/embryogenic tissues of *Dioscorea bulbifera* were cryopreserved using encapsulation-dehydration technique. The embryogenic tissues of about 1–2 mm in diameter, with a group of embryoids were encapsulated into calcium alginate beads.

These were then precultured in 0.5 M sucrose for 7 d followed by dehydration under *laminar* airflow for 4 h. High frequency (75%) of embryogenic survival was recorded after storage in LN.

The plants hence produced and transferred to field have been found to be morphologically similar to the non-treated controls.

The *diosgenin* content in the plants recovered after cryopreservation was analyzed using HPLC and the content was found to be stable. Molecular studies using RAPD analysis proved that the plants were genetically stable.

Table 6.3: Summary of cryopreservation studies on some important medicinal plant species.

Plant	Explant
Atropa belladonna	Protoplasts, cells
Anisodus acuntangulus	Suspension cultures, cells
Catharanthus roseus	Cells
Coleus blumei	Cells
Chicory	Shoot tips
Cinchona ledgeriana	Protoplasts
Datura innoxia	Protoplasts
D. stramonium	Cell suspension
Dioscorea caucasia	Organogenic callus
D. balanica	Organogenic callus
D. bulbifera	Somatic embryos
D. floribunda	Shoot tips
D. deltoidea	Cell cultures
	Shoot tips
Digitalis lanata	Cell cultures
D. thapsi	Cell cultures
Eucalyptus	Leaf
Gentiana scabra	Axillary buds
Holostemma annulare	Shoot tips
Nicotiana tabacum	Suspension cultures
N. sylvestris	Suspension cultures
N. plumbaginifolia	Suspension Cultures
Olea europe	Shoot tips
Panax ginseng	Hairy roots
	Cells
P. quinquefolium	Cell cultures
Papaver somniferum	Transformed Cells
Polygonum avuculare	Suspension cells
Trifolium repens	Shoot tips

Cryopreservation of Hairy Roots of *Panax ginseng*

The protocol for cryopreservation of hairy roots of *Panax ginseng* was developed by Yoshimatsu et al.. Hairy root *segments* including root tips were placed on to phytohormone-free halfstrength Murashige and Skoog solid medium and stored at 4°C in the dark for 4 months. The root segments resumed elongation when the temperature was raised to 25°C in the dark. For *cryopreservation*, the root tips were *precultured* with 0.1 mg l–1 2,4-D for 3 d and dehydrated with PVS_2 for 8 min before immersion in liquid nitrogen. Sixty percent survival could be obtained.

The hairy roots regenerated from cryopreserved root tips grew well and showed the same ginsenoside productivity and patterns as those of the control hairy roots cultured continuously at 25°C. The conservation of T-DNAs in the regenerated hairy roots was proved by PCR analysis.

Cryopreservation of Transformed Calli of *Papaver somniferum*

The transformed *P. somniferum* cells maintained on MS solid medium at 22°C in the dark were precultured in 50% loading solution (1 M glycerol + 0.2 M sucrose) at 20°C in the dark for 1 d, dehydrated with PVS2 at 25°C for 35 min without loading, and then cryopreserved in liquid nitrogen.

After rapid thawing and washing, the cells were precultured on MS solid medium at 22°C in the dark. All the four clones used for *cryopreservation* regenerated successfully showing the same *morphological* characteristics as the untreated cultures. To confirm the conservation of T-DNA derived from *Agrobacterium rhizogenes,* the existence of T-DNA in the regenerants was examined by PCR analysis.

Amplification of T-DNA bands was clearly observed in the regenerated cells as well as in the untreated ones. Preliminary evaluation of genetic stability using RAPD analysis was performed and no *significant* difference was observed between the untreated and *cryopreserved* cells.

Cryopreservation of Cell Cultures of *Digitalis thapsi*

Cell cultures of *Digitalis thapsi* were treated for 3 d with 0.15 M mannitol followed by treatment with a cryoprotectant solution composed of 0.5 M DMSO, 0.5 M glycerol and 1 M sucrose, slow cooled for 30 min at –20°C and freezed by rapid immersion in LN, rapid thawing and transfer of cells without washing to a standard semi-solid medium. High viability (60%) was recorded and the cultures originating from cryopreserved cells retained their capacity to accumulate carotenoids.

MONITORING GENETIC STABILITY OF REGENERANTS FROM CRYOPRESERVED GERMPLASM

It is important to consider that the plants regenerated from cryopreserved germplasm have been exposed to a range of different experimental conditions including tissue culture, pre-growth, cryoprotection, freezing-thawing, recovery (re-growth) and manipulations to enhance regeneration. All these stages have the potential to influence genetic stability. Successful post-thaw storage recovery must not only be assessed in terms of survival (viability) of plant tissues, but also the ability to regenerate and produce complete plants.

This necessitates a tissue culture regeneration system. It is therefore essential to consider the effects that *in vitro* regeneration will have on the genetic stability of the surviving plants as these may show somaclonal variation. For instance, the time taken to regenerate plants and the quality of germplasm recovered after cryopreservation are likely to be important features in maintenance of genetic stability and operation of a functional gene bank.

It is thus imperative to assess the genetic stability of plant material regenerated from cryopreserved germplasm and to determine if it is genetically identical to the mother stock (Germplasm prior to storage in LN). Studies on ginsenoside production from *cryopreserved* cells of *Panax ginseng* revealed that total amount of ginsenosides together with product remain unchanged. *Furthermore,* it was demonstrated that alternative conservation method such as preservation under mineral oil for six months as well as continuous subculturing for 14 months failed to preserve biosynthetic capacity of the cells.

Benson and Hamil reported stability in the biosynthetic capacity in transformed roots of *Beta vulgaris* after *cryopreservation*. All recovered cultures of *Coleus blumei* showed the same growth and production characteristics of rosmaric acid as controls. Stability was also *demonstrated* for recovery after different storage periods in liquid nitrogen (from 1 day to 15 months).

These works clearly show that cultures were stable even after successive cryopreservation cycles. Similar conclusions of successful application of cryopreservation can be made from freezing experiments with biotin producing callus cultures.

CONCLUSIONS

Cryopreservation of medicinal plants has multifacet advantages. The technology of cryopreservation has been refined and it enables the storage of *in vitro* cultures for the long-term conservation of medicinal plants. The retention of biosynthetic potential of the retrieved cultures amply demonstrates the use of this technology for the storage of rare, high alkaloid/secondary metabolites/medicines producing cell cultures for pharmaceutical purposes.

Medicinal plants are potential candidates for transformation as well and cryopreservation of transgenic lines is an important line of research. The successful cryopreservation of cell cultures of tobacco, *Datura, Panax, Dioscorea, Catharanthus, Anisoidus, Atropa,* etc., without any evidence of deterioration coupled with potential expected benefits warrant the extension of this technique to other species. As far as genetic stability is concerned the chances of an undesired cell selection after cryogenic storage seem to be less important than expected.

It is even more remarkable that in most of the cases the important characters of the preserved cell lines did not change. Finally, it may be concluded that routine *cryopreservation* procedures developed and applied on a large scale in medicinal plants will help in conservation of this important group of genetic resource.

7

Chapter

MOLECULAR MARKERS

Crop plants have evolved initially by incidental consequences of human gatherers, and more recently through sophisticated plant breeding programmes. While changes in cultural practices and mechanization have had significant impact on agricultural productivity, yield gains in most crops have been due to genetic improvement.

Although the gains have already achieved, further improvement of agricultural productivity and quality are demanded continuously mainly due to population growth, the increasing cost of inputs such as water, fertilizer and energy, concerns about the effects of agrochemicals on the ecosystem, and rapidly changing consumer preferences.

Plant breeding, a process being used for centuries is largely depending on selection for desirable traits. These selections often take many cycles of breeding in order to place desirable agronomic and quality characteristics from different parents into a single genotype. Recent advances in biotechnology have led to the development of a number of novel tools that offer the promise of making plant breeding more precise and faster.

Among the most promising are molecular markers, which are segments of plant DNA that breeders use to detect the presence or absence in experimental plants of specific alleles of interest and thus use them as selection tools. Such a selection of desirable plants based on linked markers is termed as Marker-Assisted Selection (MAS). By using molecular markers, breeders can by-pass traditional phenotype-based selection methods, which involve growing plants to maturity and closely observing their physical characteristics in order to infer underlying genetic make up. Several molecular marker systems have been developed and put to use. The more frequently used ones are discussed as follows.

MOLECULAR MARKERS

RFLP

Restriction fragment length polymorphisms, the first molecular marker developed by Botstein et al. are detected by the use of restriction enzymes that cut genomic DNA molecules at specific nucleotide sequences (restriction sites), thereby yielding variable size DNA fragments.

Identification of genomic DNA fragments is done by Southern blotting, a procedure whereby DNA fragments, separated by electrophoresis, are transferred to nitrocellulose or nylon filter.

In this, filterimmobilized DNA is allowed to hybridize to radioactively labeled probe DNA. RFLP is a codominant marker in which the probes are usually small (500 to 3000 bp), cloned DNA fragments (e.g. genomic or cDNA). The filter is placed against photographic film, where radioisotope disintegration from the probe results in visible bands.

RAPD

Random amplified polymorphic DNA is a dominant marker based on polymerase chain reaction (PCR). It employs a single decamer primer of arbitrary sequence, which is annealed to the template DNA typically at 37°C. The variation in RAPD profile is in the form of presence or absence of a band resulting from variation in primer binding sites.

A major limitation of this marker system is non-reproducibility due to low annealing temperature. However, utility of a desired RAPD marker can be increased by sequencing its termini and designing longer primers (e.g. 24 nucleotides) for specific amplification of markers. Such sequenced characterized amplified regions (SCARs) are similar to sequence-tagged-sites (STS) in construction and application.

CAPS

Cleaved amplified polymorphic sequences are based on the restriction enzyme site variation in the DNA fragments generated by PCR. The source of the sequence information for the primers can come from a gene bank, genomic or cDNA clones, or cloned RAPD bands. This marker is a co-dominant marker.

SSRs

Simple sequence repeats or microsatellites are ubiquitous in eukaryotes. SSR polymorphism reflects variation in the number of repeat units in a defined region of the genome. The frequency of repeats longer than 20 by has been estimated to occur every 33 kb in plants. Nucleotide sequence flanking the repeat is used to design primers to amplify different number of repeat units in different varieties.

These primers are very useful for rapid and accurate detection of polymorphic loci and the information could be used for developing a physical map based on these sequence tags. This type of polymorphism is highly reproducible.

AFLP

The amplified fragment length polymorphism markers are generated by selective amplification of DNA fragments obtained by restriction enzyme digestion. High molecular weight DNA is digested by two restriction enzymes: one hexacutter (e.g. EcoRI) and one tetracutter (e.g. Mse I). Adapter molecules are ligated to the ends of DNA fragments.

Two primers possessing sequence complementarity to the adapter as well as few extra random nucleotides at their 3' ends are used for selective amplification of fragments employing PCR. The amplified products are separated on sequencing gels or even ordinary PAGE and visualized by silver staining. Alternatively, the primers are labeled either by radioisotope or fluorescent dye so that the AFLP profile can be obtained by autoradiography or by using image analysis.

The highest number of amplified products (50-100) is produced in AFLP among all the DNA profiling systems. This increases the probability of detecting polymorphism many folds. The technique is, at present, lengthier and costlier than other PCR based techniques. It requires good quality DNA for ensuring complete digestion by enzymes. Partial digestion of DNA results in non-reproducible variation in DNA profiles.

SNP

Molecular markers are polymorphic when there is DNA sequence variation between the individuals under study. Molecular markers are, therefore, simply an indicator of sequence polymorphism. Sequence polymorphism between individuals can take many forms, for instance, it can be due to the insertion or deletion of multiple bases, or it can be due to single nucleotide polymorphisms (SNPs).

Insertions, deletions and SNPs are important in determining sequence variation between individuals. SNPs are abundant in plant genomes. They are being used for genotyping human populations for certain genetic diseases. The cost of developing SNPs is very high, since for each locus DNA has to be sequenced and suitable PCR primers designed.

The primers must then be used to amplify the corresponding fragment from all other possible genotypes. These fragments must then be sequenced and the sequences compared with one another to determine the SNPs for each haplotype. The term 'haplotype' is used in the context of SNPs instead of the term 'allele'. There are number of methods for identifying SNPs within a genetic locus namely direct sequencing, single-strand conformation polymorphism (SSCP), chemical cleavage of mismatches (CCM) and enzyme mismatch cleavage (EMC).

MOLECULAR MAPPING OF GENES OF AGRICULTURAL IMPORTANCE

Earlier, construction of genetic linkage maps using morphological markers could not be initiated in most crop plants due to lack of sufficient number of molecular markers. Map construction was highly laborious, took many years and required several mapping populations since all the morphological markers could not be obtained in a single cross.

These maps contained limited number of markers and, therefore, could not be used for efficient mapping of target genes. With the availability of a large number of molecular markers, like RFLP, RAPD, AFLP, microsatellites, etc. saturation mapping of plant genomes has become a reality. Molecular genome maps have been constructed in almost all important crop plants. The number of markers employed to construct these maps and marker density varies greatly.

Most of these maps are based on RFLP markers. The recent mapping efforts have included mostly the PCR based markers such as AFLP, STMS, RAPD, CAPS, SCAR and STS. Among the crop plants, the rice genome map is considered most saturated. The map reported by Harushima et al. contained the maximum number of markers.

Significantly, this map was made using a single F_2 population. Recently, this map has been further saturated by combining additional STS and STMS markers. In most of the crop plants F_2 population has been used since it could be generated in the shortest possible time with the least effort. However, for mapping genes, particularly those for quantitative traits, permanent mapping populations such as recombinant inbred lines (RILs) and doubled haploids are preferred since they can be maintained over years by selfing and replicated over locations and seasons.

Table 7.1: Molecular mapping of agriculturally important genes in crop plants

Crop	*Pathogen/Trait*	*Gene*	*Marker(s)*
1. Disease resistance			
Rice	*Pyricularia oryzae*	*Pi-2,4*	RFLP
		Pi11	RFLP
		Pi-5(t), Pi–7(t)	RFLP
		Pi-Z^{-6}	RFLP
		Pi-10	RFLP
		Pi-12(t)	RFLP
		Pi-18(t)	RFLP
		Pib	RFLP
		Pik^m	RFLP
		Pita-2, Pita	RFLP
		Pi-5 (t)	AFLP
		Pi20	RFLP
		Pi44	RFLP
		Pb1	RFLP
Pyricularia grisea		*Pi-1(t)*	RFLP
		QTL (1)	RFLP
	Xanthomonas oryzae pv. *oryzae* (Bacterial blight)	*Xa-1, Xa-3, Xa-4*	RFLP
		Xa-5	RFLP
		Xa-13	RFLP

(Table Contd.)

Crop	Pathogen/Trait	Gene	Marker(s)
		Xa-21	RFLP
		Xa-1	RAPD
		Xa3, Xa4, Xa5, Xa10	RFLP
		Xa13	RFLP
		Xa22(t)	RFLP
		Xa-1	RAPD
		Xa 23 (t)	SSR
	Rice yellow mottle virus	*RYMV (QTL)*	RFLP
		RYMV	RFLP & STS
	Rice stripe	*Stv-b^i*	RFLP
	Tungro	*RTSV*	RFLP
	Puccinia striiformis f. sp. tritici	*Yr5*	RGA
	Rhizoctonia solani Kuhn	*Rsb1*	RFLP, RAPD, AFLP, SSR
Wheat	*Erysiphe graminis p.v. tritici*	*Pm1, Pm2, Pm3b, Pm4a*	RFLP
		Pm1, Pm2	RFLP
		Pm2	RFLP
		Pm3b, Pm4a	RFLP
		Pm2	RFLP
		Pm1	RFLP
		Pm12	RFLP
		Pm21	RAPD
		Pm	RFLP
		Pm4b	AFLP
		Pm4a & Pm4b, Pm6	STS, RFLP
		Pm13	RFLP, RAPD, STS, DDRT-PCR
		MlG	SSR

(Table Contd.)

Crop	Pathogen/Trait	Gene	Marker(s)
	Adult plant resistance to powdery mildew	APR	STMS, RFLP
	Common bunt	*Bt-10*	RAPD
		Bt-11	RAPD
	Karnal bunt	*KB*	RFLP
	Durable stem rust	*Sr2*	RFLP
		Sr2	STS
		Sr2	RFLP
		Sr22	RFLP
	Puccinia recondite	*Lr9*	RFLP, RAPD
		Lr18	N-band
		Lr1	RFLP
		Lr9	RFLP
		Lr 19	RFLP
		Lr24	RFLP
		Lr24	RAPD
		Lr 29	RAPD
		Lr32	RFLP
		Lr34	RFLP
		Lr24	RAPD, SCAR
		Lr10	STS
		Lr10	RFLP
		Lr23	RFLP
		Lr27	RFLP
		Lr31	RFLP
		Lr34	RFLP, RAPD
		Lr34	RFLP
		Lr13	RFLP, STMS
		Lr35	PCR
		Lr28	RAPD, STS

(Table Contd.)

Crop	Pathogen/Trait	Gene	Marker(s)
		Lr3	RFLP
		Lr3	mRNA fingerprinting,
			cDNA cloning
	Stripe rust	*Yr15*	RFLP
		Yr15	RAPD, STMS
		YrH52	STMS, RFLP
	Loose smut	*T19*	Monoclonal antibody
		T10	RAPD, RFLP
	Septoria nodorum	–	RAPD
	Septoria tritici	–	AFLP
	Fusarium head blight	–	AFLP, RFLP
		–	AFLP
	Yellow rust	*YrMoro*	AFLP, STS
	Tilletia indica	*QTL (1)*	SSR, AFLP
	Wheat streak mosaic virus	*Wsm1*	*STS*, RAPD
Maize	*Heliminthosporium turcicum*	*Ht1*	RFLP
	Maize dwarf mosaic virus	*mdm1*	RFLP
	Cercospora zeamaydis	QTL (>10)	–
	Maize streak virus	*QTL (1)*	RFLP
	Maize mosaic virus	*QTL (1)*	RFLP
	Maize stripe virus	*QTL (1)*	RFLP
	Sugarcane mosaic virus	*Scm1*	RFLP, SSR
		Scmv1, Scmv2	RGA-CAPs
		Scmv1, Scmv2	AFLP, SSR
Barley	*Erysiphe graminis*	*QTL (2)*	RFLP
		Rar1	AFLP
	Rynchosporium secalis	*Rrs 13*	RFLP
	Puccinia hordei	*Rph Q*	RAPD

(Table Contd.)

Crop	Pathogen/Trait	Gene	Marker(s)
		QTL-Rphq (6)	AFLP
		Rph7.g	RFLP
	Xanthomonas campestris	QTL (3)	– pv. hordei
	Puccinia striiformis	*QTL (1)*	– f.sp. *hordei*
	Barley yellow dwarf	*Yd2*	AFLP Luteovirus
	Barley yellow mosaic	*rym5*	CAPs, SSR virus
	Cochliobolus sativus	*Vhv1*	AFLP
	Pyrenophora graminea	QTL (2)	–
Sorghum	*Sporisorium reilianum*	*Shs*	RFLP/RAPD
Tomato	*Stemphylium vesicarum*	*Sm*	RFLP
	Cladosporium fulvum	*cfa*	RFLP
	Fusarium oxysporum	*I2*	RFLP
	Pseudomonas syringae	*Pto*	RFLP
	Leveillula tourica	*Lv*	RAPD/RFLP
	Verticillium dahliae	*Ve*	RAPD
		Ve	RFLP
	Medoidogyne sp.	*Mi*	RFLP
	Psuedomonas solanacearum	*QTL (3)*	
	Oidium lycopersicum	*Ol-1*	RFLP, RAPD, SCAR
	Alternaria solani	*QTL (1)*	RFLP, RGA
Potato	Potato virus X	*Rx1*, Rx_2	RFLP
		Nb	AFLP
	Phytophthora infestans	*QTL (11)*	RFLP
		R2	AFLP
	Potato X potexvirus	*Nxphu*	RFLP
	Potato Y potyvirus	Ry_{adg}	RFLP
Soybean	*Phytophthora sojai*	*Rps 1*	RFLP
	Soybean mosaic virus	*Rsv*	RFLP/SSR

(Table Contd.)

Crop	Pathogen/Trait	Gene	Marker(s)
	Pseudomonas syringae	*Rpg 1*	RFLP
		pv. glycinea	
Common bean	*Uromyces appendiculatus*	*PI 181996*	RAPD
		Ur-9, Fin	RAPD
	Potyvirus	*I*	RAPD
	Xanthomonas campestris	*QTL (7)*	RFLP
	Colletotrichum lindemut hianum	*Co-4²*, *Co-7*	RAPD, SCAR
Pea	Pea seed borne mosaic virus	*sbm-1*	RFLP
	Ascochyta pisi	*QTL (3)*	RFLP
Tobacco	*Chalara elegans*	*Brr*	RAPD
Apple	*Venturia inaequalis*	*Vf*	RAPD
		Vf	AFLP & SCAR
Melon	*Fusarium sp.*	*Form 2*	RAPD
Mungbean	*Erysiphe polygoni*	*QTL (3)*	RFLP
Cocoa	*Phytophthora palmivora*	*QTL (5)*	AFLP
Oil palm	*Fusarium sp.*	*QTL (1)*	SSR & AFLP
Rubber	*Microcyclus ulei*	*QTL (8)*	–
	Phyllochora herberi	*Phr*	Isozyme
Sugarcane	*Puccinia melanocephala*	–	RFLP
Brassica	*Leptosphaeria maculans* (Desm.) Ces.et de Not	*QTL (10)*	–
	Plasmodiophora brassicae	Pb-Bn1, QTL (2)	–
	Sclerotinia sclerotiorum	*QTL (1)*	RFLP, AFLP, SSR, RAPD
Chick pea	*Fusarium sp.*	*Race 4*	ISSR (Inter-Simple Sequence repeat)
Pearl millet	*Puccinia substriata var. indica*	*Rr1*	RAPD,RFLP
Rose	*Diplocarpon*	*Rdr1*	RAPD, AFLP

(Table Contd.)

Crop	Pathogen/Trait	Gene	Marker(s)
Grape	Powdery mildew	*Run1*	AFLP
Pepper	Potato virus Y	*Pvr4*	RAPD, SCAR
Cassava	Cassava Mosaic Virus	*CMD2*	SSR, RFLP
Rye	Rust	*Lr26*	AFLP, RGA,
		Sr31	STS
		Yr9	SrR
Banana	Banana Streak Virus	–	AFLP
Tobacco	*Ralstonia solanacearum*	*QTL (1)*	AFLP
Lentil	*Colletotrichum truncatum*	*LCt-2*	RAPD, AFLP
2. **Nematode and insect resistance**			
Potato	*Globodera rostochiensis*	*QTL (2)*	–
	Globodera rostochiensis,	*Grp1*	AFLP, CAPs & RFLP
	G. pallida		
	G. pallida	*QTL (1)*	AFLP, SSR
		QTL (1)	AFLP
Tomato	*Globodera rostochiensis*	*Hero*	SSR
	Meloidogyne spp.	*Mi-1*	RFLP
Sorghum	Head bug	*B2/b2*	RFLP, SSR
	Schizaphids graminum	*QTL (1)*	RAPD, SSR
Soybean	*Helicoverpa zea* Boddie	*QTL (1)*	RFLP
	Heterodera glycines Ichinohe	*QTL (1)*	RFLP
Wheat	*Diuraphis noxia* Mordvilko	*Dn2*	RAPD, SCAR
		Dn2, Dn4	RFLP
		Dn4	SSR *Dn6*
	Hessian *fly*	*H23, H24*	zRFLP
		H3, H5, H6, H9- H17	RAPD
		H21	RAPD
		H6	RAPD, STS
	Pratylenchus neglectus	*Rlnn1*	AFLP, RFLP
	Cereal cyst nematode	*Cre1*	RFLP

(Table Contd.)

Crop	Pathogen/Trait	Gene	Marker(s)
	zresistance		
		Cre1	RFLP
		Ccn-D1	RAPD, RFLP
Maize	*Ostrinia nubilalis*	*QTL (1)*	RFLP, SSR
Rice	*Orseolia oryzae* (Gall midge)	*Gm2*	RFLP
		Gm4(t)	RFLP
		Gm7	AFLP, SCAR
	Brown planthopper	*Bph1*	RFLP
		Bph10	RFLP
		Bph(t)	RFLP
	Green leafhopper	*GLH*	RFLP
		Grlp3	RFLP
		Grlp11	RFLP
		Grh1	RFLP
	Whitebacked planthopper	*WBPH*	RFLP
		WBPH	RFLP
Apple	*Dysaphis devecta* Wlk.	*Sd-1*	AFLP, SSR, RFLP
3. Abiotic stresses			
Rice	Submergence tolerance	*Sub1*	RFLP
	Salt tolerance	*Salt*	RFLP
		OSA3	RFLP
	Phosphorus uptake	*QTL (1) (Pup1)*	RFLP
	Aluminium tolerance	*QTL (1)*	–
Wheat	Thermosensitive earliness *per se*	*Eps-A^{m}1 (QTL)*	RFLP
	Aluminium tolerance	*Alt2*	RFLP
		AltBH	RFLP
	Tolerance to salt stress	*Kna1*	Protein poly morphism
Barley	Aluminium tolerance	*Alt (QTL)*	AFLP, SSR

(Table Contd.)

Crop	Pathogen/Trait	Gene	Marker(s)
4. Male sterility, wide compatibility and fertility restoration			
Petunia	Restorer of fertility	*Rf*	RAPD, AFLP
Rice	Male sterility and fertility restoration	*tgms1.2*	RFLP
		tms2	RFLP
		tms3	RFLP
		tgms	RFLP
		tgms-vn1 (tms4)	RFLP
		pms1	RFLP
		pms2	RFLP
		pms3	RFLP
		ms-h(t)	RFLP
		Rf-1	RFLP
		Rf?	RFLP
		Rf2	RFLP
		Rf3	RFLP
		Rf5	RFLP
		Rfu	RFLP
		Rf?	RFLP
	Hybrid breakdown	*Hwd1, hwd2*	RFLP
	Wide compatibility	*S5*	RFLP
Wheat	Fertility restoration	*Rf4, Rf3*	RFLP
Rye	CMS	*Rfg1*	RFLP, RAPD
	Self-fertility	S1Z1S5	Isozyme, RFLP
Brassica	CMS restorer	*Rfp1*	RFLP, RAPD
Sorghum	Fertility restorer	*rf4 (QTL)*	AFLP
Sunflower	Fertility restoration	*Rf1*	RAPD, AFLP, SCAR
Cotton	CMS fertility restoration	*Rf1*	RAPD, SSR
Coffee	Pollen viability restoration	*QTL (3)*	AFLP

(Table Contd.)

Crop	*Pathogen/Trait*	*Gene*	*Marker(s)*
5. Grain quality			
Sorghum	Grain quality and yield	*QTL (6)*	RFLP, AFLP & SSR
components			
Rice	Grain aroma	*Fgr*	RFLP
	Cooked-kernel elongation	*KNE*	RFLP
	Amylose	Wx	RFLP
Wheat	Flour colour	*QTL (1)*	RFLP, AFLP
	Grain yield	*QTL (1)*	RFLP
	Red grain colour	*R3, R1*	RFLP
	High molecular weight	*Glu-D1*	PCR-based glutanin
	Grain protein content	*QTL (1)*	STMS, RFLP
		QTL (1)	SSR
	Kernel hardness	*ha*	RFLP
		ha	RFLP
	Bread making quality	*Glu-D1(1Dx5)*	PCR
	Amylose content	*Wx-B1*	RFLP
Sunflower	Grain oil content	*QTL(2)*	AFLP, SSR
6. Yield, its components and other traits			
Eucalyptus	Wood density, stem growth and stem form	*QTL (1)*	RAPD
	Lignification genes	*EgHypar* and	SSCP (Single Strand
		EgTub A1	Confirnation
Polymorphism)			
Oil palm	Fruit morphology and fertility *Sh*	AFLP	
Carnation	Flower type	*QTL (1)*	RAPD, SCAR & RFLP
Soybean	Specific leaf weight and leaf size	*QTL (1)*	RFLP

(Table Contd.)

Crop	Pathogen/Trait	Gene	Marker(s)
	Stearic acid content	*Fas*	SSR
Wheat	Plant height	*Rht-B1*	
		Rht-D1	RFLP
	Dwarfing genes	*Rht12*	STMS, RFLP
		Rht8	SSR
		Rht-B1, Rht-D1	RFLP
		Rht-B1, Rht-D1	RFLP
	Haploid formation	*QTL (1)*	AFLP
	Green plant formation	*QTL (1)*	AFLP
	Semi-dwarfing genes	*Rht-B1b (Rht-1)–*	PCR-based
	Rht-D1b (Rht2)		
	Ear emergence time,	*QTL (1)*	RFLP plant height
	Preharvest sprouting	*QTL (1)*	RFLP tolerance
		Major gene	STMS, STS
	Vernalization response	*Vrn1*	RFLP
		Vrn1	RFLP
		Vrn1	RFLP
		Vrn-A^{m}1, Vrn-A^{m}2	RFLP
		Vrn-D1	STMS
	Cadmium uptake	*Cdu1*	RAPD
	ABA production and response – RFLP		
	Coleoptile pigmentation	Rc1	RFLP
	Milling yield	–	RFLP, STMS
	Eyespot	*Pch2*	RFLP
	Tan spot	–	RFLP
	Na^+/K^+ discrimination	–	RFLP
Apple	Growth and development in juvenile apple trees	*QTL (1)*	RAPD
Potato	For foliar glycoalkaloid	*QTL (1)*	RFLP and

(Table Contd.

Crop	Pathogen/Trait	Gene	Marker(s)
			aglycones
Peach	Fruit quality	*QTL (1)*	Isozymes, RAPD, RFLP, AFLP
Rose	Recurrent blooming, double corolla, thorn density of the shoots	*QTL (1)*	AFLP
Grape	Seedlessness, berry weight	*QTL (1)*	AFLP, SSR, isozyme, RAPD, SCAR
Barley	Intermedium spike-C and	*int-c*	
	non-brittle rachis1	*btr-1 (QTL)*	AFLP
Pea	Rhizobium nodulation	*sym9, sym 10*	AFLP, RFLP
Sugarbeet	Sucrose content, yield and quality	*QTL (1)*	RFLP, AFLP
Rice	Photoperiod sensitivity	*Se1*	RFLP
	Semidwarf gene	*sd1*	RFLP
		Sdg	RFLP
	Shattering-resistance gene	*Sh2*	RFLP
		Sh4	RFLP
		Sht	RFLP
	Seed dormancy, heading date	*QTL (5)*	RFLP
	Heading date	*QTL-Hd-1,* Hd-2 & Hd-3	RFLP
	Yield	*QTL (1)*	SSR, STS
	Root morphology	*QTL (1)*	–
Pinus palustris Mill. × P. *elliottii* Engl.	Early height growth	*QTL (1)*	RAPD

(Table Contd.

Crop	Pathogen/Trait	Gene	Marker(s)
Maize	Popping explosion volume	*QTL (4)*	SSR
Cotton	Fibre strength	*QTL (2)*	SSR, RAPD
Sunflower	Agronomic traits [grain weight by plant (GWP), 100-grain weight (TGW), percentage of oil in grain (POG), sowing to flowering date (STF)]	*QTL (1) for TWP, QTL (6) for POG, QTL (2) for STF*	AFLP, SSR

Availability of molecular markers and saturated linkage maps has enabled mapping of genes for qualitative as well as for quantitative traits. The qualitative traits that are controlled by one gene show simple Mendelian pattern of monogenic inheritance such as genes controlling biotic stresses, fruit flesh colour in peach, kernel colour in corn, flower colour in Petunia, etc.

The mapping of such genes with different molecular markers is listed in Table elsewhere in this chapter. Quantitative traits such as yield, drought and cold tolerance, wood density etc. that show continuous variation are controlled by many genes. The individual genes controlling the expression of quantitative traits are now called Quantitative Trait Loci (QTL).

QTL with relatively strong effects are good targets for marker-assisted selection especially if the trait is difficult to measure. Many QTL have relatively small effects. These QTL are difficult to accurately map, especially with population sizes typical of most mapping studies. So, now the population size for QTL mapping has increased to about 300-500 individuals.

However, in order to have an understanding of what the gene does and how it interacts with other genes, it is useful to know where the genes are within the genome and in relation to other genes of interest. The number of QTL controlling a trait varies from 1 to more than 10 for different crop plants. The progress in mapping of genes of agricultural importance in some major crops is described here.

Rice

A large number of genes for qualitative and quantitative traits including disease resistance, insect resistance, cooking quality, drought and flooding tolerance etc. have been mapped using DNA markers as listed in Table elsewhere in this chapter. Mapping of disease resistance genes are of major concern for imparting stability to rice production.

Few examples of mapping such genes using molecular markers is described here. For instance, RFLP markers have been used to map gall midge resistance gene *Gm2* using recombinant inbred lines derived from a cross between 'Phalguna' (resistant variety) and 'ARC6650' (a susceptible land race). Another gall midge resistance gene, *Gm4t,* which is non-allelic to *Gm2* and is known to confer resistance against insect biotypes 1, 2, 3 and 4 has also been tagged using RAPD in combination with bulk segregant analysis of a F_3 population.

Several of the putative resistance gene analogues (RGAs) have been cloned, sequenced and found to be tightly linked to known disease resistance genes. Genetic mapping of resistance to rice tungro spherical virus (RTSV) and green leaf hopper (GLH) in ARC11554 was achieved using RAPD and RFLP markers. Yu et al. mapped a major locus *Pi-2(t)* for resistance to blast caused by the fungus *Magnaportha grisea* using RFLP markers.

Several of the major genes to the bacterial leaf blight (BLB) pathogen, *Xanthomonas oryzae pv. oryzae,* have been tagged with RFLP or RAPD markers. Two microsatellite markers tightly linked to BLB were located at approximately 2 and 18 cM from the *xa5 locus.* RFLP tagging of a gene for resistance to brown plant hopper (BPH) was reported by Mei et al..

Genes have also been mapped using RFLP markers for submergence tolerance, salt tolerance, phosphorus uptake and Al tolerance. As far as male sterility and fertility restoration is concerned, several reports have been available on mapping genes using RFLP markers. Similarly, for traits like grain aroma, cooked kernel elongation, genes have been mapped using RFLP markers. Recently, quantitative trait loci for yield have been mapped.using SSR and STS markers.

Wheat

Several reports on mapping of genes for various disease and insect resistance, abiotic stresses, grain quality and other traits are listed in Table elsewhere in this chapter. Wheat rust disease is a major concern and many successful results have been reported on gene mapping.

A sequence-tagged-site (STS) marker linked to *Lr28,* a wheat leaf rust resistance gene has been identified by Randomly Amplified Polymorphic DNA (RAPD) analysis of near isogenic lines (NILs) of *Lr28* in eight varietal backgrounds. Of the 80 primers tested, one RAPD marker distinguished the NILs and the donor parent from susceptible recurrent parent.

Comparisons between near isogenic lines (NILs) and their recurrent parents have been useful for identifying molecular markers linked to host genes showing resistance to pathogens. Inter-simple sequence repeat (ISSR primers) markers for stem rust *(SR39)* and leaf rust *(Lr 35)* resistance genes have been developed by Gold et al., which would facilitate the transfer of these genes to elite wheat lines.

Microsatellite markers have been used for detecting DNA polymorphism in yellow rust-resistance accessions of *Triticum dicoccoides.* Nine microsatellite markers were identified to be linked to striperust resistance gene YrH52. Microsatellite markers have also been used to tag several genes or QTL, including the genes *Rht8*, *Rht12* and Vrn1, and QGpc.ccsu.2D.1, a QTL for grain protein content.

The problem of pre-harvest sprouting, particularly in amber kernels, is quite common in major wheat growing regions of the world, including India. The improvement in grain protein content and its composition in bread wheat is also a difficult task and remains a major concern to plant breeders.

The QTL for pre-harvest sprouting tolerance and grain protein content have been tagged using sequence tagged multiple sites (STMS) and sequence tagged sites (STS) markers.

Brassica

Brassica juncea (Indian mustard), *B. rapa* (turnip rape) *and B. napus* (rapseed) are the major oilseed Brassicas. In this group of crops, molecular markers have been employed for mapping of genes primarily for disease resistance and oil and meal quality. Several efforts have been made to identify markers for resistance to white rust caused by the fungus *Albugo candida* (Pers.) Kuntze, which is a widespread and destructive disease in these crops with yield reductions of 30-60% in severely infested fields.

A locus *(ACA1)* controlling resistance to *A. candida* has been mapped in *B. napus* using RFLP markers. A single locus controlling resistance to *AC2* in *B. rapa* was mapped using RFLP markers and a segregating population from *Per* (resistant to both *AC2 and AC7)* × 'R500' (susceptible). A co-segregating RFLP marker (X140a) and two closely linked RFLP markers (X42 and X83) were identified which were useful for MAS and map based cloning of a single gene *(Acr)* responsible for conferring resistance to *A. candida* in *B. juncea.*

Prabhu et al. mapped a resistance gene *($Ac2_1$)* in *B. juncea* from a Russian source imparting resistance to a predominant Canadian isolate of *A. candida. B. juncea* accession BEC-144 from Poland shows resistance to the Indian isolates of the white rust pathogen. Identification of two markers linked in coupling and repulsion phases flanking the gene controlling resistance to *A. candida* in BEC-144 was reported by Mukherjee et al..

This work has been further extended to develop AFLP and CAPS markers for this gene. Moreover the CAPS marker has been validated in different populations revealing thereby its utility in marker assisted selection. Kole et al. mapped genes for resistance to white rust in *B. rapa* using a recombination inbred population and a genetic linkage map consisting of 144 RFLP markers and 3 phenotypic markers. Molecular markers have been generated for the genes conferring resistance to *Leptosphaeria maculans* in *B. napus* by various workers.

The resistance locus *LmFr1* was linked to markers cDNA 011 and cDNA 110, and localized onto the linkage group 6 (LG6); [48]. Loci pb-3 and pb-4 conferring resistance to *Plasmodiophora brassicae* in *B. oleracea* were identified and linked to RFLP and AFLP markers. Similarly, Figdore et al. also identified markers 14a on LG1, marker 48 on LG4 and 177b on LG9 linked to clubroot resistance (resistance to *Plasmodiophora brassicae* wor. Race 7) in *B. oleracea.*

A number of studies have been undertaken to generate markers for fatty acids such as linolenic acid, linoleic acid, oleic acid, palmitic acid and erucic acid. Two RAPD markers, K-011100 and 25a were generated and linked to the linolenic acid concentration. RAPD markers linked to oleic, linolenic and linoleic acids were identified in *B. napus.* RAPD marker linked to the linolenic acid content was converted to a co-dominant SCAR marker.

Markers linked to genomic regions controlling linolenic acid concentration in *B. napus* corresponding to *fad3* (omega-3-desaturase) gene in *A. thaliana* were also identified. In another study, a single QTL containing 6 markers associated with oleic, palmitic and linoleic acid content was detected in *B. rapa.* Sharma et al. recently mapped two major QTLs influencing oleic acid level in *B. juncea* using both single factor analysis of variance and interval mapping.

Erucic acid loci have been linked to molecular markers by using BSA or RFLP analysis in *B. napus.* In each of the studies, two QTL were detected. These QTL have been positioned on LG 6 and LG12 or on LG7 and LG15. In an independent study, two QTL associated with the erucic acid level in *B. napus* were detected and mapped onto two different loci termed as E1 and E2.

QTL E1 and E2 correspond to the two alleles of the β-ketoacyl-synthase (KCS) derived from *B. campestris and B. oleracea,* the two parental species of *B. napus* and encode the *Fatty acid elongation 1 (Fae1)* protein. In *B. rapa (Syn campestris)* erucic acid loci were linked to RFLP markers. The seed coat colour gene has been tagged to various RFLP and RAPD markers. The RFLP markers linked to seed coat colour in *B. napus* were identified using the Bulked Segregant Analysis (BSA) approach. Seed coat colour trait in *B. campestris* was tagged with RAPD markers using *B. campestris-oleracea* additional lines.

A 3:1 ratio of segregation of brown : yellow seed in *B. rapa* indicated a monogenic control of this trait and was mapped to LG5. Upadhyay et al. studied segregation of the trait in an F_2 population of *B. juncea* and reported duplicate dominant gene action giving a phenotypic ratio 15:1. Two RFLP markers flanking one of the interacting loci were identified. In a recent report, the seed coat colour trait was tagged using a combined approach of BSA and AFLP in *B. juncea.*

Soybean

In soybean, emphasis is laid on genetic mapping of pest and disease resistance genes. Apart from that quantitative traits such as oil quality, plant height, sprout yield etc. have been characterized using molecular maps. The soybean cyst nematode (SCN) *(Heterodera glycines* Inchinoe) is the most economically significant soybean pest.

Two SSR markers BARC-Satt 309 and BARC-Satt 168 have been reported that segregate and map 0.4 cM from *rhg1.* When these markers were used to assay lines from SCN-susceptible × SCN - resistant crosses, they proved to be highly effective in identifying lines carrying *rhg1* resistance from those carrying the allele for SCN susceptibility at the *rhg1* locus.

In another study, field resistance to SCN race 3 in soybean cv. Forrest was found conditioned by two QTLs. The underlying genes are presumed to include *rhg1* on linkage group G and *rgh4* on linkage group A2. A high density map for the intervals carrying *rhg1 and rhg4* have been developed using AFLP markers. A12-way analysis of variance showed two loci controlling SCN resistance in Essex × Forrest RILs.

Using 139 RFLPs QTLs associated with resistance to corn earworm *(Helicoverpa zea Boddie*) were identified. With the help of AFLP, four markers closely linked to soybean mosaic virus resistance gene, Rsv1 was mapped, thus demonstrating the utility of genetic mapping for generating markers tightly linked to important plant disease resistance genes. Soybean death syndrome (SDS) caused by *Fusarium solani f.* sp. *glycines* results severe yield losses.

Two QTLs for resistance to SDS were mapped in cv. Pyramid using SSR markers namely, BARC-Satt 163 and BARC-Satt 080. Similarly, a QTL was identified from cv. Douglas using SSR marker BARC-Satt 307. Njiti et al. suggested that gene pyramiding would be an effective method for developing cultivars with stable resistance to SDS. Increasing the stearic acid content to improve soybean oil quality is a desirable breeding objective for food processing applications.

Three SSR markers, Satt 070, Satt 474 and Satt 556 were identified to be associated with stearic acid content by Spencer et al.. Identification of these markers may be useful in molecular marker-assisted breeding programmes targeting modifications in soybean fatty acids. RFLP markers have also been used to identify QTLs associated with plant height, lodging and maturity.

The major locus associated with plant height was identified *as Dt1* on LG L. *Dt1* was also associated with lodging. In addition, with the help of RFLP markers, two QTLs for plant height (K007 on LG H and A516b on LG N) and one QTL for lodging (cr517 on LG J) were identified. For maturity, independent QTLs were identified in intervals between R051 and N100, and between B032 and CpTI on LG K. RFLP markers were also used for identifying QTLs associated with soybean sprout-related traits.

Four QTLs were associated with sprout-yield in the combined analysis done for two years by Lee et al.. They also found that the QTLs conditioning sprout yield were in the same genomic locations as the QTLs for seed weight. These data demonstrates MAS may be feasible for enhancing sproutyield in soybean.

Pea

Aphanomyces root rot, caused by *Aphanomyces euteiches* Drechs, is the most important disease of pea worldwide. No efficient chemicals are available to control the pathogen. Thus, to facilitate breeding for Aphanomyces root rot resistance and to better understand the inheritance of partial resistance, identified QTLs associated with the disease using DNA markers AFLPs, RFLPs, SSRs, ISSRs and STS.

The resulting genetic map consisted of 324 linked markers distributed over 13 linkage groups covering 1,094 cM. A total of seven genomic regions were associated with Aphanomyces root rot resistance. The first one was named *as Aph1,* which was considered a major QTL. Two other specific QTLs, namely *Aph2 and Aph3* were identified, which were mapped near the *r* (wrinkled/round seeds) *and af* (normal afila leaves) genes.

Four other minor QTLs were identified. The resistant alleles of *Aph3* and the two minor QTLs were derived from the susceptible parent. RAPD and SACR markers linked to genes affecting plant architecture of pea, namely, three ramosus genes *(rms2, rms3 and rms4)* and two genes conferring flowering response to photoperiod *(sn and dne)* have been reported by Rameau et al..

In another study, QTLs affecting seed weight in pea were mapped using RFLP markers. Four QTLs were identified in marker intervals on three different linkage groups.

Chickpea

Ascochyta blight is an economically important disease of chickpea caused by the fungus *Ascochyta rabiei.* Udupa and Baum identified and mapped a major locus *(ar1)* using SSRs, which confers resistance to pathotype I, and two independent recessive major loci *(ar2a),* with complementary gene action conferring resistance to pathotype II.

In another study, integration of co-dominant STMS markers improved the mapping of ascochyta resistance in chickpea. Resistance gene analogs (RGAs) of *Cicer* were isolated by different PCR approaches and mapped in an inter-specific cross, segregating for *Fusarium* wilt by RFLP and CAPS markers. A total of 13 different RGAs were isolated and classified into nine distinct classes.

This study by Huettel et al. provides a starting point for the characterization and genetic mapping of candidate resistance genes in *Cicer* that is useful for MAS and as a pool for resistance genes of *Cicer.*

Tomato

Tomato is an important vegetable crop. Extensive work has been done on genetic mapping of agriculturally important genes in this crop. Early blight (EB) caused by a devastating fungus,

Alternaria solani Sorauer causes plant defoliation, reduces yield and fruit quality, and contributes to significant crop loss. QTL mapping using 14 RFLP markers and 23 RGAs identified 10 significant QTLs for EB by Foolad et al..

Potato virus Y (PVY) is also an important disease causing organism which affects the yield of tomato. Resistance against PVY was identified in the wild tomato relative *Lycopersicon hirsutum* PI247087. The locus *pot-1* was mapped using AFLP markers to the short arm of tomato chromosome 3, in the vicinity of the recessive *py-1* locus for resistance to corky root rot.

Another important pathogen, cucumber mosaic virus (CMV) gene, *Cmr was* mapped using RFLP and isozyme markers in *L. chilense* and was located on chromosome 12. The chromosome 12 markers were found to be significantly associated with CMV resistance in both qualitative and quantitative models of inheritance. This knowledge of the map location of *Cmr* should accelerate introgression by marker-assisted selection.

Identification of tightly linked markers for the genes of importance has facilitated isolation of genes from tomato. For instance, map based cloning strategy was designed to isolate the rootknot nematode resistance gene *Mi* in tomato using PCR-based flanking markers. Fine structure mapping of recombinants with newly developed AFLP and RFLP markers from physically mapped cosmid subclones localized *Mi* to a genomic region of about 550 kb.

Two recessive mutations have been discovered in tomato that completely suppress the formation of flower and fruit pedicel abscission zones, i.e. jointless (*j*) and jointless-2 (j-2). Both the genes were tentatively localized to chromosome 11 about 30 cM apart.

However, RFLP and RAPD markers helped in correctly identifying and mapping the *j-2* locus on chromosome 12 instead of chromosome 11 that enabled map-based cloning of this gene. Improving organoleptic quality is an important but complex goal for fresh market tomato breeders. A total of 26 traits involved in organoleptic quality variation were evaluated. Physical traits included fruit weight, diameter, colour, firmness and elasticity.

Chemical traits were dry matter weight, titratable acidity, pH, and the contents of soluble solids, sugars, lycopene, carotene and 12 aroma volatiles. A total of 81 significant QTLs were detected for the 26 traits using DNA markers. RFLP mapping of 32 independent tomato loci corresponding to genes known to influence fruit ripening and/or ethylene response was reported by Giovannoni et al..

The placement of ripening and ethylene-response loci on the tomato RFLP map would facilitate both the identification of candidate gene sequences corresponding to identified single gene and QTL contributing to fruit development and ethylene response.

Potato

In vegetable crops like potato, genetic mapping has been done primarily on disease resistance. *Phytophthora infestans* is a very devastating fungus causing late blight of potato. Eleven resistance alleles (R1-R11) are known which confer race-specific resistance to this fungus. In two of the reports, R6 and R7 alleles were mapped by RFLP markers on chromosome XI similar to R3 allele and R2 allele was mapped using AFLP marker.

A study on mapping of the resistance gene of root knot nematode *(Meloidogyne chitwoodi)* derived from *Solanum bulbocastanum* in a BC_2 population using RFLP markers have been reported by Brown et al.. RFLP mapping has also been carried out for the potato virus X controlled by a single gene, *Nxphu.* Four RFLP markers CT220, TG328, CT112 and TG424 from the long arm of chromosome IX that were linked to the hypersensitive phenotype have been reported by Tommiska et al..

Mapping of QTL for resistance to potato cyst nematode *(Globodera rostochiensis)* has been reported by several researchers. In one of the studies, the nematode resistance locus *Gpa2* was mapped on chromosome 12 of potato using 733 AFLP markers. This study also showed that *Gpa2* is linked to the *Rx1 locus* conferring resistance to potato virus X.

Linkage maps using AFLP and RFLP markers were constructed and used to identify three QTLs on chromosomes V, VI and XII, respectively, for resistance against the potato cyst nematode. In a recent study by Baker at al., nine resistance gene homologues (RGHs) were identified in two diploid clones of potato with a specific primer pair based on conserved motifs in the LRR domain of the potato cyst nematode resistance gene *Gpa2* and the potato virus X resistance gene *Rx1.* AFLP marker was used to facilitate the genetic mapping of the RGHs in the four haplotypes under investigation.

Sugarcane

Sugarcane is an important cash crop and in order to analyse the inheritance of quantitative traits, extensive study on Quantitative Trait Allele (QTA) mapping was done. The first extensive QTL mapping study performed in cultivated sugarcane was reported by Hoarau et al. based on a population of 295 progenies derived from the selfing of cultivar R570, using about 1,000 AFLP markers.

The population was evaluated in a replicated trail for four basic yield components, plant height, stalk number, stalk diameter and brix, in two successive crop-cycles. Forty putative QTAs were found for the four traits of which five appeared in both years.

In another study, mapping of QTLs for sugar yield and related tarits, namely pol, stalk weight, stalk number, fiber content and ash content were done using 735 DNA markers. Fifty of the 61 mapped QTLs were clustered in 12 genomic regions of seven sugarcane homologous groups.

Forest Trees

Molecular markers have been successfully applied in tree species also which may be incorporated into existing improvement programmes in an efficient and cost efficient manner. In loblolly pine, marker-trait associations for components of radial wood density profiles had been found and verification populations have been established to confirm these associations. RAPD markers were employed to map the genome and quantitative trait loci controlling the early growth of a pine hybrid F1 tree *(Pinus palustris* Mill. × *P. elliottii* Engl.) and a recurrent slash pine tree (*P. elliottii* Engl.) in a (long leaf pine × slash pine) × slash pine BC1 family consisting of 258 progeny.

With the help of RFLP markers 13 different height increment and eight different diameter-increment QTLs were detected in loblolly pine by Kaya et al.. Similarly, chemical wood property traits were analysed for the presence of QTLs in a three-generation outbred pedigree of loblolly pine *(Pinus taeda L.)* using DNA markers. Kumar et al. reported multiple-marker mapping of wood density loci in an outbred pedigree of radiata pine using DNA markers.

The effect of locations of QTL was found to be significantly associated with the expression of wood density at different ages. These results are encouraging for the application of marker information to early selection in order to increase juvenile wood density. The single dominant gene *(R)* that confers resistance to the white pine blister rust fungus *(Cronartium ribicola Fisch.)* in *Pinus lambertiana* Dougl. has been mapped using RAPD markers.

Thirteen RAPD loci were identified by Harkins et al. that were linked to *R*. This would help in subsequent high-resolution mapping experiments to identify very tightly linked markers to facilitate the eventual cloning of *R*. RAPD markers have been used to determine the genetic location and effects of genomic regions controlling wood density, stem growth and stem formation in *Eucalyptus*.

A total of 86 and 92 markers distributed among 11 linkage groups covered 1295 cM and 1312 cM for *E. urophylla* and *E. grandis,* respectively. This application of marker information will help in early selection of hybrid trees to be vegetatively propagated for the production of clonal varieties.

MARKER-ASSISTED SELECTION (MAS)

Molecular marker-assisted selection involves scoring for the presence or absence of a desired plant phenotype indirectly based on DNA banding pattern of linked markers on a gel or on autoradiogram depending on the marker system. The rationale is that the banding pattern revealing parental origin of the bands in segregants at a given marker locus indicates presence or absence of a specific chromosomal segment which carries the desired allele. This increases the screening efficiency in breeding programmes in a number of ways such that:

(a) the segregants can be scored at the seedling stage for traits that are expressed late in plant development. This includes traits such as grain quality, male sterility and photoperiod sensitivity.

(b) it is possible to screen for traits that are extremely difficult, expensive or time consuming to score and measure such as tolerance to drought, salt, mineral deficiencies and toxicity, root morphology, resistance to nematodes or to specific races or biotypes of diseases or insects.

(c) selection can be practiced for several traits simultaneously, which is difficult or even impossible by conventional means.

(d) suppliers may cut across country boundaries. Furthermore, for most PGRFAs, the level of substitutability by other PGRFAs appears to be quite high. heterozygotes are easily identified and distinguished from either homozygotes without resorting to progeny testing. This saves time and effort.

MAS is an attractive option for improvement of certain traits of interest for which phenotypic evaluation is often expensive or unreliable. MAS provides a potential for increasing selection efficiency by allowing earlier selection and reducing plant population size during selection. Breeders can rapidly determine inheritance patterns at the genomic level by directly examining the genetic make up of experimental plants when they are still seedlings.

This is especially useful for traits that cannot be identified until the plant is mature such as fruit characteristics and for traits that are difficult to test such as disease resistance. Resistant plants are selected based on DNA markers that are linked to the gene(s) controlling the trait, instead of actually evaluating the disease resistance of the plants. Incorporating natural resistance genes into varieties is the most effective, economic and environmentally safe means of controlling the disease.

This is the response to the demand for cost-effective, "green" solutions since it eliminates the need for expensive chemicals to control diseases. It is a uniform method of scoring, tells percentage of genome from each parent, and tells which parts of each chromosome come from each parent. In addition to that as the precision in selection is increased, less unwanted side effects appear in the following generation of plants. MAS can also be used to pyramid two or more desirable genes in a new plant variety.

Some MAS Advantages in Backcrossing Breeding

There are cases where many conventional backcross programmes fail. For example, despite carefully made backcrosses to the recurrent parent, progeny derived exclusively from selfpollination due to failure of crossing, have been found in backcross programmes.

Hence, in conventional breeding, breeders are often not working on the genetic material that they assume they are, because crossing fails more often than expected. A reason why backcrossing fails is the misclassification of a plant for the presence of the donor gene (disease escape instead of disease resistance) and is used as a parent in further backcrossing.

All these problems and others like need to make time and resource consuming selfed generations to identify a recessively controlled character are avoided in marker assisted selection.

MAS Status in Different Crops

Enormous work is being carried out on marker-assisted selection in India and abroad. The progress made in some major crops is presented here.

Rice

The ongoing significant efforts on marker-assisted breeding and gene pyramiding in rice include resistance to blast, blight, gall midge, dwarfing and also drought. Bacterial blight (BB) caused by *Xanthomonas oryzae pv. oryzae (Xoo)* is one of the most destructive diseases of rice throughout the world and in some areas of Asia it can reduce crop yield by upto 50%.

The most effective approach to combat BB is the use of resistant varieties. So far, 19 resistant genes have been identified and some of these have been incorporated into modern rice varieties. However, the large-scale and long-term cultivation of varieties carrying one of the most important resistant gene *Xa-4* has resulted in significant shifts in the rice frequency of Xoo.

In many areas of Indonesia, India, China and Philippines, rice varieties with only *Xa-4* for defense against Xoo have become susceptible to the pathogen. Thus, DNA marker- assisted selection was used to pyramid four bacterial resistance genes, *Xa-4, xa-5, xa-13 and Xa-21.* Breeding lines with two, three and four resistance genes were developed and tested for resistance to bacterial blight pathogen.

The pyramid lines showed a wider spectrum and a higher level of resistance than lines with only a single gene. To speed up the gene pyramiding process and to facilitate future markeraided selection, Huang et al. developed PCR markers for two recessive genes *xa-5 and xa*13, and used these to survey a range of rice germplasm.

The results of the germplasm survey will be useful for the selection of parents in breeding programmes aimed at transferring these bacterial blight resistance genes from one varietal background to another. In India, at Punjab Agricultural University (PAU), Ludhiana, three BB resistance genes *xa-5, xa-13 and Xa-21* were pyramided in PR106 and Pusa 44 background.

After multi-location and replicated testing, two PR106 pyramid lines were identified and these have been included in All India Coordinated Testing during 2002. This is the first ever marker-assisted product reaching testing at national level. Pusa 44 pyramid lines were tested in multi-location replicated trials during 2002. Blast caused by the fungus *Magnaportha grisea* is another devastating disease of rice. The most economical and effective approach to reduce the yield loss is to breed varieties that are resistant to the disease.

However, the resistance often breaks down within a few years of cultivar release. Many genes for qualitative blast resistance have been mapped using molecular markers and some of those markers have also been tried in MAS for blast resistance. For example, RG64, a RFLP marker on chromosome 6 is tightly linked (2.8 cM) to *Pi-2(t),* a major gene for blast resistance.

The RG64 rice genome clone was sequenced and primers based on the DNA sequences were found useful in producing polymorphism between the susceptible and resistant varieties after the monomorphic PCR product was digested with restriction enzymes. The CAPS marker was then used to identify rice plants carrying *Pi-2(t)* from an F_2 population derived from the cross between CO39 and CO10151.

The effectiveness of the selection for resistant plants based on linked DNA markers was then compared with phenotyping for blast resistance through progeny testing in the F_3 families by blast inoculation. Results indicated that identification of plants carrying *Pi-2(t)* in a large segregating population is possible using one linked marker as well as flanking markers.

The accuracy of identification of homozygous resistant genotypes was 96% when RG64 marker was used. The accuracy of selection increased to 100% when two markers flanking the *Pi-2(t)* were scored simultaneously. These results illustrate that marker-assisted identification of linked target gene in a segregating population is efficient in identifying resistant genotypes.

This work has been extended futher to pyramid three major genes for blast resistance. Another objective of marker-assisted breeding in rice is to transfer resistance against gall midge *(Orseolia oryzae),* a major insect pest of rice. The resistance to gall midge biotypes is governed by single dominant genes. At national level, effort has been made to screen rice germplasm for new sources for resistance genes effective against one or more biotypes of the pest.

PCR based markers have been designed and currently are being used in markerassisted selection at different research institutions in the country. The semi-dwarf gene *(sd-1)* in rice is one of the most important single genes in the history of rice improvement. This single, recessive gene causes reduced culm length and has been widely used to confer lodging resistance, high harvest index, responsiveness to nitrogen fertilizer, and favourable plant type, in the breeding of high-yielding rice varieties. In rice, molecular mapping of *sd-1* has been reported by several workers.

Chao et al. used 20 mapped clones as probes, based on an existing rice RFLP map, and conducted experiments to establish the location of *sd-1* gene. They evaluated the efficacy of marker-assisted selection in F_2 and F_6 plants derived from the cross Milyang 23/Gihobyeo. The application of MAS for *sd-1* gene has potential to greatly improve the efficiency of the Australian rice-breeding programme. In their breeding programme, the semi-dwarf character when detected before maturity even as a heterozygote eliminates the need for progeny testing in a backcrossing programme.

Wheat

The wheat stem and leaf rusts are two major pathogens, which can potentially devastate wheat crops. Resistant cultivars have long been depended upon to control disease epidemics. Most of the genes for resistance have been mapped using molecular markers and currently being used in MAS at different national as well as international centres. In India, RAPD markers linked to *Lr19* (leaf rust) gene has been converted into SCAR markers.

Lr 28 gene was also tagged by two flanking RAPD marker S464700 and S326350. These linked molecular markers are further in use in pyramiding of rust resistance genes, which is difficult, and time consuming by conventional breeding procedures. A population of 220 BC_1F_2 plants segregating for two genes Cre1 *and Cre3* was evaluated with three molecular markers Xglk 605, Xcdo 588 and Cd 2.2 and the markers were found to provide a reliable means of gene pyramiding and selecting plants carrying the genes in wheat breeding programmes.

With overall goal of transferring new developments in genomics to wheat breeding and production, investigators at 12 public wheat-breeding and research programmes across the US including University of California, Colorado State University, Cornell University, Kansas State University, Montana State University, University of Idaho, University of Minnesota, Purdue University, University of Nebraska, USDA and Washington State University have constituted a national wheat Marker Assisted Selection (MAS) consortium that aims to use molecular markers as chromosome landmarks in MAS programme to facilitate introgression of small chromosome segments carrying the genes of interest.

Available molecular markers will be used to transfer genes for resistance to fungi, viruses and insects as well as gene variants related to improved bread, pasta and noodle quality. These genes will be incorporated into a minimum of 240 adapted cultivars or breeding lines belonging to all major market classes of US wheat, and since they are transferred by normal recombination, the resulting lines will not be classified as transgenics. These improved cultivars will transfer the value of genomic research to the wheat growers' fields.

Other Crops

Use of molecular markers in marker-assisted selection has been carried out for improvement of several other crops such as sunflower, tomato, sugar beet, barley, soybean, apple etc. throughout the world. These are briefly described crop-wise as follows.

Sunflower

MAS for two rust resistance genes in sunflower was reported using RAPD markers Ox20600 and OO04950 linked to the gene *RAdv* responsible for rust resistance in the proprietary inbred line P_2. This gene confers resistance to most of the pathotypes of *Puccinia helianthi* identified in

Australia. These RAPD markers were converted into SCAR markers and the robustness of these markers were demonstrated through the amplification in a diverse range of sunflower germplasm. This will be useful in further attempts for molecular-assisted breeding to produce durable resistance in sunflower to *P. helianthi*.

Tomato

MAS has been carried out for several traits including fruit characteristics in tomato. More recently, MAS was used to transfer the ability to accumulate acylsugars to cultivated tomato. RFLP and PCR-based markers were used through three backcross generations to select plants containing five target regions associated with acylsugar accumulation.

In another example, MAS has been demonstrated for QTL influencing blackmold resistance. Blackmold, caused by the fungus *Alternaria alternata,* is a major ripe fruit disease of processing tomatoes.

Five QTLs were selected for introgression from *Lycopersicon cheesmanii* into cultivated tomato using marker-assisted selection. RFLP and PCR-based markers flanking and within the chromosomal regions containing QTLs were used for MAS during backcross and selfing generations.

Barley

The effectiveness of molecular marker-assisted selection for malting quality trait in barley was reported by Han et al.. In this study, the flanking markers, Brz and Amy2, and WG622 and BCD402B, for two major QTL regions present on chromosomes 1 and 4 were used for MAS. The MAS for QTL1 was more effective than phenotypic selection.

It could substantially eliminate undesirable genotypes by early genotyping and keeping only desirable genotypes for later phenotypic selection. The MAS was also used for verification of yield QTL in a barley cross. The objectives of this study were to verify the value of four QTLs for selection and to compare the efficiency of alternative MAS strategies using these QTL vs. conventional phenotypic selection for grain yield. It was shown that MAS was as good as phenotypic selection.

Soybean

MAS offers the potential to reduce linkage drag and to pyramid genes with similar phenotypic effects into elite genotypes. One such example was seen in soybean breeding programme where a QTL conditioning corn earworm resistance in the accession PI229358 and a synthetic *Bacillus thuringiensis cry1Ac* transgene from the recurrent parent 'Jack-Bt' were pyramided into BC_2F_3 plants by marker-assisted selection.

Segregating individuals were genotyped at SSR markers linked to an antibiosis/antixenosis QTL on linkage group M, and were tested for the presence of *cryAc1*. MAS was used during and after the two backcrosses to develop a series of BC_2F_3 plants with or without *cryAc1* transgene and the QTL conditioning for resistance in BC_2F_3 plants that were homozygous for parental alleles at markers.

This work by Walker et al. demonstrated the usefulness of SSR for MAS in soybean, and showed that combining transgene and QTLmediated resistance to lepidopteran pests might be a viable strategy for insect control.

Apple

MAS is also a promising method to select resistant individuals in horticultural crops like apple. Molecular tools have the potential to give very early information on the genetics of apple seedlings. The aims of apple breeding such as high fruit quality, consistently high yields and durable disease and pest resistance can be achieved more efficiently.

Progress in MAS for apple breeding is being achieved mainly in the area of disease resistance especially in the durable incorporation of scab and mildew resistance. The AL07-SCAR and M18-CAPS molecular analysis in progenies, in which both parents are heterozygous for the resistance gene, made it possible to identify clearly the homozygous plants for the *Vf* gene (resistance for apple scab) and these plants showed higher level of resistance than the heterozygous one. Progenies were developed from crosses with parents that carry different resistance genes such as Vf, *Vm, Vb,* Pl1 and QTLs in different combinations.

Kentucky Bluegrass

The MAS has wide range of utility in this grass species. It was reported by Albertini et al. that MAS has helped in avoiding costly and time-consuming phenotypic progeny tests in *Poa pratensis* to study mode of reproduction. Genotypic apomixis in Kentucky bluegrass involves the pathenogenetic development of unreduced eggs from aposporic embryo sacs.

Two SCAR primer pairs were tested and identified the apomictic and sexual genotypes among progenies of sexual × apomictic crosses with low bias. Furthermore, when tested on a wide range of Italian and exotic *P. pratensis* germplasm, they were able to unequivocally distinguish sexual from apomictic genotypes. This system should, therefore, allow new selection models to be set up in this species.

FUTURE PROSPECTS OF MAS

The above review of the available literature reveals that during the last 17 years since the publication of the first paper on the use of RFLP markers for construction of linkage maps in tomato and maize in 1986, molecular markers have been extensively used for mapping and tagging of hundreds of different agriculturally important genes/QTL in various crop species. With the availability of linked markers, the first requirement for successful MAS has been fulfilled. Besides, the feasibility of MAS based on these linked markers has been demonstrated in several crops both for qualitative as well as quantitative traits as evident from the above description.

However, MAS is yet to be used routinely in plant breeding programmes. Utility of the MAS in crop plants is currently limited by factors such as recombination between the marker and the target gene, low level of polymorphism between parents with contrasting traits and lower resolution of QTLs due to interaction with the environment. With the recent developments in both structural and functional genomics, it would not be difficult to find solutions to these problems.

Availablity of high-density genetic and physical maps will enable finding markers physically closer to the target gene that would not allow failure of MAS due to genetic recombination. Moreover, cloning and characterization of the target genes, which are possible based of their position on the linkage map, would allow development of allele-specific markers.

Use of such markers would completely eliminate the possibility of breakdown of the marker-trait linkage. Besides, markers based on the sequences of the genes would facilitate allele mining in the germplasm resources, thereby leading to identification and utilization of newer alleles in crop improvement. Different alleles of a gene would differ for a number of nucleotides at different positions in their sequence that would be the basis of developing highly polymorphic single nucleotide polymorphism (SNP) markers.

The problem of low level of polymorphism in narrow crosses can thus be circumvented. Use of MAS for QTL, particularly those having little effect on trait expression and highly interacting with environment would require greater amount of research effort and newer experimental strategies. Complete integration of MAS with the conventional plant breeding programmes demands consideration of two important factors: a) size of population and b) cost.

Plant breeding experiment requires screening of large segregating populations routinely over generations. Genotyping of large number of samples manually is an extremely difficult task. MAS to be practicable should be amenable to automation that would allow handling of large number of samples.

Development and use of PCR based markers such as STS and SCAR will be a key to success of MAS in crop improvement. As the technology develops and gets modified to analyze large number of samples, the cost will automatically go down. The investment in gene tagging, and selection based on molecular markers should be weighted against the overall cost and time involved in traditional breeding program. Even though the cost of MAS is higher at present level of estimation, its integration with traditional plant breeding is desirable because of immense possibilities it offers.

8

Chapter

MODIFIED GENOMIC ORGANISMS

Genetics underwent a paradigm shift in the early 1970s. Genes available for the breeding of new varieties were no longer limited to the same genus and species. Instead, new technology allowed genes from any source to be introduced and incorporated into any genome. Termed lateral transfer, this technology is based on *enzyme* systems originally discovered in bacteria, and viruses that allowed genes to be isolated, prepared in quantity, and transferred and incorporated into the genome of any organism.

The technology has been applied in the development of new strains in agriculture, science, and medicine, giving rise to what are called *genetically* modified (GM) or genetically engineered (GE) organisms. In this article we will deal with GM plants in commercial agriculture, focusing on their adoption and acceptance in North America. By the 1990s the technologies for gene isolation and transfer developed with *microorganisms* were being applied to higher organisms.

In agriculture there was an obvious need for new methods to protect plants from foraging insects and competitive weeds. Genes with potential to deal with these problems had been identified in bacteria. The first of these genes encodes a protein that, when ingested, is toxic to insect larvae, but in a higher organism is digested without effect.

The second of these genes encodes a variant of an essential plant enzyme that makes the enzyme insensitive to the broad-spectrum herbicide glyphosate, commercially sold as Roundup. These genes were taken from the bacterial genomes that harbored them, cultured in vitro, and introduced into commercial varieties of corn, cotton, soybean, and rapeseed, a highly divergent range of plant types.

Comparing selected GM isolates with their parent strain for vigor and crop yield showed no detectable differences, while biochemical analysis revealed expression of transferred genes and, more *importantly*, field tests demonstrated strong protection from insect pest and/or herbicide damage.

These positive results allowed imaginations to soar; the possibilities appeared boundless. Weedfree, pest-free, and chemical-pesticide-free plants, in addition to enhanced yield, *enhanced* nutrition, retarded spoilage, and a longer shelf life are just some of the properties that were predicted for GM crops.

RISKS AND CONCERNS

Multiple Gene Copies and Unanticipated Genetic Results

The first lateral gene transfers into plant cells were carried out using gene guns, a method considered crude in comparison to the biological vector-based transfers systems used today. Gene guns literally shoot DNA into cells. The DNA goes everywhere and multiple copies of the transferred gene can be distributed across a recipient's genome. The ambiguity associated with this approach raised fears about the possibility of creating undetected, deleterious secondary effects.

Additional concerns about GM strains focused on the escape of laterally transferred genes into other species. Specifically, Table elsewhere in this chapter lists concerns about the adoption of lateral gene transfer technologies for the development of new plant varieties raised by US interest groups such as Greenpeace, Friends of the Earth, the Organic Consumers Association, the Union of Concerned Scientists, and the American Council on Consumer Interests.

Table 8.1: Risks and concerns about GM products.
Food allergies
Food toxicity and anti-nutrients
Antibiotic resistance Increased weediness
Horizontal gene transfer

While all of the concerns listed are valid, and were especially so in the early years of this work, improved technologies in production of strains and more exhaustive characterization of product strains have considerably reduced risks. Genetic analysis of the recombinant strains reveals the number of gene copies inserted as well as the location of the inserted genes with respect to those of the parent genome.

Such analyses are used to identify and eliminate *recombinants* with potential for alteration in expression of the parent organism's genes. Additional tests on promising, *transgenic* isolates are performed to ensure that the expression of critical parent genes is unaltered. These tests include the determination of levels of production of *indigenous* plant toxins and detection of production of proteins with allergenic potential.

Acceptance of GM products has become a matter of trust that the work characterizing the recombinant isolates genetically and the associated comprehensive risk analyses testing have been responsibly performed and are reliable.

Escape of Transferred Genes

Another set of concerns regards the escape and dissemination of laterally transferred genes. The concern divides into two areas: the escape of genes that would make foreign or weed species more competitive (such as becoming resistant to glyphosate) and the transfer of bacterial antibiotic resistance genes (which are used as selective agents in lateral gene transfers) back into the bacterial kingdom and ultimately to pathogenic species.

The first concern is a real possibility if there are genetically related species growing within pollen-transferable distance to GM crop agriculture. However, it can be argued that with the initial species used for GM agriculture (corn, cotton, soybean, and rapeseed) that in the USA and Canada there are no genetically compatible native plants.

Additionally, trans-genus gene transfer by natural mechanisms has never been observed, nor, in spite of centuries of trying, achieved. However, as GM crop development is expanded, a choice of culture habitat must be taken into account in applying the technology to additional species. For example, the culture of a glyphosate-resistant commercial variety of bent grass in North America could open the way to the spread of the resistance gene into native, wild bent grass varieties.

However, this should not be a concern for its use in lawns and golf courses in non-bent grass areas such as arid areas in the Middle East. The second concern about lateral transfer is that bacterial antibiotic resistance genes (used to aid the selective transfer of desired genes) cross kingdom lines and return to the bacterial world.

Such transfer is an extremely remote possibility and would be an unprecedented event, not possible via documented, naturally occurring mechanisms of gene transfer described below. Transmissible antibiotic resistance between bacteria was first encountered in the 1960s.

The resistance was shown to result from two factors:

(i) a gene coding antibiotic resistance, and

(ii) a DNA carrier molecule that is capable of being transferred from one cell to another.

The rapid dissemination of antibiotic resistance in bacterial populations has been and continues to be a major concern in the practice of medicine. However, antibiotic resistance genes by themselves have proven to be a highly useful tool in genetic engineering.

DNA that carries genes to be laterally transferred is joined to DNA carrying a gene that confers antibiotic resistance, and the hybrid DNA molecule is transferred to the recipient organism. Antibiotic-resistant clones are selected. Offspring that show antibiotic resistance are then tested for cotransfer of the gene of interest.

The concern that GM foods developed in this way may become an additional source for the spread of antibiotic resistance to bacteria is without foundation because the second element, an *appropriate* DNA carrier molecule required for the transfer, is absent. The transfer of DNA from one bacterial cell to another depends on *conjugative* plasmids, temperate phage or species specific transfer of naked DNA.

There is no evidence for conjugative plasmids or temperate phages that can cross between the plant and animal kingdoms, and there are only a few species of bacteria capable of taking

Table 8.2: Common plant toxins and anti-nutrients.

Toxin family	*Examples of occurrence in plants*	*Effect on humans and animals*
Cyanogenic	Sweet potatoes, stone lima beans	Gastrointestinal inflammation, glycosides fruits, inhibition of cellular respiration
Glucosinolates	Rape (canola), mustard, radish, cabbage, peanut, soybean, onion	Goiter, impaired metabolism, reduced iodine uptake, decreased protein digestion
Glycoalkaloids	Potato, tomato	Depressed nervous system, kidney inflamm-ation, carcinogenic, birth defects, reduced iron uptake
Gossypol	Cottonseed	Reduced iron uptake, spermicidal, carcinogenic
Lectins	Most cereals, soybeans, other beans, potatoes	Intestinal inflammation, decreased nutrient uptake/absorption
Oxalate	Spinach, rhubarb, tomato	Reduces solubility of calcium, iron and zinc
Phenols	Most fruits and vegetables, cereals, soybean, potato, tea, and coffee	Destroys thiamine, raises cholesterol, estrogen-mimic
Coumarins	Celery, parsley, parsnips, figs	Light-activated carcinogens, skin irritation

up and incorporating naked DNA. In these species for the uptake of DNA to occur, the DNA must carry speciesspecific signature sequences that are rarely found in other DNAs.

It is very improbable that an antibiotic resistance gene present in a plant product without the appropriate signature sequence would find its way into a bacterium. This possibility is made more remote by the fact that there are enzymes in the digestive tract that would degrade the DNA. Although there is an extremely low possibility for the transfer of antibiotic resistance genes from plants back to bacteria, the industry has not taken this possibility lightly and has adopted a precautionary approach in order to preclude any problems from this possibility.

The genes that convey antibiotic resistance chosen for use in GM crop development provide protection against antibiotics that are clinically ineffective, either by their nature, or because resistance to them is already widespread. In the meantime, a search for alternative, less controversial, selectable markers has been initiated.

Creation of Superweeds

The concern that GM plants may escape to become weeds that cannot be controlled because they are resistant to herbicides is also unlikely. The majority of *agricultural* varieties are highly selected for yield of desired product and not for *aggressive* habitation.

Most of them are poor competitors if left in the wild and are almost entirely dependent on humans for their perpetuation. Plant breeders have selected for plants that have large, starchy seeds with thin coats that stay attached to the plant longer.

They have also selected for plants that contain fewer seeds. The factors that have been bred into plants to increase their value as a food source have ultimately curtailed their ability to competitively reproduce as a seed plant. The genetic alterations that have been put into plants do not change this at all. The seeds are no more competitive or able to go out and establish themselves than they were before. A greater risk may come from a broad application of glyphosate or another herbicide.

The widespread and continued use of a single herbicide provides a real chance for the selection of spontaneous mutants in the background population of native species. Continued use of the chemical would only help the spread and eventual replacement of native flora, including weeds, by resistant variants. The appearance in 2005 and spread of glyphosate-resistant southern pigweed *(Amaranthus palmeri)* in three areas of the USA (Georgia, Arkansas, and Tennessee) shows that these concerns are real and justified.

The same argument applies to the widespread use of Bt-protected crops. When insect pest variants that are resistant to the toxin appear, a continued cultivation of Bt crops would support replacement of the sensitive parent strain(s).

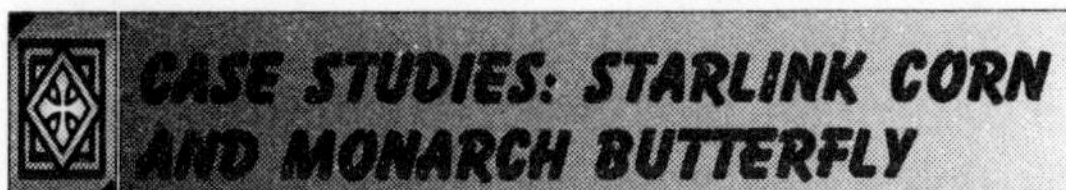

CASE STUDIES: STARLINK CORN AND MONARCH BUTTERFLY

The general public's awareness of the potential problems arising from the employment of GM products was raised by two reports that were widely circulated in the press: wind dissemination of Bt pollen upsetting the food net outside of cultivated areas, and new allergens produced by Starlink GM corn.

The report in the journal Nature states that there is a potential danger to the Monarch butterfly feeding on milkweed near fields growing corn producing *Bt-resistant* pollen. The study claims that the pollen, when ingested with the leaves, is poisonous to the insect larvae and could stunt its growth or kill it.

While subsequent studies showed that the larvae are susceptible to the Bt toxin, the likelihood of a sufficient amount of pollen falling on nearby leaves to cause larvae any harm is negligible, but the *secondary* reports gained only brief mention in the press. Another problem widely reported in the press was of people reporting allergenic reactions following ingestion of foods containing Starlink corn. Starlink corn is a variety of GM corn that is Bt-resistant and resistant to glyfosinate herbicides.

There was concern that Starlink corn might pose an allergy risk and accordingly it was not approved for human *consumption*; however, it was approved for use as animal feed. Because of miscommunication between farmers and Aventis (the company that *produced Starlink* corn) some of the GM corn was mixed in with conventional corn and used in food products highly consumed by humans, namely TacoBell and Mission foods. Twenty-eight cases of allergic reactions attributed to the ingestion of Starlink corn were reported. The CDC conducted a thorough investigation of the individuals and concluded that the individuals did experience allergic reactions; however, there was no evidence that the reactions were associated with the presence of Starlink corn in the corn products they consumed. Although these two reports of GM problems turned out not to be supported, the issue of potential problems and safety is really a case-by-case affair and concern about the technology remains unchanged.

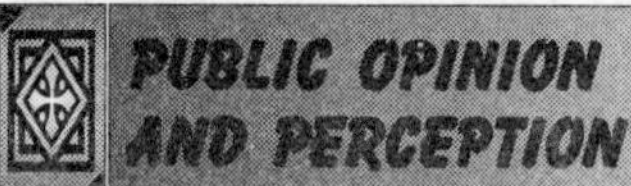

PUBLIC OPINION AND PERCEPTION

The negative publicity generated by the reports of pollen toxicity and Starlink corn allergens raised concerns about the public's acceptance of GM foods. Several surveys were carried out to gauge the public's attitude to GM products.

The results of the polls were inconsistent. Some surveys showed a 70% positive response when the questions were asked with reference to the benefits of GM products or a 75% negative response when questions were framed around the risks and concerns for the technology.

What these and other surveys showed is that the general public is ignorant of GM technology and its level of *implementation* in this country. The development of GM products is based on sophisticated genetic techniques, and the general public is *woefully* ignorant of even the basics of genetics.

For example, a 2004 survey of a representative cross-section of the American public found that only 15% of the respondents were sure that the incorporation of a catfish gene into a tomato would not produce a fishy tasting fruit and, even worse, only 9% of the respondents were confident that tomatoes contained any genes at all.

THE PRESENT AND FUTURE APPLICATIONS OF GM FOODS

GM foods were introduced commercially in 1994. Initially a limited number of farmers were willing to take the risk that GM crops would be marketable. Ultimately this concern turned out to be unfounded in the US market, as GM foods passed quietly into the food supply free from announcement or identifying labeling.

As a result, the vast majority of people in the USA consume GM foods but are unaware that they do so; the products appear *unchanged* and no problems have been detected from their consumption. GM crops provide farmers higher yields and lower costs, leading to more farmers planting more acreage each year.

Figure elsewhere in this chapter shows that for the last 7 years (1999-2005) the acreage of GM plantings in the USA and the world as a whole has been increasing exponentially.

Additionally, the world, including the USA, shows more than a doubling in the number of acres of GM crops planted over the period. The USA had a head start on the *employment* of GM agriculture and, as of 2005, still *accounted* for nearly 55% of the world's GM acreage. Genetically modified forms of three crop species, corn, cotton, and soybean are now well accepted. These three species account for 94% of the land cultivated with GM crops in the USA.

A breakdown of the statistics for the planting of these crops in the USA shows that by 2004, GM cotton had reached 76% replacement and its use was growing exponentially. Soybean and corn exhibit even greater rates of replacement, plantings extrapolate to 100% replacement by 2006 and 2008, respectively. The USA has, by default, become a giant demonstration project for the safety of GM crops. Still, when surveyed, consumers would like to know what foods contain GM products.

However, with the depth of penetration of some GM products, it may be simpler to follow the lead of pesticide-free and organically grown foods and label products as non-GM or GM-free. At this time, however, the public's general lack of *awareness* or concern about GM foods leaves the food industry without any impetus to initiate labeling beyond that required by the Department of Agriculture for nutritionally altered GM products. What is the future for GM foods?

With any new product there are always risks, and unforeseen problems can arise. With GM foods, each new construct needs to be tested thoroughly to help ensure that the desired *properties* are the only changes to the organism. However, the same argument applies for conventional breeding methods. For example, the use of conventional breeding methods to develop commercial varieties of potatoes and celery with high levels of insect resistance resulted in an increase of natural toxin production to levels that made the plants toxic to humans as well as to insects.

These results support the use of GM technology to incorporate a foreign gene with targeted toxicity as a less risky approach to restrict insect damage. Currently, two different Bt toxin genes have been introduced into cotton (Bollgard II from Monsanto) and corn (Herculex XTRA and YieldGard Plus from Pioneer) that together provide protection to rootstock as well as to top growth. Monsanto is experimenting with Roundup Ready Flex varieties of cotton that contain two glyphosate resistance genes in order to raise glyphosate tolerance in mature plants. Beyond crop protection there appear to be an unlimited number of possibilities for the incorporation of genes would add to the nutritional value of foods.

One example is "golden rice". Much of Asia suffers from a diet poor in vitamin A. With rice, the primary food source in Asia, a project was initiated to construct a strain of rice rich in vitamin A. This goal was achieved by the incorporation of vitamin A-producing genes from the daffodil. The resulting golden rice variety is now under cultivation to build sufficient seed stock, as it awaits full deregulation so that it can be put into general food production. Another example is high lysine corn. The lysine content of proteins found in corn kernels is inherently low, so low in fact that corn cannot be used as a sole protein source in the feeding of humans or livestock.

With humans, corn is often served with beans to achieve a *nutritionally* balanced diet; with animals the amino acid lysine, produced from bacterial culture, is added to the feed to bring the level of lysine to an essential level. There was hope that conventional genetics could solve the problem, especially following the discovery of a naturally occurring mutant maize strain with kernels containing protein with a significantly raised level of lysine.

However, in cultivation, the fragility of the kernels, which opens them to disease, and the still inadequate level of lysine to meet *nutritional* requirements doomed its general adoption. Recently, a GM approach was used to resolve the lysine problem, which was to introduce a bacterial gene

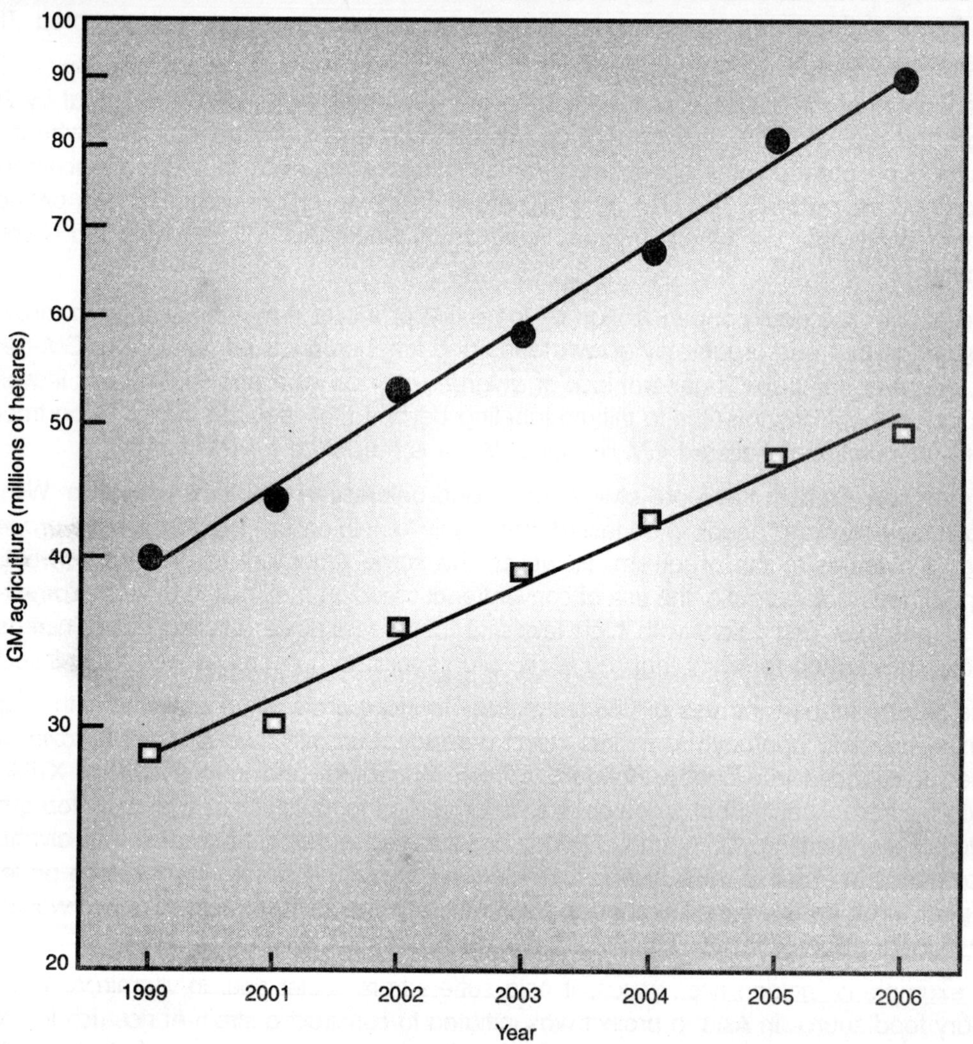

Figure 8.1: Hectares used for GM agriculture in the world and in the USA. • Values for plantings in the world, ❑ values for plantings in the USA. The world value of 81 × 10^6 ha in 2005 is one third of the world's arable land under cultivation.

into the maize genome to raise the level of lysine. Two considerations were taken into account in the construction: first the location of the bacterial gene with respect to the maize genes and second, biochemical properties of the introduced lysine *synthesizing* enzyme.

Knowledge of the maize genome permitted the placement of the bacterial gene under the control of a maize gene predominantly expressed in grain, and so did not alter the overall biochemistry of the plant. The bacterial gene for lysine production was chosen for its lack of biochemical regulation and the ability of its enzyme product to produce unrestricted amounts of lysine.

The resulting strain, termed Lysine Maize LY038, contains five times as much lysine in its kernels as standard corn, and three times as much as the high lysine mutant strain mentioned above, while neither altering the biochemistry in the rest of the plant nor *compromising* the quality of the kernels.

In fact, the level of lysine is so high in this strain that it should be possible to mix it 1 : 1 with standard corn and still meet the dietary requirements of all animals. A still greater possibility for GM technology lies in the expansion of the number and variety of plant species used in agricultural production. Today the vast majority of plant species are not acceptable for *consumption* as food because they produce and contain significant amounts of toxic compounds.

Eliminating toxin-producing genes could add valuable plant food resources and minimize the dangers of relying on too few species for our world food supply. The USA remains a verdant land for the continued development of GM *technology*.

Chapter 9

GENOMES OF LARGE CROPS

Plant genomes vary enormously in size. A part of this variation is generated by polyploidy, which is ubiquitous in the plant kingdom; however, even between closely related, ostensibly diploid species, it can still vary by an order of magnitude. A notable, but not atypical example is the contrast between rice (1 C DNA content of 0.50 pg, equivalent to 450 Mbp) and barley (5.55 pg, 5300 Mbp).

The gene content of these two species is thought to be rather similar, numbering something under 40,000, depending on the gene prediction program employed. Thus, much of the difference in DNA content is made up of nongenic DNA-in particular, retrotransposons. When large-scale genome sequencing became possible in the 1990s, the large size of the majority of the leading crop genomes was technically and financially prohibitive.

This prompted the plant research community to identify species (in particular *Arabidopsis thaliana)* with more tractably sized genomes as genomic models. Technical improvements in the efficiency of sequencing achieved the finishing of the *Arabidopsis* genome by 2000 (4 years ahead of schedule) and the sequence was released with some fanfare in *Nature*. At the time, *Arabidopsis* represented one of the first eukaryotes to be sequenced fully (along with *Saccharomyces cerevisiae,* human, and *Caenorhabditis elegans).* Its protein-encoding gene content has been estimated to be about 25,000.

In the meantime, the genome sequence of *Arabidopsis* has been joined by those of a bewildering and ever-growing list of eukaryotic and prokaryotic organisms numbering over 300 as of December 2005. Of the 40 fully sequenced eukaryotic genomes, 25 belong to simple organisms (protozoans and fungi), 7 are vertebrates, 3 are insects, 2 are nematodes, and 3 are plants (of which 2 are the *indica* and *japonica* subspecies of rice). The divergence of the monocot from the dicot clade is an ancient event, currently dated using molecular clock methods applied to the chloroplast genome at 140 to 150 MYA during the late Jurassic to early Cretaceous periods.

Independent estimates based on mitochondrial sequences have placed it somewhat earlier, at 170 to 235 MYA. Dating of the time of speciation within each clade has been attempted by applying molecular clock methodology to repetitive sequences such as retrotransposons, but sequence homology in this class of element between clades is insufficient to use this method to date the monocot–dicot divergence.

Thus, it was recognized at an early stage that the *Arabidopsis* genome sequence would probably be of only partial relevance to monocot genomes. With a genome size about three times larger than that of *Arabidopsis,* rice was rapidly identified as the donor of a suitable model monocot genome.

Before completion of the rice genome sequence, it became apparent that only a poor level of commonality in gene order existed between *Arabidopsis* and rice, thereby justifying post hoc the need for a separate model for the two major plant clades. Nevertheless, the two genomes do retain some similarity as a result of common descent. Although some 85% of *predicted Arabidopsis* proteins were found to share significant homology with those of rice, about a tenth of them show a strong level of conservation; in addition, most monocot–dicot homologs maintain exon order as expected.

Perhaps most surprisingly, in many homologs, intron number, position, and even relative size show a remarkable level of conservation. Despite the apparent disparity in gene number between the two models (25,000 vs. 40,000), it has recently been claimed that only a few hundred, or at most a few thousand, rice genes appear to lack close homologs in *Arabidopsis.* The infrastructure and efficiency of whole genome sequencing is now at a point at which it has become much more realistic to undertake on a large scale.

Current crop species targets include oat, *Brassica* spp., orange, coffee, barley, soybean, cotton, ryegrass, alfalfa, tomato, banana, bean, poplar, castor oil, sorghum, and maize. A growing number of other species has been targeted for sequencing of the gene space (ESTs or similar). If these trends continue, it is likely that within 10 years, most of the major crop genomes will have been fully sequenced.

In the meantime, species that are nodal in crop phylogenies may be chosen to serve to generate a network of submodels; a particular example of this lies behind the current proposal to sequence the grass *Brachypodium distachyon.*

This chapter attempts to take stock of model genomes' contribution to understanding of the genomes of crop species to date. Perhaps other contributors to this volume will show the lasting value that model species biology has made to crop improvement.

Dicot Models

Arabidopsis Thaliana (Thale Cress)

Arabidopsis is by far the most well developed of the crop plant models. In addition to its completed genome sequence, it is easily transformable and enjoys a huge range of genetic (mutants, mapping populations, ecotypes) and genomic (cloned genes, libraries, arrays, markers, etc.) resources and an ever expanding database relating phenotype to genotype. The closest crop relatives to *Arabidopsis* are the three diploid *Brassica* species *rapa, nigra,* and *oleracea* that carry, respectively, the A, B, and C genomes as described elsewhere in this chapter.

Although all of these represent rather minor crop species, the major contributor of *Brassica spp.* to agriculture is *B. napus* (oilseed rape or canola), which is an AC allotetraploid formed from the combination *B. oleracea* × *B. rapa.*

The lineages of *Arabidopsis* and *Brassica* are thought to have diverged from one another between 14 and 20 MYA; this divergence has included a number of distinct polyploidization events because the present-day diploid *Brassica* spp. carry multiple paralogous copies of chromosomal segments collinear with the *Arabidopsis* genome.

This copy number is most commonly three, so the inference is that the diploids must have evolved from a hexaploid ancestor. Copy number is frequently less than three, varying in 4× *B. napus* from four to seven.

Within the triplicated paralogs, a common pattern of interspersed gene loss is emerging, with the result that each paralog typically carries a slightly different spectrum of the full gene set presumably present on the progenitor segment.

A further complication is that *Arabidopsis,* as revealed from its genome sequence, is a cryptic polyploid, carrying a sufficient number of large segmental duplications for an evolutionary history of at least four different large-scale duplication events to have been proposed.

Overall, an estimated 74 translocations, fusions, deletions, or inversions separate the genomes of *Arabidopsis* and *B. napus,* of which about one half are common to A and C genomes in present-day oilseed rape.

Lotus Japonicus (Trefoil) and Medicago Truncatula (Barrel Medic)

The Fabaceae, one of the largest families of flowering plants with 650 genera and over 18,000 species, is distinguished from other dicot families by its symbiotic relationship with nitrogen-fixing *Rhizobium.* The economic and nutritional importance of nitrogen fixation has been sufficient to justify targeting a model representative, and two competitive species are currently being pursued.

Medicago truncatula has some importance in its own right as a forage crop in Australia. It has a small diploid genome (1 C DNA 0.48 pg) and a rapid generation time, is self-fertile, transformable, and is a prolific seed producer. *Lotus japonicus* is a short-life-cycle, perennial wild legume that also has a small genome size (1 C DNA 0.48 pg).

The genomes of both species are currently being sequenced. The two sequences show a high degree of similarity to one another. Collinearity between *M. truncatula* and pea at the level of coarse genetic maps appears to be encouragingly high, although there is significant sequence divergence between those of *Lotus* and the major legume crop species soybean.

In a computational approach, *Lotus, Medicago,* and *Glycine* unigenes were BLASTed against non-legume unigene sets and the rice genome sequence to define legume-specific gene motifs; this delivered some 2500 such contigs, of which less than 3% showed any homology to any previously identified legume genes. Such results underline the utility of a model legume to define sequences specific to this group of agriculturally important crop species.

***Populus Trichocarpa* (Poplar or Black Cottonwood)**

Conventional genetic approaches in trees are limited by the large size, long generation interval, and outcrossing mating system of most species. The need for a tree model reflects the importance of many traits that are not shared by an herbaceous annual plant such as *Arabidopsis*. Important among these are wood formation, longevity, seasonal growth, and hardiness.

The genus *Populus* consists of 30 to 40 species, 4 of which have significant commercial importance. Selection and hybridization programs in poplars began in North America in the 1960s, and the most commonly exploited crosses have involved *P. trichocarpa, P. deltoides, P. nigra, P. grandidentata, P. alba, P. tremuloides,* and *P. tremula*. Because the genomic resources of *P. trichocarpa* were the most developed at the time that genome sequencing was proposed, this species became the accepted tree model.

It was chosen as the first tree for genome sequencing largely because of its modest genome size (0.6 pg)—about 40 times smaller than that of pine, the most important of all forestry species. It also has a number of other advantages over potential alternative tree species specifically related to its rapid juvenile growth, which allows for phenotypic assessments to be made relatively quickly; its wellestablished transformation and regeneration protocols; and the pre-existence of a body of genetic mapping, which includes placement and tagging of a number of quantitative trait loci (QTL). The final draft sequence was scheduled for release in early 2005, but is still awaited at the time of writing.

Monocot Models

***Oryza Sativa* (Rice)**

Rice is the pre-eminent monocot model and is uniquely both a model and a crop in its own right. The particular importance of this duality lies in the much greater potential that this allows for transferring phenotype, as well as genotype, from model to crop. Rice is a tropical species and thus more likely to share pathogens and/or abiotic stresses with its tropical crop relatives such as the millets (and, to a lesser extent, maize and sorghum) than with its important temperate small-grain and pasture-grass relatives (wheat, barley, rye, oat, and ryegrass).

Nevertheless, shared morphology and crop architecture among all cereal species do allow many phenotypic connections to be made. The dicot models, in contrast, are far removed in a crop morphology, making such transfers much less predictable. The grasses belong to the Poaceae, which evolved from a common ancestor some 50 to 60 MYA; together, they provide an estimated 60% of global human calorific intake. The family includes at least 10,000 species, classified into 650 genera.

The crop species within the family fall into the three subfamilies: Pooideae (which includes the temperate cereals and ryegrass), Panicoideae (maize, sorghum, millets, sugar cane), and Bambusoideae (rice). Until the development of generic DNA technology, primarily in the 1990s, genetic research in each grass crop was conducted in isolation from that in the others.

Before this time there was no secure way of verifying what had already been suspected for some time: that because these species were related by (albeit distant) descent, they were likely to share genetic content and, at least at a basic level, genetic mechanisms. The first demonstration of what is now referred to as "comparative genetics" was carried out in the Solanaceae, where common RFLP linkage relationships in tomato and potato were uncovered using DNA probes developed from a tomato template.

The concept spread quickly to the Poaceae, and numerous crossspecies comparisons began to appear in the literature during the early to mid-1990s. These led to the construction of partial consensus maps linking maize with sorghum and wheat with barley and rye.

A synthesis of these maps was generated by relating them all to that of the rice genome. The concept of "synteny" elaborated by these cross-species comparisons of gene order reflects conservation over evolutionary time at the macroscale. Whether this was extendable to the microscale was questionable, given the large variation in genome size between individual Poaceae species.

The outcome of sequence-based comparisons in selective syntenic regions is that although gene structure and sequence are extremely well conserved between taxa, intergenic regions are highly divergent, even at the level of genotypes within a taxon. Much of this intra- and interspecific divergence is generated by retroelement activity and, in particular, helitron-like transposons composed of multiple genederived fragments.

In addition, the increasing body of evidence generated from large-scale sequence comparisons between related taxa demonstrates how synteny is also disturbed by the presence of species-specific localized duplications and other forms of genome reorganization. By the end of the 1990s, with *the Arabidopsis* genome project already well under way, rice became an increasingly attractive candidate for whole genome sequencing in the private and public sectors.

These efforts were combined to produce almost full genomic sequences of *japonica* and *indica* subspecies, along with a nearcomplete compendium of full-length cDNA sequence. The finished sequence currently covers about 95% of the genome, including most euchromatic regions and 2 (out of 12) complete centromeres. Mirroring the situation in *Arabidopsis,* the genome sequence has revealed a history of polyploidization in the evolution of modern day rice, with about half of the gene content duplicated as paralogs.

***Brachypodium spp.* (False Bromes)**

The false bromes are a group of non-cultivated grasses, mostly regarded in agriculture as weeds rather than as beneficial plants. The perennial *B. sylvaticum* (slender false brome) and the annual *B. distachyon* (purple false brome) have been suggested as intermediate models for the temperate cereals. *B. distachyon* has been claimed to have a genome size indistinguishable from that of *Arabidopsis,* but measurement of 1 C DNA content suggested that it is three times larger.

The genome size of *B. sylvaticum* is slightly higher still (0.48 pg), but both genomes are smaller than that of rice. The value of both as genomic models for the temperate grain cereals lies in their membership within the Pooideae clade and hence their much closer relationship to wheat, barley, rye, and oats than rice enjoys. The significance of this relationship has been confirmed in two recent positional cloning projects, one in wheat and the other in barley since both the quality of probe hybridization to and prediction of overall gene content in the target were superior in *Brachypodium* to that offered by rice.

Although *B. sylvaticum* has been proposed to date only as a genomic and not a biological model, *B. distachyon* does have a number of generic advantages as a functional genomic and biological model (self-fertility, in-breeding habit, short life cycle, small size [approximately 20 cm at maturity], lack of seed-head shatter, and undemanding growth requirements). At the time of writing, there is a concerted effort to develop *B. distachyon* as a fully functional genomic model, but this proposal remains controversial.

Harnessing Model Genomes for Crop Genetics and Improvement

The impact of model genomes on crop species has been felt mainly in their delivery of a strategy for gene isolation in the large genome crop species. This strategy relies on the maintenance of synteny, assuming that gene content in the model in a specific genomic region is more or less conserved in the target crop genome. The model-tocrop paradigm follows a combination of:

Mapping a trait to a defined genetic interval in the crop

Identifying the corresponding genomic region in the model via the use of common genic markers (because it is substantially only the gene content, not the nongenic, largely retrotransposon-containing, repetitive content that is conserved across clades)

Identifying a potential candidate sequence in the model on the basis of a relationship between predicted gene function (derived from the annotation of the model genome) and the target trait

Validating the crop homolog of the candidate, demonstrated by allelic association and/or mutation complementation

The first major success of the model-to-crop genomic approach in the monocots came with the isolation of the "green revolution" wheat semidwarfing genes *Rht-B1* and *Rht-D1*. Together, these two genes have been responsible for probably the most far-reaching and widespread change in the appearance of any crop worldwide.

Their incorporation into the breeding pool has generated shorter plants that enjoy an enhanced grain yield potential, thanks to the consequent increase in harvest index, and are responsive to higher application rates of fertilizer without becoming liable to straw collapse. The isolation of these genes predated the availability of the full rice or *Arabidopsis* genome sequence, but nevertheless relied heavily on genomic information from both model species.

Critical to the success of their cloning was that the physiological nature of the semidwarf variants of wheat was similar to that of previously characterized mutants in maize and *Arabidopsis*. This allowed an approach whereby the rice ortholog of *the Arabidopsis gai* gene was identified from a rice EST collection.

When this rice sequence hybridized to wheat DNA at the genomic locations of the *Rht-1* genes, the rice probe was exploited to extract the full genomic sequence of both of the wheat genes. Thereafter, the sequence and functional basis of these important semidwarf alleles were readily obtained. Finishing the genome sequences of the models enabled the model-to-crop paradigm to be tested.

A textbook illustration was provided by the recent successful cloning of the barley gene *Ppd-H1*, the major determinant of flowering time under long photoperiods. Unlike the situation with *Rht-1*, the physiological model provided by *Arabidopsis* was not informative because the candidate genes provided by *Arabidopsis* did not map to the genomic location of the barley gene target. Thus, the initial step was to fine-map *Ppd-H1* in a conventional cross between parents carrying contrasting alleles, and the linked markers thereby derived then allowed for construction of a physical contig based on the presence of key marker loci on barley bacterial artificial chromosomes (BACs).

The gene content of the homologous region in *Brachypodium sylvaticum* helped to define the matching region in rice, and the critical barley recombinants finally identified a region in the homologous rice segment that contained only a single candidate sequence.

This rice gene, *Os-PRR,* shares significant sequence homology *with Arabidopsis At-PRR7,* which, when mutated, leads to delayed flowering under long day conditions, just as the inactive form of *Ppd-H1* does in barley. *Ppd-H1* and *At-PRR* also share temporal patterns of expression. Finally, resequencing of the critical parts of *Hv-PRR* across varieties of known allelic status at *Ppd-H1* was able to demonstrate a correlation between a functional glycine to tryptophan change in a domain of the gene that is well conserved across taxa.

A more elaborate but essentially equivalent strategy was used to isolate the wheat gene responsible for determination of winter habit (vernalization requirement). Once again, a large mapping population, this time in the diploid wheat *Triticum monococcum,* was used to delineate a genetic interval of <0.1 cM containing the target.

Sequencing of the 324 kb represented by this segment identified two genes, with no additional candidates present in the homologous segments of rice or sorghum. Both candidate genes had *Arabidopsis* homologs, but only one of them, *AP1,* is required for the transition between vegetative and reproductive phases in *Arabidopsis;* the other is a floral meristem identity gene.

The association between sequence variation at *Tm-AP1* and phenotype was established by demonstration of three independent deletions distinguishing the promoter sequence of spring from winter accessions. The most recent example of positional cloning in a monocot crop that has relied on the availability of model genomes is the isolation of the *Ph1* locus in wheat.

This "gene" is responsible for the diploid-like inheritance of hexaploid wheat, and its isolation was hampered at the outset by a lack of any verifiable allelic variation. Because of this, it was not possible to generate a fine-scale genetic map as a first step to defining the target genomic region. Instead, a series of overlapping deletion mutants was generated, and phenotype (loss of diploid-like chromosome pairing) was related with loss of genic markers in the *Ph1* region, which had been derived from synteny comparisons between wheat and rice and/or *Brachypodium.*

As a result, the number of genes present in the smallest genetic interval defined was over 30, and because the effect of *Ph1* is specific to polyploids, there were no sensible leads derived from the predicted function of any of these candidates. To progress beyond this point, it was necessary to sequence a substantial tract of wheat DNA directly; the identity of the locus was finally determined through an internal comparison among the individual A, B, and D genome segments.

A reasonable level of synteny between *Arabidopsis* and *Brassica* exists, the complications of segmental duplication notwithstanding, and the finished *Arabidopsis* sequence has been available for longer than that of rice; however, gene isolation in *Brassica* has relied more on functional homology than on positional cloning.

Thus, having established function of a gene in *Arabidopsis,* primarily by mutation/complementation, homologs in *Brassica* have been extracted from genomic or cDNA libraries and function in *Brassica* established by associating variation in phenotype with polymorphism at the RFLP or sequence level. Beyond the *Brassica* spp., high rates of sequence divergence have greatly inhibited the success of orthologous cDNAs as hybridization probes against genomic DNA and restricted the applicability of the model to its immediate relatives.

PERSPECTIVE

The value of model plants in a strictly genomic context is probably ephemeral. This is primarily because large genomes are increasingly considered practical to sequence on cost or technical grounds. Within 10 years, it is likely that most of the major crops will have been sequenced, at least with respect to their gene space. At the same time, comparative genomics is showing that although gene order at the macroscale is well conserved over large taxonomic distances, the microsynteny necessary to predict sequence across species (and even, to a surprising extent, within species) at the microscale is insufficient for a small number of models to be able to serve many diverse crop species.

The cereals are exceptional in this respect, in that so many cereal crop species are clustered within a narrow taxonomic clade, but even for these, the models have their limitations. The more lasting value of models will surely lie in the insights into plant biology that they will allow.

Some of these will include the rapidly developing fields of epigenetic and micro-RNA-directed gene regulation, where *Arabidopsis* is already serving as a model organism for species well beyond the plant kingdom. Many of the more specifically plant-orientated areas of biology informed by model species are covered by other contributions to this volume.

10

Chapter

GENETIC ANALYSIS AND IMPROVEMENT

Brassica species are the closest crop relatives to the reference dicotyledon plant species, *Arabidopsis thaliana;* they contribute to a diverse range of agricultural and horticultural crops worldwide, encompassing vegetable, salad, oil, mustard, fodder and nonfood uses.

Throughout the genus, taxa are characterized by the wide range of developmental adaptations, many of which have been domesticated into crops that include oilseed rape/canola and swede (*B. napus*); cabbage, cauliflower, broccoli, Brussels sprout *(B. oleracea);* Chinese cabbage, pak choi, turnip, and oil (*B. rapa*); and the mustards and associated oils crops (*B. nigra, B. juncea, B. carinata*).

These crops contribute basic food energy, nutrients, and secondary metabolites to human and livestock diets, as well as providing a potentially increasing number of nonfood uses. In addition, brassicas can be beneficial as break crops with soil remediation properties and contribute to rotational cropping systems.

Brassica species are naturally mostly outbreeding, with a strong and well described sporophytic self-incompatibility system. Among the major challenges facing crop improvement in changing economic, market, and climate conditions is the ability to harness genetic diversity through an information-led approach that maximizes information from model species.

For brassicas, as with other crops, there is a wide range of valuable traits for which genetic variation exists, but where understanding is required.

These include improving harvest index and yields in the context of changing climate and reduced inputs, optimizing harvestable and processed product quality fit for purpose, and identifying the scope for increased added value through nutritional, prophylactic health, or nonfood use. A wide range of genetic resources are available for *Brassica*. These may be used in genetic analysis or as a source of alleles for introgression into new crop varieties. There is considerable scope to accelerate introgression by use of gene- or locusspecific genetic markers.

An increasing amount of genomic information is accumulating, together with ready access to information derived from the related Brassicaceae model plant *Arabidopsis*. There has been considerable progress in the genetic analysis of agronomic and related plant traits in *Brassica*. However, compared *with Arabidopsis,* there has been relatively slow progress in identifying and characterizing the behaviour of the underlying genes, genomic regulatory networks, and associated metabolism.

This has partly resulted from the complexities of genome organization, based on segmental chromosome duplication and divergence, existing within the diploid brassicas. These are compounded in the amphidiploid species such as *B. napus* and *B. juncea*. For effective crop-based research, it is essential to be able to navigate between trait and gene and thus integrate information from agronomy, breeding, genetics, and genomics. Genomic information is the key to exploiting knowledge gained at the level of gene expression, biochemistry, metabolism, and physiology.

To manipulate crop traits based on genomic knowledge, genes need to be located in their relevant genomic context in order to understand their regulation and any evolutionary selection pressures. In practical terms, this knowledge may then be used to develop locus-specific molecular genetic markers for use in marker-assisted selection or to introduce novel alleles via site-selected mutagenesis or transgenic modification. At the time of writing, developments within the international research communities have led to establishment of the Multinational *Brassica* Genome Project, which was initiated in 2002.

This is a long-term project, an early output of which is the Multinational *Brassica rapa* Genome Sequencing Project, which is generating contiguous sequence of gene-coding regions over the *B. rapa* "A" genome. This builds on previous initiatives to develop public-access genomic resources, which have included expressed sequence tags (ESTs), bacterial artificial chromosome (BAC) genomic libraries, BAC-based physical contigs, saturated sequence-tagged genetic maps, and associated populations.

This chapter will concentrate on providing examples of current developments in *Brassica* research that have benefited from genetic and genomic approaches, highlighting areas well positioned to capitalize on data, information, and knowledge from the model *Arabidopsis* for the benefit of crop improvement.

As such, it does not aim to provide a comprehensive review of *Brassica* research, which has a long and rich history based on diverse genetic resources and breeding systems amenable to genetic analysis. The central role of comparative genomic approaches will be highlighted, together with an assessment of current gaps in knowledge.

COMPARATIVE GENOMICS

The ability to carry out comprehensive comparative genomics depends upon the adoption of common standards and nomenclature for describing various constituent entities or objects such as linkage group, locus, gene identity, and trait. It has taken some time for the linkage group nomenclature for *Brassica* species to converge because different researchers have developed linkage maps based upon diverse sets of arbitrary markers.

The availability of sequence-tagged markers and subsequent development of maps anchored to the *Arabidopsis* genome is currently accelerating convergence in this area. The linkage group nomenclature established by Parkin et al. and Sharpe et al. for *B. napus,* by Bohuon et al. for *B. oleracea,* and by Lagercrantz et al. for *B. nigra* has now been adopted to describe other linkage maps.

This nomenclature scheme allows alignment of linkage groups for the amphidiploid genomes with their constituent diploid linkage groups. Thus, *B. rapa* (R1-R10) corresponds to the *B. napus* (N1-N10) A genome, and *B. oleracea* (O1-O9) corresponds to the *B. napus* (N11-N19) C genome. For *B. oleracea,* the linkage groups have now been aligned and oriented with respect to the karyotype.

Recently, the pattern of chromosome segments conserved within the A and C genomes of *B. napus* has been aligned with collinear regions of the *Arabidopsis* genome.

Genome Organization

The relationship between the canonical *Brassica* genomes has been characterized in the schema commonly referred to as the "triangle of U." Three distinct diploid genomes or "cytodemes" are recognized and each is represented by a type species.

The genome sizes vary and have been estimated to range from 470 Mbp *(B. nigra)* to 1540 Mbp *(B. carinata).* In contrast, *Arabidopsis* species have five chromosomes, with a genome size for *A. thaliana* of ~120 Mbp.

Collinearity and Chromosome Segmental Duplications

Different hypotheses have been proposed to account for the origin of and relationships between contemporary crucifer genomes. Based on comparison of linkage maps, Truco et al. have proposed a possible chromosome phylogenetic pathway based on an ancestral genome of at least five, and no more than seven, chromosomes. A number of studies have focused on specific regions of *Arabidopsis* and compared them with the genome organization in *Brassica* species.

For example, a comparison of *B. napus* with *A rabidopsis* Chr 5 based on sequence-tagged (RFLP) markers revealed six highly conserved copies in *B. napus* that corresponded to an 8-Mb segment from Chr 5. This included a single inversion that appeared to be the primary rearrangement that accounted for two lineages since divergence of the genera. These results were used to suggest that the constituent genomes of *B. napus* were generated from a fusion of three ancestral genomes that together had strong similarities to the contemporary *Arabidopsis* genome.

This is consistent with the hypothesis that diploid *Brassica* genomes evolved from a common hexaploid ancestor. For specific regions, a genetic distance of 1 cM in *B. napus* was found to be equivalent to 285 kb in *Arabidopsis*, although this figure may vary considerably across the genomes. Within different regions of the genome, comparisons between *Arabidopsis* and *Brassica* suggest differential patterns of divergence, with reciprocal translocations described at a genetic level in natural and resynthesized amphidiploid *B. napus.*

Table 10.1: Relationships among the Species of *Brassica*, with Chromosome Numbers (n) and Indicative Genome Sizes.

Genome	Species	n =	~Genome size
A	*B. rapa*	10	500–550 Mbp
B	*B. nigra*	8	470
C	*B. oleracea*	9	600–650
AB	*B. juncea*	18	1100–1500
AC	*B. napus*	19	1130–1240
BC	*B. carinata*	17	1540

The first evidence for this came from RFLP probes, which allowed detection of a reciprocal chromosomal transposition found in several oilseed *B. napus* genotypes, and it involved exchange of interstitial homologous regions on N7 and N16. This was confirmed by cytological analysis on synaptonemal complexes. Up to a third of the physical length of the N7 and N16 chromosomes appeared to be involved, although only a few recombination events were detected in the region.

It is interesting that this region corresponds to an inverted segmental duplication that has been described in *B. oleracea* 06, with the self-incompatibility locus (S) located at or near the junction of the duplication. Higher seed yields were associated with parental configurations of the *B. nap us* rearrangement in segregating progenies, with complete complements of homologous chromosomes from the diploid A and C genome progenitors of *B. napus.*

A more comprehensive survey of mapped RFLP probes has shown that 73% of genomic clones detect two or more sequences in the *Brassica* A and C diploid genomes. Most duplicate loci appear to be in distinct linkage groups as collinear blocks of linked loci and display a variety of rearrangements, including inversions and translocations, following duplication.

The presence of some identical rearrangements can be taken as evidence that these occurred before divergence of the two species. For some of the linkage groups, their current organization appears to be consistent with earlier centric fusion and/or fission processes that may have played an important role in the evolution of *Brassica* genomes.

Overall, it appears that at least 16 gross chromosomal rearrangements can account for differences between these two diploid genomes since divergence from a common ancestor. It also appears that there are homologous loci in the C genome for almost every mapped locus in the "A" genome.

Given current efforts to sequence the complete *B. rapa* genome, this homologous conservation should facilitate the inference of gene location in other *Brassica* genomes through comparison of saturated physical maps and the complete genome sequence of *B. rapa.*

The ability to infer information about the organization of diploid *Brassica* genomes has been complicated by the recognition that a large amount of internal duplication has taken place within the *Arabidopsis* genome. Up to 80% of this genome appears to comprise duplicated sequences with about 20% of genes tandemly duplicated. Thus, the overall pattern of rearrangements and genome organization may not represent a simple one-to-one correspondence, as recognized by Lan et al.. At a more detailed level, O'Neill and Bancroft have demonstrated that the collinearity between a neighboring set of 19 predicted gene sequences in *Arabidopsis* and the corresponding replicated l oci in *Brassica* is characterized by local rearrangements and deletions.

The *Brassica* genome can be regarded as a mosaic of segments sharing a common ancestry with *Arabidopsis.* Parkin et al. have recently compiled a comprehensive data set that allows the pattern of duplicated chromosome segments in *B. napus* to be compared with the *Arabidopsis* genome. They used evidence from 359 sequenced *Brassica* RFLP probes to detect 1232 loci in *B. napus.* Comparative sequence analysis with the *Arabidopsis* genome revealed 550 homologous sequences from which they could infer relative chromosomal position.

They were able to identify 21 blocks conserved within *Arabidopsis* that, following replication and rearrangement, can account for almost 90% of the genetic map of *B. napus.* They estimate that a minimum of 74 gross rearrangements (38 and 36 in the A and C genomes, respectively) may have separated the two lineages since their divergence 14 to 24 MYA.

Pairs or sets of genes are said to be orthologous when they have diverged following a speciation event; paralogous genes arise as a result of duplication events. The organization of the diploid *Brassica* genomes can be represented as a mosaic of segments that align in varying degrees of collinearity with corresponding orthologous regions of *the Arabidopsis* genome.

The extent of collinearity between paralogous segments may differ significantly as a result of the time elapsed since their divergence and the degree to which they may have been subject to infection by transposable elements.

Genetic Markers

The use of molecular markers in *Brassica* has recently been reviewed in detail. The emphasis here will be on developments in genetic markers that have the potential to provide additional information about genome evolution between *Brassica* and other taxa, and *Arabidopsis* in particular.

The increased availability of markers amenable to reproducible high-throughput screening, combined with locus-specific linkage to genes underlying crop traits, provides a key technology for crop improvement involving marker-assisted selection.

Marker Systems

In general, simple sequence repeats (SSRs, or microsatellites) have the advantage that they are relatively often polymorphic, locus specific, sequence tagged, and transferable between laboratories. They have been isolated and characterized from *Arabidopsis* and a range of *Brassica* species and primer sets used to develop informative and locus-specific sequence-tagged genetic markers. The process of marker development has conventionally been based upon sequencing of clones randomly selected from SSR-enriched genomic libraries. When the original clone sequences are of sufficient quality and made available, the sequence flanking the repeat can be used for comparative genomic analysis.

For *B. rapa, up* to 90% of 228 microsatellites identified in one study were found to amplify corresponding regions in other *Brassica* species, and 40% of primer sets *amplified Arabidopsis* loci. A related technology, inter simple sequence repeats (ISSRs), involves using one locus-specific primer flanking the SSR together with a nonanchored primer.

This has been used to amplify and sequence 44 fragments from cauliflower, where the majority of the internal regions of the ISSRs had homologies with known sequences (e.g., from Arabidopsis)-primarily with protein-coding genes implicated in DNA interaction and gene expression. This is one line of evidence to suggest that there are long and numerous regions conserved with the *Arabidopsis* genome.

More recently, the availability of genome survey sequence (GSS) and other data sets for the *B. oleracea* genome has allowed informatic detection of SSRs. A comparison of different sequence-tagged molecular markers has been carried out in order to establish their relative ease of development. This involved using EST sequences from *Brassica* and *Arabidopsis* to design primer pairs to amplify gene sequences using sequence characterized amplified region (SCAR), cleaved amplified polymorphic sequence (CAPS), and PCR-RF-SSCP (a combination of CAPS and single-strand conformation polymorphism) markers.

The level of polymorphism detected was assessed in two sets of *B. oleracea* breeding lines. More single genes were amplified using primer pairs from *B. oleracea* data than from *B. rapa* or *Arabidopsis.* The PCR-RF-SSCP method was the most efficient, yielding informative assays between even closely related parent lines in situations where CAPS assays were uninformative.

Gene-Specific Markers and Comparative Genomics

Considerably more information about genome evolution and organization can be obtained by developing genetic linkage maps based upon markers directly associated with coding regions. For *Brassica* species, a number of technologies have been developed that make use of genic information from *Brassica* or *Arabidopsis.* The sequence-related amplified polymorphism (S RAP) technique was developed to provide a relatively high level of information per assay. SRAP is based on a two-primer system that involves core as well as filler sequences and provides a similar level of efficiency to AFLPs, while being technically simpler.

This approach was used to develop markers from *B. oleracea* cDNAs. Of these, 169 had similarity to genes in the *Arabidopsis* genome. Orthologous and paralogous genes were identified by the clear differences in their similarity score values. In common with other comparative studies, the resulting genetic map revealed extensive collinearity between the two genomes over chromosomal segments, including many inversions and segmental indels, as well as an uneven distribution of large-scale duplications.

However, based on the particular markers used, it was found that most of the duplicated segments corresponded to *Arabidopsis* Chr 1 and Chr 5, whereas representation in *Brassica* of segments from *Arabidopsis* Chr 2 and Chr 4 was lower. SRAP has also been used to determine the level of diversity among oilseed rape (*B. napus*) cytoplasmic male sterility maintainer and restorer lines.

In this case, 118 polymorphic loci were used to calculate similarity indices from between 0.46 and 0.97. A subsequent cluster analysis was successful in grouping the lines and was in agreement with existing pedigree data. A more directed approach for gene marker development has involved development of amplified consensus genetic markers (ACGM). Based on sequence analysis and design of PCR primers, this technique allows rapid sequencing of homologous genes from species within the same phylogenetic family, as well as detection of intragenomic polymorphism.

As a demonstration, a set of 32 ACGMs has been used to amplify genes from *Arabidopsis* and *B. napus*. The polymorphism detected with ACGMs is primarily associated with intron sequences and, in this study, allowed mapping of 43 genes, together with attribution of homologs to A or C genomes within B. *napus.* A similar approach has been used with primers that amplified regions of 22 putatively orthologous functional loci in *Arabidopsis* species and *B. oleracea.*

Karyotype and Physical Map Analysis

Chromosome fluorescent *in situ* hybridization (FISH) has been successfully applied to *Brassica and Arabidopsis* to locate genes, markers, and marker loci in the context of existing karyotypes and chromosomal features. The linkage maps of *B. oleracea* have been aligned and oriented to the karyotype using a combination of RFLP-, cosmid-, and BAC-labeled probes.

Comparative fiber-FISH mapping has been demonstrated between *B. rapa* and *Arabidopsis.* This technique allows direct measurements to be made on DNA fibers isolated from chromatin and provides a valuable level of resolution between that achieved with chromosomal FISH and currently available genomic sequence or physical maps of *Brassica.*

These results supported the hypothesis that chromosomal duplications played a major role in expansion and evolution of *B. rapa,* which contrasts with the alternative of regional expansion due to accumulation of repetitive sequences in intergenic regions, such as has been found in grass genomes.

Chromosome FISH mapping of low-copy sequences from *Arabidopsis* BACs onto B. *oleracea* allows investigation of patterns of chromosomal duplication and relative physical distances. This has been used to demonstrate conserved organization of two BACs on two *B. oleracea* chromosomes.

A combination of *Arabidopsis* and *Brassica* BACs has also enabled confirmation of conservation of the order of genomic DNA between a chromosomal segment of *Arabidopsis* Chr I and two duplicated segments on B. *oleracea* linkage group O6. The FISH analysis was able to resolve the inverted duplication on O6, together with a short inversion within one duplicated copy.

This study also demonstrated that although genetic distance (frequency of recombination) between the two segments diverged, the physical distance appeared relatively conserved. Lysak et al. recently used a multicolour FISH approach to investigate the pattern of rearrangements within the Brassicaceae. They fluorescently labeled adjacent segments within an *Arabidopsis thaliana* BAC contig of ~8.7 Mb from chromosome 4 and used the visualization of three colours to trace homologous chromosome regions in 21 species.

Their data were consistent with the Brassiceae tribe being a monophyletic group, with all species analyzed descending from a common hexaploid ancestor. They confirmed the presence of three copies of the segment in the three diploid *Brassica* species and six in the amphidiploids.

Detailed analysis of the amphidiploid *B. juncea* (AB genome) and its two component species, B. *rapa* (A) and B. *nigra* (B), indicated that two of the three A genome homologs in B. *juncea* had a structure that deviated from that of *B. rapa.* This most likely occurred due to translocation and inversion event.

Genome Rearrangments

Mechanisms

One of the primary mechanisms for genomic change among plants is whole genome or segmental polyploidy. Superimposed on such variation are the ongoing processes and effects of transposition.

Transposable elements (TEs) are a major component of plant genomes and, together with chromosomal segmental duplications, are likely to account for most of the differences in genome size between *Brassica* species.

It has been possible to determine some of the patterns of 'I'B amplification, diversification and loss since divergence of 1'Bs between *A. rabidopsis* and *B. oleracea* by making use of the shotgun genomic sequence data available for *B. oleracea* and subjecting these to comparative analysis with the *Arabidopsis* genome. From this analysis, it appears that nearly all 'I'B lineages are shared between the genera, with the number of elements in each lineage larger for *B. oleracea.* In both species, Class I retroelements are the most abundant, of which LTR and nonLTR elements are the most prevalent.

Within *B. oleracea* several families of class II (DNA) elements are present in very high copy numbers and can account for much of the observed genome expansion. The TIGR plant repeat database contains a collation of repeat sequences for *Arabidopsis* and *Brassica,* and these are coded into superclasses, classes, and subclasses based on sequence and structure similarity.

The distribution of long terminal repeat (LTR) transposons has also been investigated in the *B. oleracea* genome by probing specific elements onto gridded BAC libraries. Analysis of these data provided estimates of between 90 and 320 copies of individual Tyl *(copia-like)* and *Ty3* (gypsy-like) retrotransposons per haploid genome.

This was consistent with sequence analysis of the same elements in available shotgun genome survey sequence, which indicated between 60 and 570 elements. There was minimal evidence for clustering of the two retrotransposon groups, which was also substantiated by FISH analysis that showed that each had a characteristic chromosomal distribution.

Taken together, these results suggest that preferential sites, and perhaps control mechanisms, may exist for the insertion or excision of the different retrotransposon groups. There was only evidence for a single LINE element in the BAC analysis and none from the sequence analysis.

Timing

Recent estimates of the timing of the whole-genome duplication that has occurred in *Arabidopsis* have been based on two alternative hypotheses. These are based either on the assumption that duplicated segments diverged from an autotetraploid form *(38* MYA) before divergence from *Brassica* or that the ancestor was allotetraploid and the duplication occurred less than 38 MYA, thus contributing to the *Arabidopsis–Brassica* divergence.

Duplicated blocks within the *Arabidopsis* genome have been detected using protein sequence similarity, together with estimates of the level of synonymous substitution present between duplicated genes. This allows estimates of the relative age of segments and suggests that *Arabidopsis* underwent two distinct episodes of duplication.

One of these would have occurred as a polyploidy prior to the *Arabidopsis–Brassica* divergence (14 to 40 MYA), and another, older set of duplicated blocks would have been formed following the monocotyledon–dicotyledon divergence. A comparative analysis involving the *Arabidopsis* and *Capsella rubella* genomes found that collinearity is more pronounced than that for *Arabidopsis* and different *Brassica* species. *Brassica* genomes are remarkably plastic.

In a comparison of the *B. oleracea* linkage map and *the Arabidopsis* genome sequence, Lukens et al. found strong evidence for genome duplication and rearrangement within the diploid *Brassica* species, but less evidence for triplication.

It appears that large-scale translocations combined with tetrasomic inheritance can account for some but not all genomic changes observed in the more recent amphidiploid species formed from the base diploids. Superimposed on these large-scale events at the level of chromosomal segment, transpositions and other small-scale sequence changes contribute to continuing genomic novelty.

Recent Genome Rearrangements and Breeding Introgression

The increased and widespread production of canola and rapeseed over the past 30 years has resulted from the development of modern low-glucosinolate cultivars that produce high-protein meal for animal feed. The low-glucosinolate trait was initially introduced from 'Bronowski,' and residual segments of this genotype are present in modern cultivars and may still contribute to reduced yield, poorer winter hardiness, and lower oil content. It has been found that at least 15 segments are still present in the 'Tapidor' genotype, representing about 30% of the *B. napus* genome.

Sharpe and Lydiate have shown that just three 'Bronowski' donor segments contain loci that can explain more than 90% of the variation for total seed glucosinolates. This level of linkage drag is common in breeding programs, reflecting a relatively low or discontinuous level of recombination.

With knowledge of the location of such introgressed loci, there is considerable scope for using high-resolution marker-assisted selection (MAS) to eliminate any associated deleterious alleles.

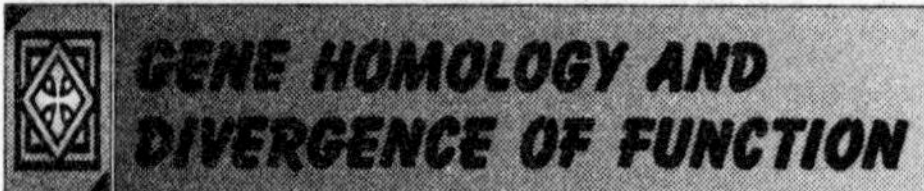

GENE HOMOLOGY AND DIVERGENCE OF FUNCTION

Detecting Orthologous and Paralogous Gene Coding Sequences

The DNA sequences in coding regions of the *Arabidopsis* and *Brassica* genomes are highly conserved and have, on average, about 85% similarity. As already described, collinearity of gene order exists between these two genomes over regions covering as much as 30 cM in *Arabidopsis.* Many studies indicate that a high proportion of loci in *Arabidopsis* are present in at least three copies in haploid *Brassica* genomes For example, Lan et al. presented a detailed comparative map of *B. oleracea* and *A. thaliana* based primarily on RFLP mapping of *Arabidopsis* expressed sequence tags (ESTs).

They found a one-to-one correspondence between linkage groups accounting for 57% of comparative loci. A similar study by Babula et al. used 110 *Arabidopsis* ESTs as molecular markers in *B. oleracea* and found that 95 were informative and suitable to use for construction of an RFLP genetic linkage map.

These EST-based markers yielded 212 new loci that covered all nine linkage groups and confirmed previous patterns of collinear organization, albeit with varying levels of sequence conservation.

By taking into account the extensive duplication within the *Arabidopsis* genome, they were able to identify long conserved regions covering entire chromosome arms in both genomes, suggesting that these are probably shared by descent. This was consistent with other studies, as was the presence of extensive rearrangements in many chromosome regions.

The difficulty associated with identifying the most closely related segments between *Arabidopsis* and *Brassica* genomes is exacerbated by the duplications present within both species. This can occasionally lead to ambiguous criteria being used to identify orthologous regions. To address this, Lukens et al. compared the positions of genetically mapped and sequenced loci in *B. oleracea* to positions of putative orthologs present within the *Arabidopsis* genome.

They defined explicit criteria to distinguish orthologous from paralogous loci and developed a conservative algorithm to identify collinear loci between the genomes, as well as using a permutation test to evaluate the significance of the regions.

This analysis enabled 34 significant *Arabidopsis* regions to be identified that were collinear with up to 30% of the *B. oleracea* genetic map. The findings were consistent with a high level of rearrangement in *B. oleracea* genome since divergence from *Arabidopsis,* probably as a result of polyploidization. Coding sequence divergence can be assessed through comparison of nonsynonymous (Ka) to synonymous (Ks) changes between coding regions.

This has been carried out for a small sample of *B. rapa* ESTs and orthologous *Arabidopsis* coding regions. Among the 218 sequences sampled, the distribution of Ka:Ks ratio was unimodal; substitution rates were more variable at nonsynonymous sites and no evidence suggested that Ka and Ks were positively correlated.

Therefore, it would appear from this sample that there was no evidence for any of the genes having evolved in response to positive selection.

As more complete data sets become available and are compared, it should become possible to relate any patterns and selective constraints to the evolutionary history of different chromosomal segments.

Consequences of Genome Duplication

Osborn et al. have addressed some of the issues relating to understanding the mechanisms associated with novel genetic variation and gene expression in polyploids. When genetic variation is achieved via the processes of gene duplication and intergenomic heterozygosity, this may increase variation in dosage-regulated gene expression and changes in regulatory interactions, as well as rapid genetic and epigenetic changes.

For crop plants, the consequences of gene duplication are likely to be manifested in greater developmental plasticity and the ability to respond to changing environments. Genetic crop improvement, therefore, may involve selection of optimal combinations of alleles at duplicated loci and the avoidance of any associated linkage drag effects.

Redundancy of Gene Function

In *Arabidopsis,* many insertion mutants have no obvious phenotypic effect, which in part may be due to redundancy of function among duplicated genes. An example of this is found with the shatterproof genes SHP1 and SHP2, which are MADS box transcriptional regulators that must both be simultaneously down-regulated or removed to generate a silique nondehiscence phenotype.

These two genes are located within a chromosomal block that appears to have duplicated about 100 MYA. A comparative study involving another related MADS box gene in *Arabidopsis* and *Brassica* has also indicated possible redundancy of gene function. The *ap1/cal* double mutant in *Arabidopsis* displays an arrest at inflorescence initiation and forms a cauliflower-like curd.

Such lines are also apetalous. In contrast, B. *oleracea* cauliflowers that have recessive loci that contain the orthologous genes *Bocal-a/Boap1a* arrest at the same stage, but have wild-type flowers. Because *B. oleracea* contains multiple loci of each of these genes, it appears this could represent differential function during ontogeny consistent with the duplication–degeneration–complementation (DDC) model.

This model predicts that degenerative mutations in regulatory elements tend to increase the probability of duplicate gene preservation rather than the converse. It also predicts that the normal mechanism associated with preservation of duplicate genes involves partitioning of ancestral functions rather than the evolution of new functions.

Divergence of Gene Function

Studies of genes in isolation can provide useful insights into the processes driving functional divergence and specificity. However, it is important to understand such changes within their genomic context because this may provide information about mechanisms underlying phenomena such as heterosis and epistasis.

For some classes of genes, especially those involved in recognition pathways, there is often a requirement for rapid divergence of specificity, as is the case with pathogen resistance recognition genes.

Glycine-rich pollen surface proteins (GRPs) have diverged substantially between *Arabidopsis* and *B. oleracea,* and this makes identification of homologous genes impossible, although they are only separated by ~20 million years.

Fiebig et al. have sequenced eight members of the GRP cluster in four related crucifer species, as well as 11 flanking genes. They found that GRP genes change more rapidly than neighboring genes, are more repetitive, and have undergone more insertion and deletion events. However, this variation occurs concurrently with conservation of the repeat amino acid composition.

The sequence analysis provided evidence that genes flanking the cluster had undergone strong purifying selection, with relaxed selective constraints in the first exon of the GRP. As a result, these researchers conclude that rapid GRP evolution was due primarily to the processes of duplication and deletion, together with divergence of repetitive sequences.

A comparison of sequence and gene content has been carried out in a targeted region *(ABI1-Rps2-CK1)* in *Arabidopsis* and *B. oleracea. BoRps2* is present in single copy, with a segment containing an additional *N-myr* gene between *Rps2* and *CK1.* The *Arabidopsis* homologs for this gene are on different chromosomes, where they are linked with additional homologs for *CK1.*

There are high levels of sequence identity for coding sequences of all genes. However, in most cases, *Brassica* has larger intra- and intergenic noncoding spacers than does *Arabidopsis.* As has been found in some other studies, the promoters of these genes were poorly conserved, except for several sequences of a few nucleotides.

Comparison of duplicate copies of caseine kinase-like (CK1) genes in *Arabidopsis* and *B. oleracea* enabled separation into two major groups based on exon number and sequence identity and thus assignment to orthology and parology.

A more thorough examination has been carried out for recent (10,000 years for *B. napus)* and ancient *(~20* MYA *from Arabidopsis)* divergence of blocks of genes with an intermediate set (A genome vs. C genome: *4* million years). This was based on a comparison of *Arabidopsis* genome sequence and clones from BAC libraries. There appears to have been extensive divergence of gene content between B. *rapa* paralogous segments and homologous segments *in Arabidopsis.*

It also appears that the pattern of gene loss in *B. rapa* and *B. oleracea* is similar, with a small number of species-specific rearrangements. From such comparisons, it can be concluded that the evolution of genome microstructure is an ongoing process, although there has been little or no change in microstructure as a consequence of the hybridization event forming modern *B. napus* amphidiploids.

ACCOUNTING FOR TRAIT VARIATION IN BRASSICA CROPS

Brassica species display a notable phenotypic plasticity in terms of morphology and developmental adaptation, which is apparent in the diversity of crop morphotypes. To understand the genetic and genomic basis underlying phenotypic variation and its relevance to crop improvement fully, it is important to be able to assign variance among different components of environmental effects, as well as components of genetic variation.

Comparative studies between *Arabidopsis* and *Brassica* can also be particularly informative, especially when a similar range of phenotypes may be observed. Forward genetic approaches involving analysis of natural or chemically induced allelic variation have been widely used both in model and crop.

Much of the current understanding of genes underlying plant development initially arose from mutational and segregation analysis in *Arabidopsis.* Within *Brassica,* progress in identifying major and quantitative gene effects has primarily relied on access to a range of different segregating populations and associated linkage maps based on molecular markers.

Beyond this, progress has been made through candidate gene and some map-based cloning efforts facilitated by development of large-insert BAC libraries and associated physical contigs anchored to the *Arabidopsis* genome.

The ease of transformation and establishment of the complete genomic sequence in

Arabidopsis have led to an array of reverse genetic approaches, which have provided deep insights into gene function. There are large collections of T-DNA insertion lines and transposon knock-out lines, as well as constitutive and inducible RNAi lines that saturate the *Arabidopsis* genome.

When combined with the ready ability to up- and down-regulate specific genes and complement mutations, as well as more sophisticated resources such as enhancer traps and cell lineage markers, a previously unforeseen level of detail is being accumulated.

Although many of these resources and much of the information can be transferred to the related *Brassica* crop species, there is a need to understand the most appropriate use of comparative analyses. This is particularly relevant in the context of the extent of genome-wide duplication events that characterize the *Brassica* species.

QTL Analysis

Among the *Brassica* crops, many agronomic traits are comprised of components having a polygenic inheritance, with varying degrees of genotype times environment (G × E) interaction. The availability of segregating populations with associated linkage maps provides the opportunity to detect quantitative trait loci (QTL).

QTL may be detected in F2 or derivative populations, although the estimates of variance components associated with G × E interaction will not be as reliable as those obtained from replicated populations of recombinant inbred or doubled haploid populations of homozygous lines. In *Arabidopsis,* QTL studies using recombinant inbred (RI) populations have increasingly been used to detect the genetic basis of more complex traits.

Although the resolving power of QTL analysis is limited by the number of recombinants available in a population or set of populations, for *Arabidopsis,* any QTL detected can usually be readily resolved further through the screening of additional near isogenic lines (NILs) or STepped Aligned Inbred Recombinant (STAIR) lines.

This then allows resolution to a relatively small number of candidate genes, which may then be functionally analyzed using knock-out mutants, RNAi, or other resources. For *Brassica* species, the availability of reference doubled haploid mapping populations with well-covered linkage maps has allowed initial detection and comparison of many QTL effects.

However, the lack of resolving power, in terms of number of recombinants, presents a major limitation to widespread use of genomic information in *Brassica* crops. Further resolution can be achieved through use of substitution lines, which provide similar resolving power to NILs.

To date, little use has been made of association genetic approaches and linkage disequilibrium to detect or resolve trait loci in *Brassica,* although such approaches could readily be applied to the wide range of genetic resources that represent the domesticated gene pool. This would provide a sound basis for extending the role of MAS in crop breeding by enabling rapid screening of germplasm for novel variation, followed by introgression of specific alleles into breeding lines.

Mutational Analysis

TILLING (targeted induced local lesions in genomes) allows screening pools of PCR products from plants that have been chemically mutagenized, and it has successfully been applied to

Arabidopsis. The methodology allows identification and isolation of mis-sense and non-sense mutant alleles within target (candidate) genes and is often more suitable than transgenic reverse genetic approaches for crop plants.

In particular, genomes containing multiple duplicated copies of genes may be able to withstand a high mutant load and thus require relatively smaller numbers of lines to be screened. In addition, TILLING allows identification of an allelic or paralogous series of mutations with which to study gene dosage effects and epistatic interactions.

As well as providing a tool for inducing variation and thereby identifying gene function in model and crop plants, there is considerable potential for using this approach in generating and identifying natural variation. EcoTILLING has been adopted to identify natural genetic variants that can provide considerable information about gene function.

The technique can also be useful for association mapping and linkage disequilibrium analysis and is likely to become an important tool for crop improvement by allowing rapid identification of allelic variation that can then be introgressed into new varieties using MAS. In *Arabidopsis,* Henikoff and Comai were able to detect small deletions, insertions, and microsatellite polymorphisms, as well as single nucleotide polymorphisms (SNPs). EcoTILLING can also be used to establish the level of heterozygosity within a gene, which could be important in screening crop lines to assess uniformity or the basis of hybrid vigor.

Earlier studies of EMS-induced mutation in *B. napus* generated plants with increased and decreased flowering times. In these experiments, there were no changes in seedling emergence with EMS concentrations between 0 to 1%. To date, no reports of TILLING applied to *Brassica* species have been published, although a number of screening programs are under way.

Other mutagens have been used effectively in *Brassica* to induce genetic or epigenetic variation. Dunnemann and Grunewaldt used N-nitroso-N-methyl-urea (NMU), which is an alkylating agent that acts on DNA to affect base pairing and thus induces SNPs. They observed a range of developmental phenotypes at a rate of 14 to 28% with 20 nM NMU. Epigenetic mutations may be induced by treatment with 5-azacytidine, which incorporates into DNA during replication in place of cytidine and appears to inhibit the subsequent action of methyltransferase, thus effecting a reduction of 5-methylcytosine (^{5m}C), the major methylated nucleotide in eukaryote genomes.

By treating imbibing seeds of *B. oleracea,* a range of developmental variants can be generated at high frequency. Although mostly transmitted through mitosis and not meiosis, the phenotypes observed were similar in type to those observed as a result of somaclonal variation in tissue culture or with mutation by NMU.

CHARACTERIZING SPECIFIC GENE FAMILIES

Comparative genomic approaches based on candidate genes *from Arabidopsis* have been applied to characterizing a range of traits in *Brassica* crop species. A number of examples relating to different areas of biology and crop traits are sufficient to illustrate the progress that has been made and to highlight complexities that can emerge as a result of paralogous genes arising from historical segmental duplications.

The relevance of such information for crop improvement lies in the ability to determine the underlying genetic complexity and inter-relationships of particular traits and to understand the scope for selecting allelic combinations likely to provide predictable crop phenotypes in particular growing conditions.

Development

Flowering

QTLs controlling flowering time have been identified in *Brassica* species, some of which have been associated with genomic regions that contain orthologs of *CONSTANS,* a regulator that plays a key role in the photoperiodic flowering pathway.

CONSTANS is controlled by the circadian clock and in *Arabidopsis* promotes flowering in long days. Four orthologs from homologous loci have been isolated from *B. napus* lines that had different flowering times.

These *BnCO* genes all appear to be expressed in *B. napus,* and the functional conservation of one allele *(BnCOa1)* has been confirmed by complementation of the *Arabidopsis co-2* mutation in a dosage-dependent manner.

Two different alleles of *CONSTANS* that possess identical DNA coding sequence have also been obtained from *B. nigra (BniCOa),* although they were isolated from early flowering and late flowering lines. Because these did not show any differential effect on flowering time, Osterberg et al. deduced that the variation influencing flowering time must be outside the coding region (cis-regulation or another gene).

Further investigation showed that a *B. nigra* ortholog of the *Arabidopsis CONSTANS-like1* gene *(BniCOL1)* was located 3.5 kb upstream of *BniCOa,* and this did display sequence divergence among alleles of early and late flowering lines.

It was also found that a single indel polymorphism in the *BniCOL1* coding region was present in several natural populations of *B. nigra* and, in most cases, had a significant association with flowering time.

More detailed analysis showed that the intergenic sequence between *BniCOL1* and *BniCOa* had a prominent peak of divergence 1 kb downstream of the *BniCOL1* coding region and may in fact contain regulatory elements for the downstream *BnCOa* gene. Further comparison of the indel among 41 sequences of complete *BniCOL1* revealed a moderate rate of within-population recombination, with no evidence for selection.

This is an exemplary example of the care and attention to detail required to interpret candidate- and map-based cloning studies, especially in complex crop genomes.

Vernalization

Several QTLs have been detected that account for vernalization in *Brassica* species.VFR2 is a major QTL for vernalization-responsive flowering time in *B. rapa.* The chromosome region in which VFR2 is located is syntenous with a region of *B. napus* that controls the same trait, as well as a region of *Arabidopsis* Chr 5 that contains several flowering time loci.

Kole et al. have backcrossed the late allele into an early flowering line and detected an additive effect attributable to the late allele. FLC is a type II MADS-box repressor of flowering and is down-regulated in response to exposure to cold temperatures.

This stable epigenetic switch is required for the winter-annual habit of late flowering ecotypes of *Arabidopsis.* It also appears to provide an explanation for cold vernalization in biennial brassicas and other species.

The expression patterns in *B. rapa* are consistent with those in *Arabidopsis,* and an RFLP detected by the *Arabidopsis FLC* sequence was found to cosegregate exactly with the VFR2 QTL in 414 gametes. In the *B. rapa* biennial parent, *BrFLC* RNA is up-regulated, whereas under cold treatment it is down-regulated.

Four *BrFLC* orthologs have been cloned and sequenced. There appears to be no evidence for differential rates of evolution, with the Ka:Ks ratios of nonsynonymous to synonymous substitutions suggesting they are not under strong purifying selection.

The *BrFLC1-3* gene has been mapped to regions that are collinear with the top of *Arabidopsis* Chr 5, which is consistent with a polyploid origin.

Another paralog, *BrFLC5,* maps near a junction of two collinear regions of *Arabidopsis.* One of these contains an FLG-like gene *(AGL31).* However, all *BrFLC* sequences appear more closely related to *FLC* than *AGL31.* Kole et al. have concluded that the duplicated *BrFLC* genes appear to have a similar function and to interact in an additive manner to modulate flowering time.

Thus, one of the consequences of segmental chromosomal duplication appears to have been to increase the sensitivity and adaptive range with respect to changes in location and environment. Environmental and endogenous flowering signals in *Arabidopsis* are integrated by a range of transcriptional regulators.

These include the MADS-box gene AGL20, which appears to be a flowering activator downstream of FLC. Knockouts of *AGL20* have a late flowering phenotype, whereas when activated, it promotes early flowering even in the presence of strong expression of FLC.

The role of AGL20 appears to have been conserved in the Brassicaceae, with the orthologs from *B. rapa (BrAGL20* genes) at least 94% identical.

When the *BrAGL20* genes were constitutively expressed in *Cardamine felxuosa* (a long-day Brassicaceae that does not respond to vernalization), it was found that although some transgenic plants flowered very early, other antisense plants had delayed flowering.

Floral Development

It has been suggested that oilseed *B. napus* genotypes with reduced or no petals would possess greater photosynthetic efficiency and activity. To manipulate this trait, hairpin (hnRNA) gene silencing has been used to silence the B-type MADS-box floral organ identity genes *APETALA3* and *PISTILLATA* in *Arabidopsis* and in *B. napus.*

This engineering approach made use of an *AP1* promoter that regulates transcription in a second whorl-specific manner. The transgenic *Arabidopsis* plants had male fertile flowers in which the petals were converted into sepals.

The corresponding transgenic *B. napus* plants also had male fertile flowers and sepaloid petals. In both cases, the phenotypes were stable and heritable, underlying the conservation of function. This study also underlines the experimental value of focusing on key transcriptional regulators to understand and manipulate crop plant development.

It should be noted that natural variation present within the *Brassica* gene pool can also result in a reduction or absence of petal tissue.

For example, Fray et al. have identified cosegregation of two loci *(STAP)* that control the production of stamenoid petals in homologous positions in *B. napus,* and have isolated orthologs of the *Arabidopsis CURLY LEAF (CLF)* gene from the same genetic loci.

They considered *CLF,* which pleiotropically affects leaf and flower development, to be a candidate gene for STAP. More recently, Jiang et al. have established that the apetalous phenotype of a mutant *B. napus* line (ap-Tengbe) was regulated by cytoplasmic genes interacting with two pairs of nuclear genes.

Gene Identification in Model and Crop

Such investigations highlight the need for parallel experimental approaches that make use of the functional genomic resources for a model plant, as well as detailed and exhaustive study of natural variation present at multiple loci within a crop plant.

Simple knockout experiments often cannot provide full insight into the functioning of key regulatory genes that may exist in two or more copies in the genome. The use of gene-silencing approaches such as RNAi may be effective in such studies, but will not always reveal the subtlety of regulation or interactions among multiple loci.

A valuable demonstration of the suitability of crop plants for isolating key regulatory genes that provide additional information about model systems is illustrated by the characterization of the *BABY BOOM (BBM)* gene. This was isolated following subtractive hybridization of RNA from embryogenesis-induced microspores of *B. napus,* against a nonembryogenic sample.

Such an approach would have been challenging in *Arabidopsis* because no optimized procedure is currently available for production of microspore-generated embryos. BBM is preferentially expressed in developing embryos and seeds and is similar to the AP2/ERF family of transcription factors.

The *Brassica (BnBBM)* and *Arabidopsis (AtBBM)* sequences have a high level of similarity (85%) and conserved intron–exon boundaries. When the *BBM* gene was ectopically expressed in *Arabidopsis* and *Brassica,* it gave rise to spontaneous formation of somatic embryos and cotyledon-like structures on seedlings, as well as similar ranges of pleiotropic effects.

The preceding examples highlight the importance of understanding the function of paralogous loci in their genomic context, as well as the need to identify and sequence multiple copies of gene-coding sequences from *Brassica* based on candidate genes in *Arabidopsis.*

Testing the conservation of gene action over a wider taxonomic range can also be informative. For example, the *OsMADS1* gene from rice is functional across angiosprem subclasses and has been successfully introduced into *B. rapa* under a constitutive promoter.

The transgenes appeared to be expressed, with one line notably involving homeotic replacement of a carpel with another flower.

Oil and Fatty Acid Pathways

Fatty acid synthesis and metabolism is of primary importance for production and genetic improvement of *Brassica* oil crops. Understanding the basis of harvestable yield (oil content) and quality (fatty acid profile) has considerable potential to enable these traits to be manipulated by transgenic or conventional means.

Triacylglyerols and proteins are used as the major seed storage reserves in the developing embryos of the Brassicaceae. The synthetic and modification pathways and products are well conserved, and considerable information is now available about the genes coding for key enzymatic steps in synthesis, elongation, and modification steps.

The accumulation of knowledge about plant lipid metabolism, including fatty acid synthesis and modification, has benefited from availability of information and functional genomic resources in *Arabidopsis,* as well as genetic variation in *Brassica* species. Considerable progress has been made by mining the *Arabidopsis* genome sequence.

This has already demonstrated that most of the genes encoding enzymes involved in lipid biosynthesis are represented by gene families that include a diverse array of isoform functions. Quantitative genetic studies of oil synthesis in *Arabidopsis* have so far revealed a QTL that can account for some of the variation in seed oil content and fatty acid composition.

This study was based on a recombinant inbred population between Landsberg *erecta* and Cape Verdi Islands ecotypes, where the QTLs included two major and two minor loci accounting for 42% of variation in oil content. As with other areas of metabolism, there is interest in knowing whether some or all of the genetic variation observed can be accounted for by specific candidate enzyme-coding genes or whether regulatory genes make major contributions.

In this study, significant QTLs for linoleic acid and linolenic acid appeared to colocate with the fatty acid desaturase 3 (FAD3) locus in *Arabidopsis,* and one for oleic acid with FAD2. Within the plastid, synthesis of fatty acids requires carboxylation of acetyl-CoA, which is catalyzed by acetyl-CoA carboxylase (ACCase). The same enzyme is also used in a number of biosynthetic pathways within the cytosol for fatty acid elongation.

Two genes located in a tandem 25-kbp duplicated region near the centromere of *Arabidopsis* Chr1 code for two multifunctional ACCase isoforms. Fatty acid desaturases important in plant lipid metabolism are located in the endoplasmic reticulum (ER) and other subcell compartments. *B. juncea* plants have been transformed with ADS1, which is an *Arabidopsis* homolog of yeast and mammalian acyl-CoA Delta9 desaturases.

These had a significant decrease in the level of seed saturated fatty acids. These preliminary data suggested that *Arabidopsis* ADS1 encoded a Delta desaturase. However, as well as altering the level of saturated fatty acids in *Brassica,* it also affected the levels of monounsaturated fatty acids.

This highlights some of the complexities associated with regulation of fatty acid metabolism that may be compounded in amphidiploid *Brassica* species. Substitution lines developed between two varieties of *B. napus* have been used to identify a range of QTLs for oil content and fatty acid profile. Of 13 QTLs detected that affected fatty acid composition, 7 also affected total seed oil content.

The approach of using substitution lines appeared to substantiate and provide additional resolution over previous results from segregating DH populations. Erucic acid is the main component of storage fatty acids in oilseed *Brassica,* with the levels controlled by activity of the fatty acid elongation 1 (FAE1) gene. Fatty acid elongases have been cloned from *Arabidopsis* and *Brassica,* including *B. rapa* and *B. juncea,* and the gene sequences compared with corresponding loci in high- and low-erucic acid lines of *B. rapa* as well as from *B. oleracea* and *B. napus.* FAE1 sequences from *Brassica* and *Arabidopsis* have a high level of nucleotide and amino acid sequence conservation.

This study concluded that differences in 13 amino acid positions in the central part of the protein were responsible for differences in erucic acid levels between low- and high-erucic acid lines. QTL analysis in *B. juncea* identified loci on two linkage groups; SNP markers to FAE1.1 and FAE1.3 cosegregated with the QTLs that accounted for 60 and 38% of the total phenotypic variance, respectively.

Quantitative transcriptional analysis has shown that mRNAs of various components involved in lipid biosynthesis are expressed in a coordinated manner during *B. napus* embryogenesis and are present in constant molar stoichometric ratios.

For biotin carboxylase, similar amounts of RNA were found in *Brassica* embryos and *Arabidopsis* siliques. Although Girke et al. found similar levels of mRNAs for fatty acid synthase components between leaves and seeds using a cDNA microarray, this was not substantiated by O'Hara et al.. Using quantitative northern analysis and RT-PCR they found that embryos accumulated between 3- and 15-fold more transcripts per unit total RNA than young leaf tissue.

In cases where there appears to be divergence of expression pattern, it is necessary to distinguish between inherent behaviour of the promoter and ancillary effects of transgene insertion site or transitory silencing. Oil within the developing embryo is accumulated in oil bodies that are small droplets containing mostly triacylglycerol and are surrounded by a phospholipid/oleosin annulus.

As the major protein component of developing embryos, oleosins have been suggested to play a structural role in stabilizing the lipid body during dessication of the seed by preventing coalescence of the oil. In *Arabidopsis,* pollen-specific oleosin-like proteins (olleopollenin) genes are located in a tandemly repeated cluster. Comparative analysis between *Arabidopsis* and *Brassica* of the complete set of oleosin genes confirmed that they were subject to rapid evolution, including whole gene duplication and loss events, as well as a high rate of nonsynonymous mutations and indels in coding sequence.

Evidence suggested that lineages leading to *Arabidopsis* and *Brassica* arose from independent duplications, consistent with the overall pattern of variation deduced from collinearity studies. Based on the Ka:Ks ratios of nonsynonymous to synonymous divergence, this class of gene appears to be among the most rapidly evolving. A survey of the *Arabidopsis* gene pool for seed oil content, very long chain fatty acids, and polyunsaturated fatty acids has demonstrated extensive natural allelic variation.

A core collection derived from the original 360 accessions has been selected and should be valuable for gene identification, as well as more detailed dissection of the genetic regulation of seed lipid traits. Comparative studies using similar approaches are likely to be very effective, especially when applied to the wider gene pools beyond relatively modern mono- or oligophyletic oilseed crops such as *B. napus.*

Interest in production of crop plants that possess enhanced or optimized nutritional capacity is increasing. One trait of particular interest is the accumulation of omega3 very long chain polyunsaturated fatty acids (VLC-PUFAs). These have been shown to help prevent cardiovascular disease, metabolic syndrome, and progression towards other prevalent Western pathologies such as type-2 diabetes and obesity.

Unfortunately, VLC-PUFAs are not normally present in the oils of higher plants. However, it has been demonstrated that, by introducing genes encoding key enzymes originating from marine microalgal, algal, and fungal species, heterologous reconstitution of VLC-PUFA biosynthesis can be achieved in transgenic plants.

This work was carried out in *Arabidopsis* and tobacco, as well as the oilseed crop linseed, demonstrating proof-of-concept accumulation (to low levels) of VLC-PUFAs. Contrary to earlier expectation, recent reports have demonstrated that some genotypes of *Brassica juncea* have the capacity to accumulate higher significant amounts of these valuable fatty acids.

This highly significant finding indicates that there may be considerable scope to engineer *Brassica* germplasm metabolically to produce economically viable amounts of VLC-PUFAs. At present, there is no knowledge of the inferred intrinsic species-specific variation in terms of combinations of naturally occurring alleles that may contribute to expression and regulation of the relevant pathways. This would provide information to guide genetic improvement through prebreeding of lines amenable to wide-scale production.

Glucosinolate Pathways

The genes in the Brassicaceae uniquely enable production of glucosinolates (GLS), which break down to isothiocyanates such as sulphoraphane that are known to provide some protection against a range of human cancers. This has been ascribed to the ability of isothiocyanates to induce phase 1 and 2 detoxification enzymes in mammalian cells, which can then lead to reduction in the rate of tumor development.

Glucosinolates also have a potentially significant role in some aspects of herbivore defense and signaling mechanisms, as well as in phytoremediation through interactions with soil microorganisms. The role of comparative studies *with Arabidopsis* in elucidating the various genes involved in glucosinolate synthetic, modification, and breakdown pathways has been well covered in recent reviews.

Different glucosinolate products vary considerably as a result of variation in activity of different steps in the synthetic and side-chain modification pathways. This variation occurs throughout the *Brassica* genepool and arises from mutations in loci controlling such steps. Genes involved in synthesis of glucosinolates have been isolated from *Arabidopsis* and *Brassica.*

Those that regulate side-chain length of *GSL-PRO* result in three carbon glucosinolates, whereas *GSL-ELONG* results in four-carbon glucosinolates. Segregation of *GSL-PRO* and *GSL-ELONG* has been shown to be independent in *B. oleracea,* and double recessive plants produce only trace amounts of aliphatic glucosinolate. Candidate genes from the *Arabidopsis* genome sequence have been used to clone several *GLS* genes, including *BoGSL-ELONG.*

Comparative sequence analysis has provided a view of the conservation of gene order between *Arabidopsis* and *B. oleracea* in a region containing several GLS genes. By comparing a region from *Arabidopsis* Chr 4, Gao et al. demonstrated that a high level of collinearity existed, with 23 out of 37 genes present and in the same orientation in *B. oleracea.*

However, whereas three 2-oxoglutarate-dependent dioxygenase *(AOP)* genes are present in the *Arabidopsis* region, together with an additional *AOP* pseudogene, in *B. oleracea* two of these genes are locally duplicated and a third *(AOP3)* was not present. This arrangement was also conserved between different *B. oleracea* crop types (collard and broccoli).

A more extensive survey did indicate that the copy number and sequence of the *Brassica* AOP2 gene varied across the gene pool. QTL studies in *B. napus* have accounted for much of the phenotypic variation for seed glucosinolate, with three QTLs detected in common between two populations were located in homologous regions.

This suggested that seed glucosinolate accumulation is controlled by duplicated genes. In *B. juncea,* QTLs have been identified associated with 3-butenyl that were consistent across years and experimental sites. Other QTLs for 2-propenyl and total GLS were detected, although some of these were highly inconsistent in different environments.

GSL-ALK affects desaturation of side-chains. In *Arabidopsis* and *B. oleracea,* it cosegregates with the *GSL-OH* responsible for side-chain hydroxylation. An ortholog cloned from *B. oleracea* has been transformed into *Arabidopsis.* This resulted in detectable transcriptional activity and associated changes in the glucosinolate profiles of leaf and seed tissues.

Myrosinase is the only known S-glycosidase in plants and contributes to degradation of glucosinolates to isothiocyanates and nitriles. Thangstad et al. have carried out promoter-fusion experiments to determine tissue specificity of myrosinase expression in *Brassica, Arabidopsis,* and *Nicotiana.* By fusing myrosinase promoters from *B. napus* and *A. thaliana* to a GUS reporter gene and transforming into *A. thaliana* and *B. napus* as well as *Nicotiana,* Thangstad and colleagues were able to determine the cell types in which they were expressed.

They found that the *Arabidopsis* TGG1 promoter directed expression within guard cells and phloem myrosin cell idioblasts of the transgenic *A. thaliana* plants, whereas the *B. napus* Myr1.Bn1 promoter resulted in cell-specific expression in idioblast myrosin cells of immature and mature seeds, as well as the myrosin cells of phloem in *B. napus.*

The *B. napus* promoter resulted in an expression pattern similar to TGG1 in the guard cells. This differential pattern of expression may result from locus-specific divergence of cis-acting factors in the amphidiploid *B. napus* because only one promoter sequence was tested in this study.

In practical terms, advances have been made in manipulation of glucosinolate pathways and contents in *Brassica* harvestable products. For example, Faulkener et al. describe a broccoli hybrid with a tenfold increase in the level of 4methylsulphinylbutyl glucosinolate. Tissue from this was able to induce more than a 100-fold increase in quinone reductase within Hepa 1c1c7 cell lines.

POLLENT-STIGMA INTERACTIONS AND SELF-INCOMPATIBILITY

Brassica species have for many decades provided the model for understanding sporophytic pollen self-incompatibility (SI) systems. In diploid brassicas, selfincompatibility is controlled by genes at a single locus (S), with the transmembrane receptor kinase *(SRK)* gene expressed in the stigma and the S-locus cysteine-rich *SCR* ligand gene is expressed in the pollen coat.

A considerable body of work has been carried out to characterize the allelic variation and function of the S-locus, particularly in *B. oleracea* and *B. rapa*. In the case of self-incompatibility, relatively few functional insights have been achieved through experimental work *within Arabidopsis.*

This is primarily *because Arabidopsis thaliana* is self-compatible and has no functional S-locus, although a functioning and orthologous S-locus system with similar levels of allelic variability as *Brassica* is present within the perennial species *A. lyrata*. The lack of a functional S-locus in *A. thaliana* has enabled the genetic basis of the Brassicaceae SI system to be confirmed through transfer of the relevant components between *Arabidopsis* species.

Nasrallah et al. isolated the *SRK* and *SCR* genes from one S-locus haplotype of *A. lyrata* and were able to demonstrate by complementation that these genes alone are sufficient to confer a self-incompatible phenotype upon self-fertile *A. thaliana*. From this, they concluded that all the other components of the relevant signaling cascade had been conserved within *A. thaliana*.

This key experiment provided the impetus to analyze other aspects of self-incompatibility with the relevant forward and reverse functional genomics tools available for *A. thaliana*. It is apparent that the transition from inbreeding to outbreeding can occur rapidly during evolution and crop domestication. This has been demonstrated from studies in *Brassica* and *Arabidopsis*. Ekuere et al. analyzed the genetic control of self-incompatibility in populations derived from crosses between resynthesized lines of *B. napus* and oilseed rape cultivars.

They were able to detect evidence for latent S-alleles in at least two *B. napus* rapeseed cultivars and also demonstrated that the S-phenotype was masked by an unlinked suppressor system common to oilseed rape. Based on this analysis, they suggested that similar latent S-alleles may be widespread throughout the domesticated rapeseed gene pool and that, moreover, they may be associated with the highly conserved C-genome S-locus of these crop types.

HOST RESISTANCE TO PATHOGENS

Brassica crops are subject to attack by a range of parasitic organisms, including viruses, bacterial, fungi, oomycetes, and various insect pests. This section will focus on a small number of examples in which knowledge of resistance mechanisms determined in *Arabidopsis* is likely to provide insights into crop-based resistance.

There are, however, several caveats to a comparative approach that will lead to genetically determined durable field resistance. These include the different life history and ecological context of *Arabidopsis* compared with *Brassica* crops, as well as global issues relating to rate of change and spread of pathogen populations as a result of international trade in seed and crop products.

The major *Brassica* pathogens include viruses such as turnip (TuMV) and cauliflower (CaMV) mosaics; bacteria such as *Xanthomonas campestris pv. campestris (Xcc)* and *Pseudomonas* spp.; fungi such as *Leptosphaeria maculans, Pyrenopeziza brassicae, Altenaria brassicae,* and *Fusarium oxysporum;* and the oomycetes *Hyaloperonospora peronospora* (formerly *Peronospora parasitica)* and *Albugo candidans,* as well as the protozoan clubroot pathogen *Plasmodiophora brassicae.* For a number of these pathogens, a large research effort has focused on the same species and their interaction with the model plant *Arabidopsis.* This has enabled considerable progress to be made in elucidating the detailed mechanisms of host–parasite recognition, signal transduction, and response. However, as Hammond-Kosack and Parker have pointed out, most attempts to harness this knowledge to engineer improved disease resistance in crops have so far failed.

In terms of transgenic approaches that introduce components of resistance mechanisms, commercial exploitation has not been possible because of detrimental effects on plant growth, development, and crop yield. Crop improvement that makes use of biotechnological approaches is increasingly focused on marker-assisted breeding, as well as a more targeted use of transgenes that involves vectors containing highly regulated transgenes able to confer resistance in several distinct ways.

To develop comprehensive and effective marker-assisted strategies for crop improvement, it is necessary to understand the underlying genetic and functional mechanisms of host–parasite interaction. In terms of host-specific resistance, it has become apparent that plants are able to resist pathogen attack by eliciting an active defense response that mostly leads to cell death or hypersensitive response.

This involves dramatic cellular reprogramming with activation of a signal transduction cascade mediated by plant disease resistance recognition (R) genes. These have been well characterized and classified in *Arabidopsis* and other model species. In a compatible response, avirulence genes encode parasite elicitor molecules that interact directly or indirectly with the corresponding plant host R gene product through R gene recognition of elicitors in a ligand–receptor interaction.

Depending upon the class of R gene, there is then an interaction with different component pathways in the signal transduction cascade. To date, very little research has been published that focuses on isolating the different components of resistance directly from *Brassica* crop species. Although there was initial progress in using rapid cycling *Brassicas* as a tool for characterizing resistance genes, in recent years the potential of these valuable resources has unfortunately not been fully realized.

Major pathotype-specific resistance gene loci have been mapped in *Brassica* for a number of pathogens. These include TuMV and *Xanthomonas,* as well as markers to *Albugo* resistance. Kole et al. made use of comparative mapping in their characterization of *Albugo* white rust resistance loci in *B. rapa* and *B. napus.*

By comparing map positions of resistance genes in these two species, they were able to identify loci where additional resistance loci may be located. Alignment of the *Brassica* maps to the physical map of the *Arabidopsis* genome identified regions to target for comparative fine mapping.

As in other crops, a number of studies have focused on identifying resistance gene analogs (RGAs) in *Brassica* species. Such information can assist in understanding the level of conservation *between Arabidopsis* and *Brassica* genomes either in terms of genome location or gene sequence. Most studies to date have focused on the NBS-LRR class of recognition genes rather than genes further down the signal transduction pathway.

A combination of 103 *Arabidopsis thaliana* ESTs homologous to cloned plant R genes and 36 *Brassica* R-gene homologs has been mapped in *Brassica napus* to identify candidate R-gene loci and explore collinearity with *Arabidopsis*. These results indicated no apparent rapid divergence of the R-gene containing loci between the two genomes.

As with many recognition genes, NBS-LRR genes are highly variable, although some conserved motifs are commonly used to amplify RGAs from genomic DNA. Vicente and King used this approach to isolate RGAs from *B. oleracea,* the sequences of which were highly variable, although most of them showed similarity to known disease-resistant genes. Brassica-specific primers were then used to amplify and map five groups of RGAs, and four locus-specific sequences were confirmed as being expressed.

Probing the specific gene sequences onto a BAC library indicated that these specific genes may only be present in low copy number. In another study, *44 B. napus* RGAs were identified, of which a third were expressed from a subset of 29. The sequence specificity allowed discrimination of each genotype within a *B. napus* collection. Although examples of conservation of specific resistance recognition genes within the Brassicaceae exist, the pattern of changes in local organization and relative rate of evolution of resistance loci is likely to be highly complex.

Grant et al. investigated the conservation associated with homologs of the *Pseudomonas syringae* pathovar *maculicola (RPM1)* bacterial resistance gene, which is completely absent in *Arabidopsis thaliana* accessions that lack RPM1 function. Collinearity of genes flanking *RPM1* is conserved between *B. napus* and *Arabidopsis,* with four additional *B. napus* loci in which the flanking marker synteny is maintained, but *RPM1 is* absent.

The *B. napus* rpm1-null loci have no detectable nucleotide similarity to the *Arabidopsis* rpm1-null allele; thus, it appears that RPM1 evolved before the divergence of the Brassicaceae and has been deleted independently in the *Brassica* and *Arabidopsis* lineages. The general conclusion from such results is that functional polymorphism at R gene loci can arise from gene deletions.

Although much is now known of the diversity and interactions between plant host and parasites from work on *Arabidopsis* and genomic information about resistance recognition genes is being accumulated, at present little progress has been made on matching the molecular diversity to known functional resistances in crop brassicas.

For example, Malvas et al. investigated homologs of the *RPS2* disease resistance gene and found that 2.5-kb fragments were conserved at a level of 95 to 98% homology among *Brassica* species and that the homolog was constitutively expressed in *Brassica oleracea.* However, they found no linkage between the gene and resistance to blackrot caused by *Xanthomonas campestris* pv. *campestris.* This underlines the need for careful genetic analysis and gene isolation in wellcharacterized resistant lines.

Given the progress being made in *Brassica* comparative genomics and the ability to map-base clone genes rapidly, a large number of major resistance gene alleles should soon be isolated. This will provide information to understand the relative range and rate of resistance recognition gene variation in the domesticated *Brassica* species compared to *Arabidopsis,* as well as "in-gene" molecular markers to assist in breeding introgression and resistance gene pyrimiding strategies.

However, although major gene resistance is valuable in some situations, the understanding of more durable "field" resistance is one of the ultimate goals in terms of integrated crop disease management. In this area, relatively little progress has been made through use of the model species *Arabidopsis,* primarily because of differences in terms of its ephemeral life history and unsuitability to the study of crop field pathogenesis and epidemiology.

UNDERSTANDING GENE REGULATION

Gene Expression and Use of Transcriptional Arrays

Transcriptional analysis (transcriptomics) based on gridded arrays of gene sequences has provided a powerful approach to understand the coordinated expression of genes and deduce the regulatory networks associated with different cell types, stages of development, and responses to changing environments. *For Arabidopsis,* a range of technological platforms exists, including those based on ESTs, amplified gene probes, and short or long oligonucleotides.

This has been made possible due to the comprehensive whole-genome and EST data sets available for the model species. For *Brassica,* transcriptomics has to date been less widely used, although the increasing availability of sequence data and validation of *Arabidopsis* resources will enable more detailed analysis. There have been concerns that the presence of multiple copies of closely related paralogous gene sequences may provide equivocal results.

However, in any transcriptomic analysis, the initial identification of up- or downregulated transcripts does require detailed quantitative verification, which requires development of locus-specific PCR assays. Amagai et al. have used an *Arabidopsis* cDNA macroarray to detect signals from hybridization with cDNA isolated from anthers and pistils of *B. oleracea.* This resulted in identification of 53 putative anther-specific genes, including a number that had already been characterized.

A third of the clones had RT-PCR expression patterns consistent with that detected in the *Arabidopsis* macroarray. More recently, Lee et al. carried out a sensitivity analysis by comparing cDNAs and corresponding 60- to 70-mer oligonucleotides for 192 *Arabidopsis* genes.

In addition to demonstrating that the sources of variation were similar for *Arabidopsis* and *B. oleracea* RNA, they showed that cDNA and oligonucleotides were similar in their ability to detect changes in expression, with a common subset of significant genes.

Characterization of Cis-Regulatory Regions

The behaviour of promoters exchanged between *Arabidopsis* and *Brassica* provides a useful tool to investigate functional motifs and properties. In many cases, a very similar range of tissue or temporal expression is observed. For example, the promoter of an *Arabidopsis* late embryogenesis abundant protein (AtEm1), when fused to a GUS reporter, was found to be highly active in vascular tissues of *B. napus* embryo and pollen grains and was also active in other late developing floral organs.

In other cases, the function of cis-acting regions across species can provide considerable insights into the relationships between sequence divergence and specificity of action. The promoter of *AGAMOUS (AG)* from *Arabidopsis* has been introduced into *B. napus* and shown to have an expression pattern limited primarily to the reproductive organs and nectarium. However, the tissue-specific pattern in this case was not conserved between species. For example, the AG *cis* elements did not express in the ovules of *B. napus,* although the pattern was temporally similar to that observed during early development of *Arabidopsis.*

Pylatuik et al. conclude from these experiments that the regulatory factors controlling the generalized local expression of AG were conserved between these species, although those that control the temporal and tissue-specific expression have not been conserved.

Future enhancer traps lines may be able to provide the relevant level of information on the subtlety of cell-specific control associated with cis-regulatory regions from paralogous copies within *Brassica* genomes. To date, a number of different approaches have been used to isolate cis-regulatory regions from orthologous copies of candidate genes. Based on the *Apetala3* gene, an ortholog was isolated from *B. napus (BnAP3)* and the 5' region of the cDNA, then retrieved by rapid amplification of cDNA ends (RACE).

Expression analysis of the multiple alleles from *B. napus* demonstrated that they are expressed in floral as well as nonfloral tissues. Further information was obtained by transforming *BnAP3* into wild-type *Arabidopsis* and *B. napus* under the control of a reproductive organ specific promoter.

Consistent with the behaviour of the *Arabidopsis* ortholog, the transgenic plants had carpels converted to stamens. As well as complementing the *ap3 Arabidopsis* mutant, the *BnAP3* gene also directed the conversion of carpels to stamens in the fourth whorl. A plant-wide survey searching for conserved regulatory sequence elements has been carried out by Hulzink et al..

This was based on the relative importance of 5' untranslated regions (UTR) of several pollen transcripts and the conservation of genetic programs in pollen. By analyzing 5'-UTR sequences of pollen and sporophytic expressed genes, they identified several pollen-specific elements containing various consensus sequences, several of which were preferentially associated with genes from dicotyledons, wet-type stigma plants, or plants with bicellular pollen.

The analysis included three sequence elements that were preferentially found in the 5'-UTR of pollen-expressed genes in *Arabidopsis* and *B. napus.* This approach has generic application for identifying functional motifs in cis-regulatory regions through identification of consensus sequence or by bioinformatic comparison of sequencedetermined physical properties that might affect binding of trans-acting regulators. Margulies et al. have outlined a rationale for identifying multispecies conserved sequences (MCSs) based on analyses between vertebrate genomes.

Although they found that about 70% of the bases within MCSs are located within noncoding regions, they deduced that most of the sequences had no known function. Many MCSs corresponded to clusters of transcription factor-binding sites, noncoding RNA transcripts, and other candidate functional elements in vertebrates.

Although it should be noted that vertebrate genomes have a different distribution of coding and noncoding sequence than dicotyledonous plants, Colinas et al. used a similar comparative bioinformatic approach to analyze the degree of conservation between upstream noncoding regions of *B. oleracea* and *Arabidopsis.*

This demonstrated that there is likely to be significant conservation of promoter regions between *Arabidopsis* and *Brassica* and that such sequence comparisons between two species at this level of divergence could reveal functional cis-regulatory elements. Another approach has been suggested by Gilchrist and Haughn, who speculated that by identifying blocks of conserved sequence within relatively unconserved noncoding regions, TILLING could help identify regulatory domains. This would have considerable application in the comparative analysis of *Arabidopsis* and Brassica.

FUTURE DEVELOPMENTS AND APPLICATIONS

As outlined in the examples given in the previous sections, there has been considerable progress in transferring information and experimental approaches from the model *Arabidopsis* to the crop *Brassica* species. A number of experimental areas still need additional focus of effort to exploit current knowledge fully.

These include the ability to identify and resolve gene effects, to distinguish between locus-specific gene sequences, and to understand fully the interactions that may occur in complex genomes comprising multiple duplicated chromosomal segments. Crop improvement has additional requirements to identify and characterize sources of genetic diversity and to understand the constraints and control of recombination.

Gene identification and haplotype reconstruction will benefit from developments in highly parallel locusspecific SNP detection, which should also allow high-throughput MAS for breeding. Recombinational resolution currently limits inferences that can be made from conventional genetic analysis.

This may be addressed through concerted development of overlapping sets of near-isogenic, substitution, or STAIR lines. A resolution gap then remains between trait loci (major gene or QTL) and existing BAC-based physical maps or the emerging genomic sequences.

As more gene sequences are anchored in the context of *Brassica* genetic maps, so we will understand and be able to reconstruct the detailed relationships with the *Arabidopsis* genome. For practical purposes this will allow more rapid interpolation and identification of additional candidates within any given region.

Understanding the adaptive significance of segmental and whole-chromosome polyploidy, especially with regard to interactions with the environment, is a further challenge. These may be mediated via variation in cis-acting regulatory sequences, with epigenetic mechanisms superimposed that confer locus or cell-specific variation in local or global patterns of gene regulation.

Assessing and Exploiting Genetic Diversity for Crop Improvement

The ability of genetic improvements in *Brassica* and other crops to meet complex and changing requirements of production and market conditions will be constrained by the ability to generate and select breeding lines that contain optimal combinations of alleles.

As an increasing amount of information becomes available that contributes to understanding the interactions between genotype to phenotype, so will the ability to utilize a wider range of allelic variation within the gene pool.

This places an emphasis on conserving, accessing, and understanding genetic diversity at the genomic level. There is considerable scope to use combinations of locus-specific markers to explore the extensive genetic diversity within the *Brassica* gene pool. Although development of informative SNP markers can be problematic in some *Brassica* material because of the requirement to distinguish paralogs from homologous loci, such approaches are likely to be increasingly valuable in addressing the relationships and limitations of natural and domesticated variation.

The success of association genetics and linkage disequilibrium studies in other species indicates that this should be a powerful approach for identifying alleles relevant for crop improvement. Development of "graphical genotypes" that display DNA marker data to create a graphical image for the genomic constitution of an individual will provide an important link for breeding pedigrees, allelic variation, and crop phenotypes.

In general, modern crops have a relatively narrow genetic base that does not reflect the existence of extensive allelic variation within the wider gene pool or germplasm collections. For *Brassica* crops, existing natural variation has been exploited for particular target traits such as resistances, low glucosinolates, or modified oil profiles.

Despite this, a considerable degree of genetic erosion has occurred compared with traditional land-race germplasm. *Ex situ* genetic resource collections for *Brassica* exist throughout the world, and core collections have been developed from these to screen for novel sources of fungal or other resistance traits. Although they are valuable in identifying hotspots of variation within the relevant gene pools, such collections consist of heterogeneous and heterozygous material. This limits their long-term use for correlating detailed genetic studies.

As a result, the concept of diversity fixed foundation sets (DFFSs) has been devised. These are defined as *an informative set of genetically fixed lines representing a structured sampling of diversity across a gene pool;* they provide the advantage of being suitable for replicated, coordinated, or distributed analysis at molecular and trait levels.

DNA and seed of fixed lines are being made available in the public domain to provide a common reference resource for *B. oleracea,* related C-genome species, and *B. napus,* with additional sets being prepared for *B. rapa.*

INFORMATION MANAGEMENT AND BIOINFORMATICS

The continuing development of crop genetics and genomics is increasing the requirement for access to a wide range of relevant information and data. The establishment of international community resources and generation of large amounts of interconnected and persistent data provide great opportunities for interdisciplinary research and involvement of stakeholders.

A range of databases and other information resources exists for *Brassica,* but require integration. These include genetic resources, genetic mapping, genomic and functional genomics, as well as disparate sources of legacy trait and pedigree data. In addition, the close relationship with *Arabidopsis* provides a large amount of functional and reference information of direct relevance to the *Brassica* community.

This is starting to be collated in a number of public-domain systems. In all situations, it is important to be able to assess and determine provenance, status, and information content of data sets. The need for stable nomenclature for objects and entities such as chromosomes and linkage groups is becoming increasingly important as data integration progresses.

For crop genetics in general, there are opportunities to establish community-wide acceptance of trait and other ontologies. Throughout this chapter, the emphasis has been on the complexity of genome organization that underlies the plasticity of *Brassica* crop phenotypes.

Developing a deeper understanding of the ramifications and constraints arising from this will be important in the context of identifying and characterizing relevant genetic variation that can then be exploited in future breeding programs.

11

Chapter

GRASS GENOMES

The releases of the genome sequences of the dicot model plant *Arabidopsis thaliana (Arabidopsis)* and the monocot crop rice represent a revolutionary advance in plant biology. These sequences have provided a potent source of *information* for genetics, evolution, development, and all other fields of plant biology at a resolution that was previously unavailable.

The choice of these species as templates for genomic sequencing was based on their small genomes. Furthermore, *Arabidopsis* has long served as a tool for exploration of basic plant processes, and rice is a member of the most important plant family (Poaceae) from the standpoint of human subsistence.

The success of the *Arabidopsis* and rice genome sequencing projects, coupled with advances in strategies and technologies, has led to efforts to sequence all or parts of the genomes of many other plant species *representing* diverse plant families.

Given the rapid progress in genome sequencing of crops, one question that emerges is whether future investment in new model species to serve as surrogates for crops is necessary.

The answer to this question is largely dependent upon the crop or crops that would be represented by such a new model species. In this regard, a cogent argument in favour of a model crop species can be made for the cool-season grass crops.

Although it is clear that the rice genome sequence has been profitably exploited by researchers studying wheat, barley, and rye, it is equally clear that the rice genome sequence has *limitations* as a template for use in isolating genes from the cool-season cereals.

This is largely due to the nearly 50 million years of evolution that separate rice from the cool-season grasses, which is reflected in profound differences in phenology, morphology, physiology, and biotic and abiotic stress susceptibility and tolerance.

These differences impose limitations on the use of rice as a model for exploring the gene structure-function interface in the coolseason grasses. An alternative model species that, like rice, is diploid, possesses a small genome, and is transformable, but more closely related to the cool-season grasses, can be expected to find a place in the laboratories of scientists interested in cool-season grass crop improvement.

The species *Brachypodium distachyon* possesses all of these characteristics and is emerging as a new model species. The intent of this review is to provide an introduction to evolutionary, genetic, genomic, and *morphological* attributes of *B. distachyon* that make it such an attractive new model plant system.

EVOLUTIONARY RELATIONSHIP BETWEEN BRACHYPODIUM DISTACHYON AND OTHER GRASS CROPS

The grass family Poaceae (Gramineae) comprises approximately 10,000 species, including most of the world's most important crops such as wheat *(Triticum spp.)*, rice *(Oryza sativa)*, maize *(Zea mays)*, barley *(Hordeum vulgare)*, sorghum *(Sorghum bicolour)*, and many forage and turf species. The members of this large plant family are distributed around the world in highly diverse environments. Results developed from the joint efforts of a consortium of grass systematicists support the presence of eight subfamilies within the Poaceae.

The largest of these subfamilies is the Pooideae which, depending upon the authority, includes about 10 tribes and over 3000 species. This one subfamily includes most of the important cool-season grain crops, forage grasses, and turfgrasses.

The genus *Brachypodium* belongs to the *subfamily* Pooideae as well and is considered distinct enough from other members of the Pooideae that it has been placed within its own tribe, the Brachypoideae. Presumably because of the *economic* importance of the *subfamily*, many studies have been undertaken to resolve evolutionary relationships between the tribes in the Pooideae, with some including *Brachypodium* in analyses.

A recent analysis based on chloroplast gene *(ndhF)* sequence data indicated that the tribe Brachypoideae is sister to the evolutionary lineage that subsequently gave rise to the tribes Triticeae, Aveneae, Bromeae, and Poeae. As such, of the thousands of species comprising the entire family Poaceae (Gramineae), those of the genus *Brachypodium* are considered to be the closest ancestors to these four "core pooid" tribes that encompass most of the major cool-season grasses.

Thus, simply based on evolutionary considerations, we can predict that members of the genus *Brachypodium will* possess a genome structure and composition more similar to the "core pooids" than other genera in the entire grass family. This is one feature that makes *Brachypodium* attractive as a genus from which to select a possible model species for the coolseason grasses.

The systematics of *Brachypodium* has been examined a number of times in an attempt to resolve the evolutionary relationships among the species in this genus. The genus has not been unambiguously delineated, but it is likely to include 12 or more species. Several members of the genus are endemic to Europe, and one or more different species are found in other parts of the world including regions of Africa, Central and South America, and Asia.

Two unusual characteristics of the genus are that the different species of *Brachypodium* exhibit significant variation in base chromosome number, including 5, 7, 8, or 9, and that polyploid series are encountered within species in the genus, so ecotypes in a given *Brachypodium* species may range from diploid *(2n = 2x)* to octaploid *(2n =* 8x) in some cases. The species within the genus *Brachypodium* are perennial in nature with one exception: the annual species *B. distachyon,* often referred to as purple false brome.

This species appears to have diverged from the other perennial *Brachypodium* species early in the evolution of the genus. The *evolutionary* divergence of *B. distachyon* from other *Brachypodium* species is bolstered by results of experimental hybridizations between *B. distachyon* and other *Brachypodium* species. Although these crosses can produce F, progeny, they are sterile and exhibit abnormal meiosis.

This contrasts with results from crosses among perennial European *Brachypodium* species, which generally produce fertile F, hybrids. With one exception, *B. distachyon* is the only species in the genus with a base chromosome number of 5. The observed ploidy levels reported in *B. distachyon* ecotypes include diploidy *(2n = 2x =* 10), tetraploidy *(2n = 4x = 20),* and hexaploidy *(2n* = 6x = 30).

This species is considered to be endemic to regions surrounding the Mediterranean Sea; however, it has spread to other regions of the world, where it can be a weed. In terms of its breeding system, *B. distachyon* is characterized as highly self-compatible with a floral morphology that encourages inbreeding. An annual habit, self-compatibility, and availability of diploid ecotypes make *B. distachyon* the most desirable of the *Brachypodium* species from the standpoint of broad potential as a model species for functional and structural genomics research.

The first publication that touted *B. distachyon* as a possible model species was that of Bablak et al., in which it was noted that various parallels between *B. distachyon* and *Arabidopsis* might support research into the former species as a new model system.

STRUCTURAL FEATURES OF BRACHYPODIUM DISTACHYON GENOME

In diploid ecotypes of *B. distachyon,* the 5 chromosomes, while small, can be distinguished morphologically, and the major ribosomal gene tandem repeat loci (5S and 45S) have been localized to their respective chromosomes by *in situ* hybridization. Several publications have reported size estimates of the *B. distachyon* genome.

Using flow cytometry, Draper et al. measured the genome size of diploid ecotypes of the species and reported a c-value of 0.21 pg that they considered equal to 172 megabase pairs-a size approximately the same as what they obtained for *Arabidopsis.* However, this value is in conflict with other reports. Shi et a]. report that diploid *B. distachyon* has a c-value of 0.3 pg, which is similar to recent results of Bennett and Leitch, who reported a c-value of 0.36 pg.

These latter values are similar to the average c-value of 0.39 pg obtained from three separate analyses by flow cytometry for a reference inbred diploid line of *B. distachyon* that we have developed from the accession PI 185134 and is deposited in the United States Department of Agriculture's National Plant Germplasm System (NPGS).

It is unclear why the estimates of Draper et al. are not congruent with other reports because they used various plant species as calibration standards in that report. At this time, it is *recommended* that the c-value estimate of 0.36 provided by Bennett and Leitch be accepted as the best estimate for the genome size of diploid *B. distachyon.*

This indicates that the genome of diploid ecotypes of this species is approximately 6.5% the size of barley and approximately 2% the size of the hexaploid bread wheat genome. Thus, the entire genome of diploid *B. distachyon* is slightly smaller than an "average" chromosome arm in barley or wheat, is smaller than the genome of rice, and is only approximately twice as large as the genome of *Arabidopsis.*

The extremely small genome of diploid *B. distachyon* is another important feature desired in a model plant species. The derivation of polyploidy in *B. distachyon* is an interesting topic worth discussing.

Given that the base chromosome number of the diploid ecotypes is 5 and that this doubled or tripled in the tetraploid and hexaploid ecotypes, respectively, one might assume that this is the signature of autopolyploidy rather than allopolyploidy.

This assumption would be supported by the fact that no other European species of *Brachypodium* with a base chromosome number of 5 are known to exist. Thus, this rules out the likelihood that the polyploids are in fact allopolyploids derived from interspecific hybridization between diploids with a base number of 5, followed by chromosome doubling in a manner analogous to the evolution of tetraploid durum wheat *(T turgidum)* and hexaploid bread wheat.

However, the observation that 15 bivalents form during meiosis in hexaploid ecotypes of *B. distachyon* is contrary to expectations in an autopolyploid, in which multivalent pairing for some of the chromosomes is expected between some proportion of the chromosomes.

Furthermore, our recent observations that a large set of putative hexaploid ecotypes of *B. distachyon* from the NPGS appear to exhibit c-values approximately two and not three times the size of diploid ecotypes are not concordant with an autopolyploid origin of the polyploid ecotypes.

A recent publication reported on similarities and differences in chromosome karyotypes, location, and abundance of ribosomal DNA loci, and results of genomic *in situ* hybridization between ecotypes representing the three ploidy levels.

From these results, the authors postulated that the tetraploid ecotypes may actually represent a distinct diploid subtype (perhaps even a different species) with a base chromosome number of 10 and with chromosomes more similar to those of the related perennial species *B. sylvaticum* (slender false brome).

They also proposed that the hexaploid ecotypes may represent the product of hybridization and chromosome doubling between diploid *B. distachyon* and a plant similar to the putative tetraploid that they proposed as a distinct subtype. As such, the polyploid ecotypes of *B. distachyon* may well have value as a system for exploring the impact of polyploidy on genome *organization-a* topic of great interest to wheat researchers because of the polyploid derivation of that crop.

GENOME COMPOSITION

The emergence of *B. distachyon* as a new model grass species is quite recent and thus information on the genome composition of the species is scarce. Nonetheless, data gleaned from available studies help to shed some light on this topic. In an early molecular phylogenetic analysis of the genus *Brachypodium,* Catalan et al. produced a HindIII library of genomic DNA from a diploid *B. distachyon* ecotype to obtain probes for RFLP analysis.

Hybridization of labeled genomic DNA against a series of randomly selected clones from this library revealed that less than 12% of them gave a strong hybridization signal indicative of the presence of high copy genomic sequences. Hybridization of several of these high copy clones to *Brachypodium* species DNA and DNA from other diverse grass species revealed that they hybridized exclusively to *Brachypodium* DNA.

Furthermore, the hybridization patterns obtained from the different high copy probes suggested the presence of tandemly repeated DNA as well as interspersed repetitive DNA. In contrast, a set of "low copy" *B. distachyon* clones hybridized to DNA of most other grass species included in the hybridizations, suggesting conservation of such sequences across the Poaceae. These results are consistent with those obtained in *B. sylvaticum* (slender false brome).

Because of its small genome size, Moore et al. proposed that this member of the genus *Brachypodium* may be a useful model genome for studying the organization of grass genomes in general. Moore et al. found that 100% of the short (<2 kb) clones from *B. sylvaticum HpaII* and *HindIII* genomic libraries that were tested cross-hybridized to wheat genomic DNA, and 60% of these clones gave simple hybridization patterns on wheat genomic DNA.

In contrast, a set of comparably sized single copy rice *PstI* genomic clones hybridized to wheat just 64% of the time, with less than 60% of the hybridizing sequences revealing simple hybridization patterns on wheat genomic DNA. Furthermore, their results indicated that the genome of *B. sylvaticum* is less methylated than those of wheat and barley.

Also, a presumed centromere repeat sequence (CCS1) isolated from *B. sylvaticum* was found to hybridize to different grass species with different intensities; hybridization was strongest in wheat and rye, while maize and rice showed hybridization intensities less than 10% of that seen in the former species.

Taken as a whole, the results obtained from these analyses of the genomes of *B. distachyon* and its close relative *B. sylvaticum* support the notion that the small genome of *B. distachyon* harbors significantly less repetitive DNA than large genome grasses such as wheat and that the genomes of rice and *B. distachyon* are similar in complexity. However, the *B. distachyon* genome exhibits higher sequence similarity to the Triticeae than does rice for low-copy clones and at least one repetitive sequence.

GENE CONTENT AND ORGANIZATION VERSUS THOSE OF OTHER GRASSES

One of the principal interests in *B. distachyon* is its potential as a compact grass genome model that can be exploited to isolate genes from larger grass crop genomes because of conservation of genome organization and gene order. Various research groups are exploring this topic, though results have yet to be published.

In one study, the structure and abundance of C-repeat binding factor genes *(CBFs),* which encode transcription factors implicated in conditioning cold tolerance, were examined in *B. distachyon* with the purpose of comparison to the organization of the *CBF* gene family on chromosome 5H of barley. In barley, approximately ten *CBF* genes are present in a tightly linked gene cluster on chromosome 5H coincident with QTLs for cold tolerance, and they are estimated to be present at a density of less than one gene per 15 kb.

In contrast, seven *CBF* genes in a diploid *B. distachyon* line derived from NPGS accession PI 185134, which include presumed orthologs and paralogs of many sequenced barley 5H *CBF* genes based on phylogenetic analysis, have been localized to just two lambda phage clones spanning 33 kb. Thus, the *CBF* gene density in the region of the *B. distachyon* genome presumed to be syntenic to the *CBF* cluster on barley chromosome 5H is at least three times higher.

These results also are in agreement with the expectation that significant compression of coding sequences takes place in the small genome of *B. distachyon* because of a reduced amount of repetitive DNA, but not a large scale loss of coding sequences relative to its large genome cool-season relatives. The phylogenetic analysis of barley and *B. distachyon CBF* genes mentioned earlier not only provides interesting insights into the evolution of the *CBF* gene family in the grasses, but also offers evidence that orthologs of genes of agronomic interest in small grain cereal crops such as barley are present in the *B. distachyon* genome.

This contention is further bolstered by the observation that a presumed ortholog of the Q gene, which played an important role in wheat domestication because it confers the free-threshing character, was identified in *B. sylvaticum.* Finally, the genes Al, *X1, X2,* and *Sh2* have been used to examine the evolutionary dynamics of genome size and organization in many grass crops spanning different subfamilies.

To date, the presence of orthologs of the first three of these has been identified in a diploid inbred *B. distachyon* line derived from the NPGS accession PI 185134 (unpublished data). Efforts are currently under way to identify the *B. distachyon* ortholog of *Sh2* so that the organization of these genes can be compared to that in other grasses. The limited information available on comparative gene content and genome organization between *B. distachyon* and other grasses is supplemented with results obtained from its sister species *B. sylvaticum.*

Foote et al. reported construction and characterization of a bacterial artificial chromosome (BAC) library of *B. sylvaticum* and then used this library to explore synteny relationships among rice, wheat, and *B. sylvaticum.* They screened the BAC library with 48 probes derived from different grass species, and 36 were found to hybridize to one or more BAC clones.

Of these, 33 were postulated to exhibit the same gene order found in rice chromosome 9 and/or the Triticeae group 5 chromosomes. The largest *B. sylvaticum* BAC contig spanned 367 kb, and 9 of 11 genes/probes examined on this contig were present as expected in the syntenic region of Triticeae group 5 chromosomes.

Furthermore, they sequenced a 163-kb *B. sylvaticum* BAC clone from this library that revealed a minimum of 17 hypothetical genes. This corresponds to a gene density exceeding one gene per

10 kb of genome sequence. Thus, results from these studies suggest that we should have confidence that orthologs of genes of broad interest to crop productivity will be present in *B. distachyon.*

Furthermore, if we accept that the genome of *B. distachyon* will be highly similar in gene content and order to that of *B. sylvaticum-aside* from changes associated with major structural rearrangements, we can also be confident that *B. distachyon* has a genome amenable to service as a surrogate genome to accelerate gene discovery efforts in wheat, barley, and other large genome grasses.

FUNCTIONAL GENOMICS

The utility of *B. distachyon* to serve as a model grass genome is not limited to serving simply as a static "nucleotide roadmap" to help researchers navigate to genes of interest in grass genomes less tractable to physical analysis. Because *B. distachyon* is expected to have essentially the same ensemble of genes found in crops within the subfamily Pooideae, it may also be a potent resource for accelerating investigations of the role of particular genes in biological processes-for instance, diverse developmental or biochemical pathways and responses to biotic or abiotic stressesthrough functional genomics methods.

Initial steps have been taken in this direction. Comparative functional analysis of molecular responses of *B. distachyon* to *Magnaporthe grisea,* the fungal pathogen causing the disease rice blast, has been reported. Microscopic analyses of host responses in the resistant ecotype strongly resembled those of rice.

The responses observed included localized cell death likely due to oxidative stress and the rapid induction of pathogenesis-related (PR) gene expression. Interestingly, ecotypes of diploid *B. distachyon* were found to exhibit differential resistance to the pathogen; this will permit the use of these ecotypes for comparative functional analysis of resistance and susceptibility to this pathogen.

Although this example illustrates that *B. distachyon* can serve as a model for functional genomic analysis of this host-pathogen system in rice, it would be desirable to extend similar opportunities to pathogens of relevance in the cool-season grass crops.

For instance, the pathogens of greatest impact worldwide in wheat are the rusts belonging to the genus *Puccinia,* including leaf rust *(P. triticina),* stem rust *(P. graminis),* and stripe or yellow rust *(P. striiformis).* Draper et al. inoculated *B. distachyon* with the leaf and stripe rust pathogens to determine its response to these pathogens.

In both instances, host responses ranged from flecking (presumably indicative of a resistance response) to no visible response. In no case was an ecotype found to exhibit a typical susceptible response, though stripe rust uredinia were observed within heavily necrotic regions in two ecotypes, indicating successful colonization by the pathogen.

In the same publication, reaction of *B. distachyon* ecotypes to *Blumeria graminis,* which causes powdery mildew, was investigated and particular host resistance responses, including the formation of papillae, were found. These initial studies suggest that *B. distachyon* holds promise for functional genomic analysis of disease resistance in cool-season crops.

However, beyond these studies, there is a dearth of information reporting *B. distachyon* as a

functional genomics tool. Nonetheless, as this species becomes adopted as a model system, this is expected to change rapidly.

TRANSFORMATION AND MUTAGENESIS

The ability to transform a plant is an important attribute for a model plant species. One important aspect of transformation competency is the regeneration of plants from tissue culture. Bablak et al. reported the first successful regeneration of *B. distachyon* from tissue culture. Three diploid ecotypes were used in this study.

A high frequency of callus induction was observed from cultured seeds, and significant differences in callus induction were present between the ecotypes. The percentage of embryogenic callus obtained with the appropriate culture medium was as high as 60% in some experiments, and the number of regenerants per gram of callus was as high as 26.

Subsequently, Draper et al. confirmed the high level of embryogenic callus that could be obtained and the high level of plant regeneration for one of the same diploid ecotypes studied by Bablak et al.. More recently, Christiansen et al. found that it was possible to obtain embryogenic callus from immature embryos of diploid and tetraploid *B. distachyon* ecotypes, but the quality of this callus varied considerably.

Nonetheless, they found that the percentage of cultured embryos that produced embryogenic callus for the two diploid ecotypes they examined was approximately 50%, similar to the results of Draper et al.. The frequency of plant regeneration from the callus of their diploids approached 100%, but was dependent on the age of the callus. Thus, it appears that *B. distachyon* is very amenable to regeneration from tissue culture.

Draper et al. and Christiansen et al. also undertook biolistic transformation experiments on *B. distachyon*. Results of transformation for a single hexaploid ecotype were reported by Draper et al..

Callus of this hexaploid ecotype was subjected to microprojectile bombardment with a plasmid carrying the hygromycin resistance gene and a glucuronidase (GUS) marker, resulting in recovery of an average of seven independent hygromycin-resistant calli per gram of bombarded callus.

The authors reported a success rate in regenerating plants from the resistant calli of approximately 70%. These plants were subsequently found to exhibit GUS activity. However, transformation of diploid ecotypes was not reported. Similarly, Christiansen et al. undertook biolistic transformation of diploid and tetraploid ecotypes.

In this study, dramatic differences in transformation efficiency between the two diploid ecotypes included were observed. Staining with GUS to examine transient transformation revealed that one diploid ecotype exhibited no apparent GUS-stained sectors in calli after microprojectile bombardment, and the other diploid exhibited an average of nearly 2000 GUS-positive sectors per bombardment experiment. This latter result was roughly similar to those obtained for the two tetraploids included in the experiments.

Plant regeneration efficiencies (percent of bombarded calli that produced a transgenic plant) from these experiments were estimated at 5% for the diploid ecotype that exhibited GUS staining of calli and approximately 4% for the two tetraploids. However, individual experiments yielded efficiencies of between 9 and 14% depending upon the ecotype.

An alternative method for plant transformation is through the use of *Agrobacterium tumafaciens,* which has the advantage of producing lower copy and more stable genome integration events than particle bombardment. Vogel et al. studied *Agrobacterium-mediated* transformation efficiency of *B. distachyon* lines derived from accessions deposited in the NPGS and found a wide range of regeneration and transformation efficiencies among lines, which included diploid and polyploid ecotypes.

For the three diploids evaluated (derived from the NPGS accessions PI 185133, PI 185134, and PI 254867), the regeneration efficiency ranged from 4 to nearly 11%. However, as reported for microjectile-mediated transformation by Christiansen et al., significant differences in transformation efficiency between diploid lines using *Agrobacterium* were found.

For two of the diploid *B. distachyon lines,* no transformed plants were generated, while the other diploid line derived from PI 254867 exhibited an average plant regeneration frequency of over 3% across different experiments. In contrast, the regeneration frequency for polyploid lines ranged from 0 to 15%, though it should be noted that significantly more polyploids were examined in the experiments.

Thus, it is clear that *B. distachyon* is transformable by two different and common methods and that certain diploid ecotypes can be transformed by one or the other method. Presumably, in the future *B. distachyon* transformation can be optimized by building upon existing information provided by these earlier studies.

Mutagenesis provides a strategy to introduce new molecular variation into a genome. In doing so, it generates novel variants of genes that can be used for dissecting the molecular basis of traits of interest.

As such, mutagenesis is an important technique in the repertoire of a model species. In *Brachypodium,* little mutagenesis research has been conducted to date. Draper et al. have subjected *B. distachyon* to gamma-irradiation, but no mutation frequencies were reported and no descriptions of the mutants obtained were provided.

The only other report of mutagenesis in *B. distachyon* is that of Engvild. In this report, three diploid accessions of *B. distachyon* were used to assess the relative efficacy of different common mutagens and mutagen concentrations to induce mutations in *B. distachyon.* The mutagens included sodium azide and ethyl methanesulfonate (EMS).

The sodium azide treatment appeared to be far more effective in inducing mutant phenotypes (based on the frequency of chlorophyll mutations) in the M1 generation than the EMS treatments. Furthermore, the relative efficacy of the mutagen treatments in inducing mutations appeared to vary depending upon the diploid accession used.

Additional attempts to establish conditions that permit EMS or alternative mutagens to be used as effective mutagens in *B. distachyon* are strongly warranted so that mutant pools can be developed to identify mutations of interest based on phenotype or to identify lesions in genes of interest using methods such as TILLING.

GENETIC STOCKS AND GENETIC VARIATION

For any model plant species it is desirable to have a large and diverse collection of ecotypes exhibiting variation in phenotype and variation at the molecular level. Broad variation is particularly useful for successful development of segregating populations to serve as the basis for genetic map construction and for positional cloning endeavors.

Furthermore, a diverse collection of ecotypes may be useful for studying the genetic and molecular basis of important crop traits by studying the same traits within the model system. The NPGS has in its collection approximately 30 accessions of *B. distachyon* collected from various regions of the world over the course of the last several decades. The collection list can be viewed at the NPGS Website.

We used flow cytometry to discriminate between different ploidy levels in lines derived from 27 NPGS accessions. Five of these lines (from PI 170218, PI 185133, PI 185134, PI 245730, and PI 254867) were identified as diploids. These were collected in Turkey and Iraq.

In addition, a collection of over 50 accessions of *B. distachyon* is maintained by a company established by the University of Wales. The Brachyomics collection includes many lines collected in Europe and Asia, and elsewhere in the world. It will be important to cross-reference collections of *B. distachyon* maintained by different entities to provide the scientific community clear information on geographic origin and other passport information associated with each ecotype.

To date, the number of unique diploid ecotypes of *B. distachyon* is limited to perhaps ten. Therefore, it would be highly desirable to expand this number further by examining *B. distachyon* germplasm collections in other countries for additional novel diploid ecotypes, as well making collecting trips to regions to which this species is indigenous.

Although the number of diploid *B. distachyon* accessions/ecotypes is limited, it is promising to note that even among the five putative diploid accessions of *B. distachyon* deposited in the NPGS, significant phenotypic variation is evident. For instance, a cursory evaluation revealed clear differences in vernalization requirement, flowering date, plant height, pubescence, shattering, and seed size among these accessions (PI 170218, PI 185133, PI 185134, PI 245730, and PI 254867).

These morphological differences are likely a reflection of molecular variation through the genome of *B. distachyon*. Diploid *B. distachyon* held by Brachyomics can be expected to contribute even more genetic diversity than that available in these few NPGS diploids.

Additionally, although little emphasis has been placed on the issue, it may be expected that molecular variation within each of these accessions exists because they very likely are composed of seeds collected from multiple plants in a wild population.

In fact, support for this prediction is found in the publication of Christiansen et al., which alludes to evidence for molecular heterogeneity between plants from the same accession as revealed by AFLP analysis. Because diploid *B. distachyon* is highly self-compatible and rarely exerts anthers, it is likely that individual plants are inbred and thus homozygous for most loci.

However, this cannot be assumed for certain; genetic homogeneity with original accessions also cannot be assumed. Thus, reference inbred genetic stocks of *B. distachyon will* be a valuable resource to the *Brachypodium* research community.

During the past 3 years, we have developed single seed descent-derived inbred lines from 27 different *B. distachyon* accessions obtained from the NPGS, including five putative diploids (PI 170218, PI 185133, PI 185134, PI 245730, and PI 254867).

In most cases, we have developed two independent inbred sister lines from each accession to capture additional molecular variation that may reside in each accession, and in some of these inbred sister lines we have observed phenotypic differences, suggesting some success toward this end. We are making these inbred lines freely available to interested parties.

These inbred genetic stocks will be of benefit for various reasons, particularly at this early stage of research with *B. distachyon*. For instance, the use of common inbred genetic stocks for research will reduce incongruous results that may be due to genetic variation in source materials studied.

Also, the use of common inbred lines will improve the process of streamlining and integrating information emerging from research undertaken by different laboratories. The use of inbred lines of *B. distachyon* to develop segregating populations will result in yet another highly useful resource. Despite the small size of the florets, populations appropriate for genetic mapping purposes will be developed unless unanticipated incompatibilities exist between the inbred lines.

The only known report of a segregating population developed in *B. distachyon* is that of Routledge et al., who developed F_2 populations from crosses between two diploid ecotypes and used them to assess the inheritance of resistance to *M. grisea*. Similarly, we have been working to develop segregating populations derived from crosses among multiple diverse inbred lines.

Crosses among all pairwise combinations of four of our inbred diploid *B. distachyon* lines have been completed. Interestingly, we found that diploids appear to possess two anthers per floret, in contrast to reports in authoritative references that indicate that the species has three anthers.

This character, therefore, could differentiate all diploids from polyploids. The putative hybrids we now have will serve as the basis of recombinant inbred line development, resulting in a diverse set of segregating populations that can be used for a range of purposes.

GROWTH CHARACTERISTICS

During our program to develop inbred *B. distachyon* lines, we noted that diploid *B. distachyon* plants are more petite than the polyploids during early vegetative growth and PI 245730) appear to be unresponsive to long days and require significant vernalization periods to flower in a synchronized fashion. The capacity to cycle *B. distachyon* rapidly is highly desired in a model species.

At the same time, the capacity to retain these same genotypes in a vegetative state simply by growing them under short days is a useful attribute if one is interested in using the species as a model for forage or turf species. Furthermore, growing plants under conditions that encourage vegetative proliferation, followed by conditions that induce flowering, permits the recovery of large numbers of seeds from single plants.

In our experience, we have been able to recover over 500 seeds from plants in this manner. Seed size varies between lines, but some of the inbred diploids we have developed produce seeds that have a mass approximately 15% that of wheat seeds, thus indicating a rather large endosperm that may also be beneficial for studies focusing on seed development.

CONCLUSIONS

The rice genome sequence is a resource that has great value for structural genomics research in the grasses. However, *B. distachyon* is poised to become a potent model system that complements rice because it also possesses attributes desired in a model system for functional genomics, including a small physical size, a growth habit that can be regulated to control generation times, and, perhaps most importantly, a close evolutionary affinity to the cool-season grain, turf, and forage species in the grass subfamily Pooideae.

There is much to be learned about *B. distachyon* before it can be fully exploited for research purposes. If we use *Arabidopsis* as a template to help determine where energies need to be directed to establish *B. distachyon* firmly as a new model species, this would include assembling genetic resources (genetic stocks, segregating populations), molecular resources (traditional genomic and cDNA libraries, as well as large insert libraries), obtaining sequence information (ESTs, genomic sequence data), optimizing transformation systems, and developing mutagenesis protocols and mutant pools.

Although this review reveals deficiencies in this list that need to be addressed, more importantly it also highlights the significant strides forward that have been made in these areas in the short period of time that has passed since *B. distachyon* was proposed as a potential model grass species. This review also highlights the fact that a significant amount of research now is under way in *B. distachyon.* Such research efforts are a tacit acknowledgment by the grass research community that a new model species for the cool-season grasses is indeed desirable and that *B. distachyon* possesses the attributes to serve in this capacity.

12

Chapter

GENOMIC ANALYSIS

Most important Gram-negative plant pathogens are extracellular and rely on a type III secretion system (TTSS) to secrete proteins into the extracellular host environment and directly into the host cytoplasm. Pathogens that are deficient in type III secretion are unable to grow *in planta* or to cause disease, suggesting that the proteins secreted by the TTSS are essential virulence factors.

Some authors make a distinction between type III-secreted proteins that are predicted—based on their predicted enzymatic activity—to be targeted to the extra-cellular host environment and those that are targeted to the host cytoplasm.

They call the first kind "helper proteins" and the latter ones "effectors". Because no evidence exists thus far for differential secretion of type III-secreted proteins in plant pathogens, all type III-secreted proteins will be called effectors in this chapter.

There is one prominent group of effectors of plant pathogens that have a striking phenotype: they can dominantly confer to virulent pathogens the inability to cause disease.

On certain hosts, these effectors induce a resistance response that is usually accompanied by a type of programmed cell death called the hypersensitive response (HR). When the dose of the bacterial inoculum is high enough, the HR is macroscopically visible as a total leaf collapse and can easily be scored by eye in controlled infections.

Because such effectors turn a virulent pathogen into one that is "avirulent," these effectors are called "avirulence" (Avr) proteins encoded by avirulence *(avr)* genes. An individual Avr protein usually is recognized by an individual plant Resistance protein coded for by a resistance (R) gene that segregates as a single locus. The concept of cognate avr–R pairs is known as the "gene-for-gene" relationship and was first described by Flor.

The first *avr* genes were identified by constructing genomic DNA libraries of a strain avirulent on one host and transforming this library "en masse" into a virulent strain on the same host. Avirulent transformants were subsequently screened by individual inoculations on plants. Library clones that conferred avirulence were sequenced and the individual *avr* gene was identified.

Among others, this clever technique allowed the molecular identification of the first *avr* gene. Other effector genes were identified by their proximity to the gene cluster coding for the TTSS components. Any mutation in a TTSS component important for the actual secretion process eliminates the ability of a pathogen to cause disease in the case of a virulent pathogen and to elicit an HR in the case of an avirulent pathogen.

The genes coding for TTSS components are therefore called *hrp* (HR and pathogenicity) genes. The *hrp* genes are always clustered and are localized either on the bacterial chromosome or on a plasmid. In *Pseudomonas syringae* the *hrp* cluster is flanked on both sides by effector genes. On one side, the effectors are conserved among many *Pe syringae* strains and this cluster was therefore called the conserved effector locus (CEL). The other side of the TTSS contains effectors that are not as well conserved among strains and was therefore called exchangeable effector locus (EEL). Note that although some of the genes in the CEL and EEL are *avr* genes, others are not and that not all *avr* genes are located in the CEL or the EEL.

Effectors in the animal pathogen *Yersinia* and some other animal pathogens are efficiently secreted into the culture medium under certain conditions. None of the plant pathogens efficiently secretes effectors in culture, but *P. syringae* secretes some effectors in sufficient amounts in culture to be sequenced and some in fact were.

Yuan and He identified HrpA, a structural component of the TTSS, and HrpZ by sequencing proteins from the culture supernatant of *P. syringae*. Because many effectors are not secreted efficiently in culture, this approach is limited. Individual effector knockouts in plant pathogens usually have subtle effects on virulence.

Why this is so has not been answered satisfactorily yet, but it is believed to be mainly to the result of redundancy between effectors. This chapter describes two methods for identifying effectors in plant pathogens independently of any knowledge about their Avr activity, location, ability to be secreted in culture, or knockout phenotype. The first method is based on two findings:

1. The AvrRpt2 effector has two distinct regions, an N-terminal region that is important for secretion, and a C-terminal effector region that harbors the avirulence activity and that is sufficient to induce an HR in *Arabidopsis thaliana* upon recognition by the plant R protein Rps2.
2. The effector region of AvrRpt2 can be secreted from *P. syringae* when fused to the heterologous secretion region of the AvrRpm1 effector.

On the basis of these two findings, we developed an in vivo screen using the effector region of AvrRpt2 as reporter. We constructed a minitransposon that carries an origin of replication for *Escherichia coli*, an origin of transfer, and the tn5 *tranposase* gene on the vector backbone. The DNA coding for the effector region of AvrRpt2 (amino acids) and an antibiotic resistance marker are located between the tn5 insertion sequences. This construct can be transferred from *E. coli* to *P. syringae* by triparental mating. When the individual construct enters an individual *P. syringae* cell, the transposase is activated and can insert the DNA coding for AvRpt2 and the antibiotic resistance marker randomly into the *P. syringae* genome.

Because the construct has no origin of replication functional in *P. syringae,* it is lost after cell division. Furthermore, because the tranposase gene is not included between the insertion sequence elements, it is lost together with the construct and a stable insertion line is created.

When by chance the minitransposon is inserted in-frame downstream of the secretion region of an effector gene, a fusion between this secretion region and AvRpt2 is created. The resulting fusion is secreted into plant cells where it elicits an HR upon the interaction between AvrRpt2 and Rps2.

Because this HR is very strong, one HR-eliciting strain mixed with seven non-HR eliciting strains is enough to cause leaf collapse. Pools of eight insertion strains can therefore be infiltrated into leaves and positive pools can then be deconvoluted to identify the "culprit."

Once an HR-eliciting strain is identified, the sequence of the gene into which the transposon inserted has to be determined. Then, the TTSS-dependent secretion of the fusion, and the Rps2 dependence of the HR have to be verified.

These controls are needed to rule out the possibility that the fusion is secreted through a different kind of secretion system and that the observed cell death is caused by the toxicity of the fusion product and not by recognition of AvRpt2 by Rps2.

We anticipate that instead of the AvrRpt2 reporter in *P. syringae* infections of *A. thaliana,* any other effector of any plant pathogen that elicits a strong HR on any plant could be used in a similar screen. The second method for effector identification described in this chapter consists of a bioinformatic approach to effector prediction.

It is based on the fact that effectors in *P. syringae* and in other plant pathogens have amino acid biases that distinguish them from other proteins and that in many plant pathogens effector genes are preceded by conserved sequences in their promoters.

In the case of *P. syringae,* effector proteins are richer in serine than noneffector proteins and effector genes (or operons) are preceded by the conserved *hrp-box* promoter element. Other plant pathogens have similar biases and similar or different promoter elements. *Erwinia amylovora* effectors for example have also hrp-boxes, whereas *Ralstonia solanacearum* and *Xanthomonas* sp. effectors are preceded by a so-called PIP box, also called hrpII box.

Once effector candidates have been identified by this approach, they can be validated using the AvrRpt2 reporter, for example.

MATERIALS

Mating the Minitransposon into P. sringae

1. Bacterial strains: *E. coli* VPE42 (kan^r, tet^r) containing the mini-transposon vector pDSG50 (amp^r, kan^r) or similar mini-transposon vector, helper strain *E. coli* RK600 (cm^r), a *P. syringae* strain of choice and a TTSS-deficient strain, derivative of the same.

2. LB medium: 1 L of water, 10 g of bacto-tryptone, 5 g of bacto-yeast extract, and 10 g of NaCl.
3. KB medium: 1 L of water, 10 g of bacto-proteose peptone, and 1.5 g of K_2HPO_4. After autoclaving and cooling to at least 65°C add 3.2 mL of autoclaved 1 M $MgSO_4$ and 25 mL of autocolaved 20% glycerol.
4. All agar plates contain 15 g/L of agar.
5. Antibiotics: streptomycin, kanamycin, and nitrofurantoin (toxic).
6. For replica plating, a commercial or homemade "replicator" is used in combination with velvets (available from fabric shops or from laboratory supply companies).

Table 12.1: Sequences of Primers Used.

Primer name	Sequence
p1	CCTTTGTTCCGTCTCACGCACGTTC
p2	GGAATCGGAAGCCACGCTCGAACTATC
p3	CGGCCGCACTTGTGTATAA
p4	TAATTCCGCGAACCCCAGAG
p5	CGGCCTAGGCGGCCAGAT
p6	GAAGGCGATAGAAGGCGATG
v1	GAGAGGCGGTTTGCGTATTG
v2	ATGCTTCCGGCTCGTATGTT

Plant Growth

1. Potting soil.
2. Seed of A. *thaliana* ecotype "Columbia" and the *rps2* mutant line.

Plant Infections and HR Ealuation

1. Toothpicks.
2. *P. syringae* mini-transposon insertion strains are grown in 96-well growth blocks with 2-mL wells (reusable if bleached and washed after each use).
3. $MgSO_4$ is autoclaved as a 1 M stock solution and diluted when needed in sterile water to 10 mM.
4. 1-mL Blunt end syringes for plant infections are available from medical or laboratory supply companies.

Sequencing Flanking Regions

1. Glycerol.
2. Cryogenic vials.
3. Restriction enzymes (BsaAI, Bsp1286I, FspI, MspI, NcoI, and *SacI* when using pDSG50)

and a thermostable polymerase.

4. Primers for inverted polymerase chain reaction (I-PCR) when using pDSG50 are listed in Table elsewhere in this chapter.
5. Sequencing of PCR products can be outsourced to a sequencing center.
6. Custom primers can be ordered from many biotech companies.
7. Agarose, TBE, or TAE buffer.

In Vio Effector Verification (Tpe III Dependence Test)

1. Emzymes for PCR amplification and ligation can be purchased from any major molecular biology company.
2. The plasmid pBAV208 is available from the authors.
3. Electro-competent or chemical-competent *E. coli* DH5a cells.

In Silico Screening

1. Bioperl software available as a free download.
2. "Amino acid bias" script written by Gregory Kettler downloadable for free.
3. Any web browser to access online databases.

METHODS

We describe here how to use the minitransposon pDSG50 that carries the AvRpt2 reporter in *P. syringae* on *A. thaliana*. This construct was used in the effector screen described in Guttman et al.. We describe in *Note 5* how to substitute the AvRpt2 reporter with other reporters to apply the in vivo screen to other pathogens on other plants.

Mating the Minitransposon Into *P. sringae*

1. Grow the *E. coli* strain VPE42 containing pDSG50, the helper strain *E. coli* RK600 and the receiving *P. syringae* strain separately overnight at 30°C in 5 mL of liquid medium each (using LB for the *E. coli* strains and KB for *P. syringae* plus respective antibiotics).
2. Dilute the *E. coli* cultures 1:50 in the morning and *P. syringae* 1:10. When the *P. syringae* culture has reached an OD600 of approx 1.0 (this is after 3 to 5 h depending on the *P. syringae* strain used), spin down 1 mL of each of the cultures at 2000g in a tabletop microcentrifuge for 5 min.
3. Resuspend each of the pellets in 100 µL of sterile 10 mM MgSO4 by vortexing at medium speed, combine the three strains in one tube, and vortex at medium speed for a few seconds.
4. Spread the mixture on KB plates without antibiotics so that the strains can mate. Always plate a series of different volumes because the mating efficiency is pretty variable. Plate 10, 50, and 100 µL (each volume is spread on several plates to obtain at least 800 well-separated colonies in total).
5. Plates have to be incubated for approx 1 d at 28 to 30°C.

6. Replica plates onto KB plates with selection for *P. syringae* (streptomycin in the case of PmaES4326; *see Note 7*) and for the transposon (kanamycin in the case of pDSG50) using a replicator (*see Note 8*).
7. After another 2 to 3 d at 28 to 30°C colonies should become visible and after another day they should be big enough to be picked into liquid cultures for plant infections (*see Note 9*).

Plant Growth

1. Soak A. *thaliana* ecotype "Columbia" seed in 0.1% agar.
2. Vernalize the seed for 3 to 10 d at 4°C (*see Note 10* for hints on planting and growing *Arabidopsis)*.
3. Plant seeds using a Pasteur pipet and rubber bulb placing individual seeds on the surface of wet soil.
4. Plants are grown under long day conditions (16 h light) at 20°C and 50% humidity. The plants are ready for infection when they are between 3 and 4 wk old (this corresponds to the week that precedes bolting).

Plant Infections and HR Ealuation

1. Fill a 96-well growth block that has 2-mL large square wells with 1 mL of KB medium per well.
2. Using toothpicks, pick up as many bacteria as you can from each colony obtained in the matings (*see Note 11* for hints on growing *P. syringae* in 96-well growth blocks).
3. Use also one colony containing *P. syringae* expressing full-length AvrRpt2 as positive control in one well of one block for each eight blocks you fill.
4. Take the toothpicks out and grow the blocks in a shaker at 30°C for approx 20 h at the highest speed possible that does not cause the cultures to spill from well to well.
5. After 20 h the cultures should be saturated. You now prepare a growth block containing 1 mL of 10 mM MgSO4/well.
6. Measure the average OD600 of your overnight cultures
7. Using a multichannel pipet, add to each row of the 10 mM $MgSO_4$ block an entire *P. syringae* growth block combining the 8 rows of one entire block with overnight cultures of *P. syringae* into one row of the $MgSO_4$ block. Add as much culture as needed to obtain a final OD600 between 0.3 and 0.5. This corresponds more or less to adding 5 to 10 µL of each of eight saturated cultures (40 to 80 µL total) to 1 mL of $MgSO_4$.
8. Keep the overnight growth blocks until the next day (room temperature is fine). You need the individual cultures in order to identify any individual positive colony of a possible HR eliciting positive pool the next day!
9. Infect one block of eight pooled overnight growth blocks on one flat of 96 plants. Infiltrate the three largest leaves/plant with a blunt-end 1-mL syringe pressing the syringe against the lower side of the leaf after marking the leaves you chose to infiltrate with a marker

pen. Do not water the plants between infection and scoring.

10. Sixteen to eighteen hours later, score the plants for an HR.
11. If a plant looks like it has an HR, go back to the corresponding growth block and stamp the eight cultures forming the putative positive pool onto a KB plate containing the appropriate antibiotics.

Identification and Verification of Positie Insertion Strain

1. If only one or a few putative positive pools were obtained, you can test the putative individual positives by growing them up individually in 5 mL of KB in 15-mL test tubes and infecting them individually on plants. If there are many putative positive pools, the individual strains composing the positive pools can be grown up again in a growth bock and then be reinfected in pools of four. If a pool of four is positive, the individual strains forming that pool should then be tested individually.
2. Once you have identified an individual HR eliciting strain, it must be streaked out to a

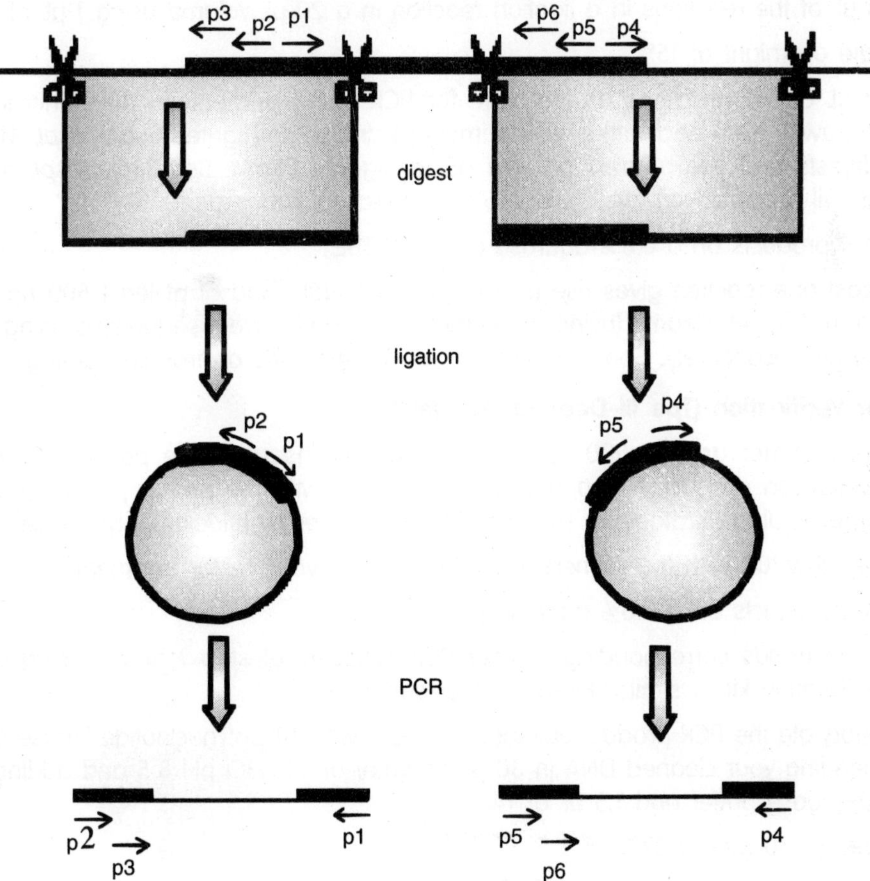

Figure 12.1: Inverted polymerase chain reaction schematic.

single colony and a few single colonies are tested again on plants to confirm the individual positive colony. At this point, *rps2* mutant plants should also be infected to make sure that the positive strain elicits an RPS2-dependent HR is caused by the recognition of the reporter and not by the toxicity of the fusion product.

Sequencing Flanking Regions

1. Once an individual positive colony has been isolated, it is grown up overnight to saturation.
2. Make a glyercol stock adding 300 µL of sterile 50% glycerol to 700 µL of saturated culture and store in a cryogenic vial at –75 to –85°C.
3. Extract genomic DNA using the Gentra Puregene kit or similar kit from other suppliers.
4. Dilute the genomic DNA to a final concentration of 50 ng/µL to use it for I-PCR.
5. Use 2 µL of DNA in a total reaction volume of 20 µL to digest DNA for at least 2 h. When using pDSG50 digest DNA separately with BsaAI, Bsp1286I, *FspI, MspI, NcoI,* and *SacI.*
6. Inactivate the restriction enzymes by heating the reactions at 80°C for 10 min.
7. Use 3 µL of the reactions in a ligation reaction in a 20-µL volume using 1 µL of ligase.
8. Incubate overnight at 15°C.
9. Use 2 µL of the reactions as template for PCR using primers on the minitransposon pointing away from each other. Use primers p1 and p2 for ligated BsaAI, *FspI, MspI,* and *SacI* digests and use primers p4 and p5 for ligated BsaAI, Bsp1286I, *MspI,* and *NcoI* digests. All primers work well with a 58°C annealing temperature.
10. Run PCR products on a 0.8% agarose gel in 1X TBE.
11. If at least one reaction gives rise to a clean individual band of at least 500 bp, cut the band from the gel, clean it (using a commercial kit) and have it sequenced using primers p3 and p6, respectively. See also Note 16 for more advice on troubleshooting I-PCR.

In Vio Effector Verification (Tpe III-Dependence Test)

1. Design a primer approx 500 bp upstream of the insertion site pointing toward the minitransposon and use this primer in combination with the primer p5 in a PCR using Pfu Turbo or Ultra (Stratagene, La Jolla, CA) or another highfidelity polymerase.
2. Amplify pBAV208 with the primers v1 and v2. This is your vector fragment.
3. Run PCR products on a 0.8% agarose gel.
4. Clean the bands corresponding to your PCR products of step 1 and 2 using Quiagen PCR purification kit or similar kit from other suppliers.
5. Phosphorylate the PCR product obtained in step 1 with T4 polynucleotide kinase (PNK) by resuspending your cleaned DNA in 30 µL of water or Tris-HCl pH 8.5 and adding 3.5 µL of a 10X ligase buffer and 1.5 µL of T4 PNK.
6. Incubate for 30 min at 37°C.
7. Clean the phosphorylation reaction using Quiagen PCR purification kit or similar kit from other supplier.

8. Ligate phosphorylated PCR product from step 5 with cleaned PCR product from step 2 using a molar ration of 5 to 1 and incubating overnight at room temperature.
9. Transform *E. coli* DH5a with your ligation and select for kanamycin-resistant colonies.
10. Extract DNA and verify the correct insert using primers on your gene.
11. Mate the resulting constructs back into the wild-type *P. syringae* strain used in the screen and into a TTSS-deficient derivative as described elsewhere in this chapter.
12. Infect two colonies from each mating, your wild-type strain and the original insertion strain on plants. If the fusion gives an HR only when expressed in the wild-type strain, but not when expressed in the TTSS-deficient strain, the fusion is secreted by the TTSS.

In Silico Screening

Identification of Open Reading Frames with an Amino Acid Bias

1. If you do not have your own sequence data, download the sequence files from public databases in which you want to search for effector candidates.
2. Save the sequences as simple text files.
3. In the case of *P. syringae,* also look for the absence of aspartate in the first 15 amino acids of the proteins you identified because aspartate is rarely found in this region.

Open Reading Frame Verification and Open Reading Frame Finding on Opposite Strand

1. Verify that the predicted effectors could actually be real genes by looking for open reading frames (ORFs) on the opposite strand. If there is an ORF on the opposite strand with homology to a known gene, your serine rich ORF is probably not a gene.
2. Look if there is a more likely START codon upstream or downstream of the predicted serine rich ORF. You do this by looking for Shine Dalgarno sequences.
3. Do a blastp search with your predicted effector at www.ncbi.nih.gov/BLAST/. If there is a confirmed effector homologous to your predicted effector, the START of the homolog is probably the START of your predicted effector.

Table 12.2: *hrp* Box Sequences.
hrp Box consensus
GGAACC 15/16N CCAC
GGAACT 15/16N CCAC
GGAATT 15/16N CCAC
GGAACC 15/17N ACAC

4. In the blastp search also look for homology of your predicted effector to eucaryotic genes. If you find such homology, this is supporting information that your predicted effector acts inside the eukaryotic host cell.
5. In the blastp search also look for homology to bacterial proteins with a known function

inside the bacterial cell. If you find such homology, you probably do not have an effector although it is serine rich. This is especially true when homology extends through the N-terminus.

hrp Box Identification

1. In the case of *P. syringae,* you now look for the *hrp* box promoter element. The *hrp* box has one of the sequences listed in Table elsewhere in this chapter and is most likely located between 30 and 200 bp upstream of the START of your effector.
2. It is relatively straightforward to look for *hrp* boxes by eye or by simply searching for either of the two conserved sequences that make up the *hrp* box using the search function of your sequence or text editor of choice and then looking for the other half of the *hrp* box at a distance of 15 to 16 nucleotides by eye.
3. Look if your *hrp* box is within a known gene on either strand. If this is the case, your *hrp* box is most likely spurious.

Effector Verification

Once you have a predicted effector you can clone it and fuse it to your reporter of choice. You can clone your predicted effector either using restriction enzymes or using the Gateway recombinational cloning system (Invitrogen, Carlsbad, CA). With the latter method, you can take advantage of a series of Gateway compatible cloning vectors for effector confirmation and characterization that are available from the authors of this chapter.

NOTES

1. *E. coli* VPE42 is a derivative of the broad host range RP4 conjugal donor strain *E. coli SM10 lambda pir.* Other *lambda pir* derivatives can be used if different drug resistances are required, for example S17-1 (strep^r, spec^r). *E. coli* RK600 is a so called helper strain and provides *mob* and *tra* genes for mobilization and transfer of the mini-transposon to *P. syringae.*
2. *P. syringae* pv *maculicola (Pma)* ES4326 (Strep^r) was the *P. syringae* strain used in Guttman et al. (10). This chapter describes the in vivo screening method when using PmaES4326. Other *P. syringae* strains and even other species can be used for the screen as well. A type III secretion-deficient derivative of PmaES4326 is available from the authors.
3. Most *P. syringae* strains are resistant to nitrofurantoin at 100 µg/mL (toxic). A nitrofurantoin stock solution is made in dimethylsulfoxide at 100 mg/mL and stored at −20°C. Nitrofurantoin stock solution is added to KB agar before pouring plates. It is important to stir KB after adding nitrofurantoin for another minute or two since it does not dissolve immediately.
4. Toothpicks can be reused many times when autoclaved after each use.
5. Tn5 transposons are functional in all Gram-negative bacteria. The pDSG50 minitransposon can therefore be used in any plant pathogen. The *avrRpt2* reporter sequence can be changed to other reporters. pDSG50 is derived from the pBSL118 minitransposon by cloning

avRpt2 into the multiple cloning site and then removing 5' upstream sequence because a STOP codon was present upstream of and in frame with avRpt2. We can provide you with pDSG50 and the DNA sequence up- and downstream of *avRpt2* so that you can replace *avRpt2* with your reporter of choice or you can request the original minitransposon with a series of useful cloning sites from the Netherlands Culture Collection of Bacteria with the catalog number 3379. Any reporter has to be well characterized before use. Make sure that the reporter is sufficient to elicit an HR inside the plant cell. Create defined effector::reporter fusions to make sure fusions to your reporter elicit a strong HR on your host plant. Also perform setup experiments to identify the optimal infection dose, plant cultivar/ecotype, age of plants, and growth conditions to reach the highest possible sensitivity in your screen. It is also useful to add an epitope tag to your reporter if you do not have an antibody against it.

6. The vector pDSG50 or similar minitransposon vector can be transferred efficiently to *P. syringae* by triparental mating. Because of its RP4 origin of replication, pDSG50 can only replicate in *lambda pir* strains like *E. coli* VPE42, but not in DH5alpha, and needs a helper strain like *E. coli* RK600 to provide mobilization and chromosomal transfer genes *(tra* and *mob)* to be mobilized and transferred to *P. syringae.*

7. In case your *P. syringae* strain is rifampicin-resistant or has no antibiotic resistance, use nitrofurantoin as selection for *P. syringae. E. coli* easily acquires spontaneous rifampicin resistance and rifampicin is therefore useless as a selective marker in matings.

8. Wash velvets immediately after each use in water and soap, rinse in water, let dry, and autoclave for 30 min.

9. An efficient mating leads to more than 100 colonies per plate. You should make sure that at least some of the colonies actually produce fusion proteins by doing Western blots on a few dozen insertion strains using an antibody against your reporter (if available) or against an epitope tag that you should have fused to your reporter. If the genome of the bacterial strain you are using were 100% transcribed, you would expect one in six strains to produce a fusion protein. Because there are non-coding regions and not all genes are expressed in culture medium, you can expect approx 1 in 24 strains to produce a fusion with your reporter. It is also useful to mate at least once a full-length reporter under the control of its own promoter into your strain and to make sure that colonies obtained from this mating give a strong HR (we can provide the construct pDSG49 containing full-length *avrRpt2* under the control of its own promoter for this purpose).

10. Do not use seed that has been vernalized longer than 2 wk because the germination rate will decrease after 2 wk and even the seedlings that look normal at germination may turn red later and not give a good HR. Plants can be grown in a 96-well grid on standard flats containing 48 cells planting two plants diagonally in each cell. To make sure to have 96 plants you can plant four seeds per insert and then thin out to two plants after 2 wk. Make also sure plants are not under or overwatered while growing. Only "happy" plants give a good HR.

11. Use as much inoculum as you can. *P. syringae* is not *E. coli* and a nonvisible amount of bacteria will take for ever to grow. Use a blob of cells of at least 1 mm in diameter to

start your culture with. You cannot use too much. If even after using a good-sized inoculum, *P. syringae* cultures do not saturate within 20 h, it is most likely because of a too-slow shaking speed. Also, residual bleach in the wells may interfere with growth. To attach the growth blocks to your shaker, you can build your own growth block holder. Screw a sturdy cardboard box to the shaking platform and squeeze Styrofoam blocks between the growth blocks and the box to keep the growth blocks from moving around. Always cover growth blocks with plastic lids and tape the lids to the blocks or fasten with rubber bands.

12. If some bacteria accumulated at the bottom of the wells, vortex until they resuspend completely before pipetting. You do not have to change tips between samples since the small quantity of cross-contamination will not give you false-positives.
13. The best time to score the HR will depend on your plant growth conditions, your *P. syringae* strain, and the OD600 of the inoculum. The first few times you do the screen check the plants several times between 16 and 24 h. The best time to score the HR is when the plants infected with negative pools still look healthy but are about to wilt. Note that positive pools may give a weaker or a stronger HR than the pool containing the positive control.
14. I-PCR is a technique that allows you to obtain PCR products of unknown genomic regions flanking a known DNA sequence, in our case the unknown sequences flanking the transposon insertion.
15. The enzymes BsaAI, FspI, *MspI,* and *SacI* cut pDSG50 downstream of the primer sites p1 and p2. These digests will be used to obtain PCR products to sequence the region upstream of the transposon insertion. The enzymes *BsaAI,* Bsp1286I, *MspI,* and NcoI cut the minitransposon upstream of the primer sites p4 and p5. These digests will be used to obtain PCR products to sequence the region downstream of the insertion site. When using a minitransposon different from pDSG50, you have to find restriction sites in similar positions regarding to the primer sites you will use.
16. If you obtain a strong band with a background smear or additional weaker bands, the strongest band should be cut out, cleaned, and diluted 1:100 (you may try a dilution series of 1:10 to 1:100) and used as template for a second PCR reaction using the forward primer from the first PCR with a more internal second primer p3 or p6. A 2-μL aliquot of this PCR should be run on a gel and if a clean band is now obtained the rest of the PCR can now be cleaned and sequenced. If no bands are obtained with any restriction enzyme, a different polymerase can be used. Pfu or other similar high-fidelity enzymes can amplify longer DNA fragments compared with Taq. Thus, using such polymerases increases the chance of getting an I-PCR product. In any case I-PCR may not lead to the identification of the complete effector and promoter sequence of every effector found in the screen. If this is the goal, a genomic library of the strain used in the screen should be constructed and hybridized to the DNA fragments obtained by I-PCR. Positive clones can then be sequenced to extend the sequences surrounding the transposon insertions.
17. To confirm that the identified effector::AvRpt2 fusions are secreted in a TTSSdependent manner, you amplify by PCR at least 500 bp of the DNA sequence upstream of the

insertions together with the AvrRpt2 reporter and the kanamycin resistance gene of the minitransposon and clone this whole fragment into a highcopy number cloning vector that has no origin of replication for *P. syringae.*

18. Pfu creates blunt DNA fragments that are unphosphorylated. A Pfu product that is used as insert needs therefore to be phosphorylated. A Pfu product that is used as vector is ready to go since it is already unphsophorylated.
19. The construct will integrate at the effector locus corresponding to the effector fragment it contains and the effector::AvrRpt2 fusions will therefore be expressed from the native promoter of the effector.
20. If no TTSS-deficient derivative of the *P. syringae* strain used in your screen is available, you can sequence the full-length effector gene you found and clone it downstream of a constitutive promoter and fuse it to the AvrRpt2 reporter on a non-integrating vector. pBAV178 is a Gateway (Invitrogen) compatible vector developed for this purpose and can be requested from the authors. You can then mate this plasmid into any TTSS proficient and deficient *P. syringae* strain to test TTSS-dependent secretion.
21. A different algorithm for effector prediction is described elsewhere in this chapter. The protocol described here can also be done in a different order, for example, looking first for genes downstream of conserved promoter elements and then looking for genes with amino acid biases.
22. In the case of *P. syringae,* also look for a high proline content instead of a high serine content in the first 50 amino acids and for an overall serine content of at least 10% over the whole protein. *P. syringae* effectors also often have cluster of serines or serines and prolines in the first 50 amino acids. *See also* Greenberg and Vinatzer for more background information.
23. Shine-Dalgarno sequences are bacterial ribosome binding sites with the consensus AGGAGG located four to seven nucleotides up-stream of a gene's START codon.
24. However, there are mistakes in the databases and you should always be critical about any published annotation you find.
25. Sometimes transposase or insertion sequences are found between effectors and their *hrp* box increasing the distance between *hrp* box and effector. An effectcor may also be in an operon with non-effector encoding genes.
26. In case of *Ralstonia solanacearum* and *Xanthomonas you will* look for the PIP box or *hrpII* promote element.

ANALSIS OF GENE FUNCTION IN RICE THROUGH VIRUS-INDUCED GENE SILENCING

In the past few years, virus-induced gene silencing (VIGS) has been used to determine gene function in plants. The mechanism of VIGS is not fully understood; however, it is known to target RNA molecules in a sequencespecific manner. Although VIGS originally was a term used to describe the recovery of the host from virus infection, it now is used most often to describe the downregulation of host gene expression via a virus vector. Doublestranded RNA (dsRNA), produced by fold back of the single-stranded viral RNA or during virus replication through the activity of a viral RNA-dependent RNA polymerase, triggers silencing. The dsRNA is then cleaved into small interfering RNAs (siRNA; 21 to 25 nucleotides in length) by an RNase III-type dicer-like endonuclease.

The resulting siRNAs are believed to provide a single strand of RNA for incorporation into the RNA-induced silencing complex. The RNA-induced silencing complex then identifies and cleaves additional single-stranded target RNA sequences complementary to the "captured" siRNA sequence.

Of interest, cleavage of additional target RNA through this pathway may not be the only method of controlling virus accumulation. It is possible that small RNAs derived from viral single-stranded RNA stem-loops, similar in structure to the microRNAs derived from host RNAs, inhibit virus accumulation through translational repression.

With time, the mechanism of VIGS will be determined, but this lack of basic information does not prevent its practical use to study gene function. To date, multiple RNA plant virus vectors (e.g. *Tobacco mosaic virus, Potato virus X,* and *Tobacco rattle virus)* are available for VIGS in dicotyledonous plant species and one RNA virus vector, *Barley stripe mosaic virus,* is available for VIGS in two monocotyledonous species (barley and wheat).

Here, we review the use of a plant virus vector newly created from a strain of *Brome mosaic virus* (BMV), to silence genes in various monocotyledons, including rice, through VIGS. To construct this BMV vector, we isolated a virus from an infected Tall Fescue plant (*Festuca arundinacea* Schreb.) and purified viral RNA from the virus capsid. Previous analysis of the capsid, through electron microscopy, and viral RNA, through Northern blot analysis, indicated that this virus was a strain of BMV, which we named the fescue strain of BMV (F-BMV; Ding, X. S. and Nelson, R. S., to be submitted).

Full-length reverse transcription polymerase chain reaction (RT-PCR) products of the three viral genomic RNAs were synthesized, gel purified, and ligated individually into the pGEM T-Easy vector (Promega, Madison, WI).

The ligation products were transformed individually into *Escherichia coli* JM109-competent cells. The plasmids containing the viral complementary DNAs (cDNAs) were named pF1-11, pF2-2, and pF3-5 (for F-BMV cDNA representing BMV genomic RNAs 1, 2, and 3, respectively). Three RNAs from the Russian strain of BMV (R-BMV) also were cloned (pB1-26 for RNA1, pB2-4 for RNA2, and pB3-3 for RNA3) using the same technique as described for the F-BMV.

We determined that the viral sequence necessary for infecting rice did not reside in RNA 3 of BMV). This information allowed us to use the clone representing RNA 3 from R-BMV, which has a more convenient restriction site for inserting host sequences than does the cDNA representing F-BMV RNA 3, in our VIGS vector.

Co-inoculation of rice plants with transcripts from pF1-11, pF2-2, and pB3-3 harboring a fragment of phytoene desaturase from maize *(Zea mays)* caused silencing of this gene in rice. VIGS has several advantages over other functional genomics approaches, such as no transformation of the host plant is required and the function of individual gene family members can be studied.

An additional major advantage is the rapidity with which a researcher can observe a silencing phenotype during VIGS. A silencing phenotype can be observed within 3 wk after inserting a host sequence into the BMV vector and inoculating plants. For important monocotyledonous crops such as rice and maize, for which no rapid gene knockout technology exists, VIGS with the F-BMV vector was effective in silencing genes.

This chapter provides procedures to execute the following:

1. Insert plant sequences into bacterial plasmids and transfer them to the BMV vector.
2. Synthesize the BMV vector transcripts.
3. Inoculate plants with transcripts.
4. Analyze virus infection and gene silencing in the plants.

It is important that all plants inoculated with BMV or BMV harboring a foreign insert be kept inside a greenhouse or growth chambers. Infected plants and materials used to grow these plants (e.g., soil, pots, and trays) should be autoclaved before disposal to prevent release of the virus into the environment, as per regulations from the US Department of Agriculture governing the use of pathogens and transgenic viral sequences.

MATERIALS

Insertion of Foreign Sequences Into the BMV Vector

1. pB3-3 (plasmid containing cDNA of R-BMV RNA 3).
2. pGEM–T Easy vector (Promega; Madison, WI).
3. JM109 competent cells (10^8 cfu; Promega).
4. SOC medium: tryptone 2% (w/v), 8.6 mM NaCl, 2.5 mM KCl, 20 mM $MgSO_4$, 20 mM glucose (Sigma; St. Louis, MO).
5. 2% X-gal (5-bromo-4-chloro-3-indolyl-(3-D-galactoside): 20 mg in 1 mL of dimethylformamide (Promega)
6. 20% Isopropyl-(-D-thio-galactopyranoside (IPTG), 200 mg in 1 mL of H_2O (Promega)
7. Luria broth (LB) medium: 25 LB medium capsules per liter of distilled H_2O (Q-BIOgene; Irvine, CA).
8. LB agar medium: 16 LB agar medium capsules per 400 mL of distilled H_2O (Q-BIOgene).
9. LB liquid medium containing ampicillin (Sigma): autoclave LB liquid medium at 121°C for 20 min in 1-L autoclavable bottle. Remove the bottle from the autoclave after the pressure is fully released. Cool the medium to approx 50°C and add ampicillin (100 μg/mL medium). Mix well and store the medium at 4°C before use.
10. LB medium plate containing ampicillin (Sigma): autoclave LB agar medium at 121°C for 20 min. Remove the bottle from the autoclave after the pressure is fully released. Cool the medium to approx 50°C and add ampicillin (100 μg/mL medium). Mix well and pour the medium into 20 Petri dishes (100 × 15 mm). Store plates at 4°C before use.
11. LB medium plate containing ampicillin, X-gal and IPTG: Mix 20 μL of 2% X-gal solution and 30 μL sterilized H2O with 50 μL of 20% IPTG. Spread the mix onto the surface of an

LB medium plate containing ampicillin. Incubate the plate at 37°C for 1 h and then store the plate at 4°C before use.

12. T4 DNA ligase with 10X buffer (NEB; Ipswich, MA).
13. Nuclease-free water.
14. 1-kb DNA ladder (NEB).
15. *HindIII* restriction enzyme with 10X reaction buffer (NEB).
16. QIAquick PCR Purification kit (Qiagen; Valencia, CA).
17. QIAquick Gel Extraction kit (Qiagen).
18. Buffer P1 (Qiagen).
19. Buffer P2 (Qiagen).

Preparation of BMV RNA Transcripts and Inoculation of Plant

1. BMV clones (pF1-11, pF2-2, pB3-3, and pB3-3 containing a plant gene sequence).
2. Phenol/chloroform/IAA solution (Ambion; Austin, TX).
3. 3 M Sodium acetate, pH 5.2.
4. mMESSAGE mMACHINE T3 kit (Ambion).
5. SpeI restriction enzyme with 10X buffer (NEB).
6. PshAI restriction enzyme with 10X buffer (NEB).

Analysis of Virus Infection and Gene Silencing in Plants

1. Polyclonal antibody against BMV coat protein.
2. Primers to allow specific detection of host insert sequence in virus genome, virus genome, and host transcript targeted for silencing.
3. SuperScript reverse transcriptase with 5X first strand buffer and 0.1 M dithiothreitol (DTT; Invitrogen, Carlsbad, CA).
4. Taq DNA polymerase in buffer B with 10X PCR buffer and 25 mM MgCl2 (Promega).
5. SUPERase In™ (Ambion).
6. 10 mM dNTP Mix: Mix 10 μL of 100 mM dATP, 10 μL of 100 mM dCTP, 10 μL of 100 mM dGTP, and 10 μL of 100 mM dTTP with 60 μL of nuclease-free H_2O.
7. Recombinant RNasin (40 U/μL; Promega).
8. TRIzol Reagent (Invitrogen).
9. DNaseI (RNase free; Ambion).
10. Chloroform.
11. Isopropanol.
12. Ethanol.
13. Agarose (Shelton Scientific, Shelton, CA).

14. Mortars and pestles: bake mortars and pestles at 180°C for 3 h before use.
15. 0.6-mL Microfuge tube (ISC Bioexpress, Kaysville, UT).

METHODS

Insertion of Foreign Sequences Into the BMV Vector

To silence a plant gene of interest through VIGS, DNA representing a portion of the gene should be inserted into the *HindIII* restriction site within the pB3-3 construct. Inserts can be 150 to 250 nucleotides in length for this vector. A gene fragment is amplified from total cellular RNA isolated from plant tissue through RT-PCR using primers specific for the gene.

The amplified gene fragment is then ligated into the T-Easy vector. After purifying the T-Easy vector containing the inserted host-gene fragment, the fragment is released through digestion with *HindIII* and gel purified. The gene fragment then is ligated into the pB3-3 vector predigested with *HindIII*.

Synthesis and Ligation of a Target Host-Gene Fragment Into T-Easy Vector

1. Amplify a fragment (150 to 250 nucleotides in length) from a host mRNA representing a gene of interest using specific primers under predetermined RT-PCR conditions. An approx 100-μL PCR reaction mix is needed for each fragment.
2. Begin purifying the PCR product by mixing it with 500 μL of PB buffer (Qiagen).
3. Load the mixed solution onto a QIAquick spin column inserted into a 2-mL collection tube. Spin the column-collection tube unit at 15,000*g* for 1 min and discard the flow-through.
4. Add 0.7 mL of PE buffer (Qiagen) to the column and spin the column with a collection tube at 15,000*g* for 1 min. Discard the flow-through and spin the column with collection tube again at 15,000*g* for 1 min.
5. Place the column into a clean 1.7-mL microfuge tube. Add 50 μL of EB buffer (Qiagen) to the column and spin the column-collection tube unit at 15,000*g* for 1 min.
6. Concentrate the PCR product by spinning the tube in a microfuge under vacuum until the final volume of the PCR product is approx 5 μL.
7. Set up a ligation reaction by mixing 3 μL of PCR product (approx 200 ng of DNA) with 0.5 μL of pGEM-T Easy Vector (50 ng), 5 μL of 2X ligation buffer, 1 μL of nuclease-free H_2O, and 1 μL of T4 DNA ligase (Promega) in a microfuge tube.
8. Mix the contents by flicking the tube several times. Spin the tube briefly to collect the ligation mixture at the bottom of the tube and incubate the tube overnight at 4°C.

Transformation of Competent E. coli Cells With Plasmid Containing the Host Sequence

1. Place JM 109 competent cells from –80°C freezer on ice until just thawed (5 to 10 min).
2. Transfer 25 μL of just-thawed competent cells into a prechilled 14-mL Falcon tube (Becton Dickinson, Franklin Lakes, NJ). Add 2 μL of ligation mixture into the tube and mix gently by flicking the tube several times.

3. Incubate the tube on ice for 20 min and then subject the tube to a heat-shock treatment by holding it in a 42°C water bath for 40 s followed by a 2-min incubation on ice.
4. Add 250 μL of SOC medium into the tube and incubate the tube in a 37°C shaker set at 150 rpm for 1 h.
5. Plate 100 μL of transformed cell culture onto a LB medium plate containing ampicillin, X-gal and IPTG. If a greater number of colonies are desired, spin the transformed competent cell culture at 1000g for 2 min, resuspend the pellet in 100 μL of fresh LB liquid medium, and plate the cells on one LB medium plate containing ampicillin, X-gal, and IPTG.
6. Incubate the plate overnight in a 37°C incubator. After incubation, pick two to three white colonies from the plate with toothpicks and inoculate each colony to individual Falcon tubes containing 2.5 mL of LB liquid medium with ampicillin.
7. Incubate the Falcon tubes in a 37°C shaker set at 250 rpm overnight.

Isolation of T-Easy Vector With Plant-Gene Insert

1. Before isolating plasmid DNA from the overnight cell culture, take 0.5 mL of the overnight culture and mix with 0.5 mL of sterile 50% glycerol in a sterile 1.7-mL microfuge tube. Store the tube at –70°C for future use.
2. Transfer 1.7 mL of the remaining overnight cell culture into a microfuge tube. Pellet cells by centrifuging the microfuge tube at 15,000g in a microfuge at room temperature (RT) for 5 min.
3. Discard the supernatant. Resuspend the pellet in 0.3 mL of buffer P1 with RNase A (Qiagen) by pipetting the cells up and down until the pellet is fully dissolved. Incubate the microfuge tube at RT for 5 min.
4. Add 0.3 mL of lysis buffer P2 (Qiagen) into the microfuge tube and mix by inversion five or six times. Incubate the tube on ice for 5 min.
5. Add 0.3 mL of 3 M potassium acetate, pH 5.5, to the microfuge tube and mix by inversion two to three times. Incubate the microfuge tube on ice for 10 min.
6. At RT, centrifuge the mixture at 15,000g for 10 min in a microfuge. Pour the supernatant into a clean microfuge tube and centrifuge again at 15,000g for 5 min.
7. Pour the supernatant (approx 0.9 mL) into a clean microfuge tube. Add 0.65 mL of isopropanol into the microfuge tube, mix and incubate the mixture on ice for 20 min.
8. At RT, centrifuge the mixture at 15,000g for 15 min to pellet plasmid DNA. Discard the supernatant. To wash the DNA, carefully add 0.7 mL of 75% ethanol to the tube, flicking the tube several times, and centrifuge the tube at 15,000g for 5 min.
9. Carefully remove the supernatant without dislodging the DNA pellet. Dry the remaining pellet by centrifugation in a vacuum centrifuge (e.g., Savant SC110A, Cambridge Scientific Products, Cambridge, MA) for 5 min.
10. Resuspend the pellet in 20 μL of nuclease-free H2O. Estimate the concentration of the plasmid DNA preparation by electrophoresis in a 1% agarose gel containing 0.1 μg of ethidium bromide per milliliter of agarose solution and comparison with differing known

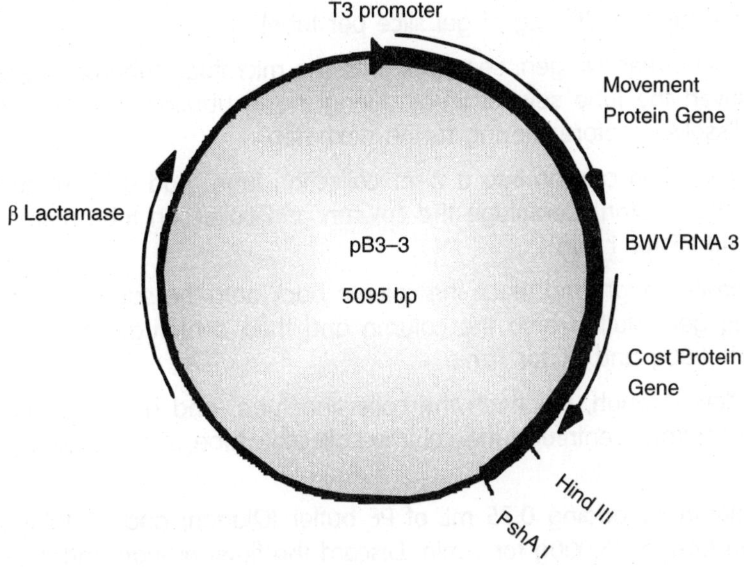

	nucleotides
T3 RNA polymerase Promoter	16–33
BMV (Brome mosaci virus) (RNA3)	33–2135
BMV Movemenbt Protgein Gene	124–1018
BMV Coat Protein Gene	1275–1840
pGEM T Easy Vector	2141–5095
β Lactamase (Ampicillin resistance) Gene	3401–4262

Figure 12.2: Schematic drawing of pB3-3 showing restriction sites for inserting foreign deoxyribonucleic acid sequence or linearizing the plasmid before in vitro transcription.

concentrations of 1-kb DNA ladder loaded in adjacent lanes. Alternatively, determine the plasmid concentration with a spectrophotometer.

Transfer of Plant-Gene Sequence From pGEM T-Easy to pB3-3

1. To release the plant-gene sequence from the T-Easy plasmid, mix 5 µL of the purified T-Easy plasmid containing the plant sequence (approx 2 µg of DNA) with 3 µL of 10X *HindIII* reaction buffer (NEB), 21 µL of nuclease-free H_2O and 1.5 µL of *HindIII* restriction enzyme (NEB) in a microfuge tube. Incubate the tube at 37°C for 2 h.
2. After digestion, mix the DNA sample with a loading dye and load the sample onto a 1% agarose gel containing 0.1 µg of ethidium bromide per milliliter of agarose solution. Electrophorese to separate the DNA fragment containing the plant gene sequence from the T-Easy vector DNA. Also load a DNA ladder as a size marker.
3. Visualize DNA bands with an ultraviolet transilluminator. Excise the gel fragment containing the plant gene sequence using a clean razor blade and transfer the gel slice into a

microfuge tube (approx 300 μg of gel slice per tube).

4. Add 1 mL of QG buffer (Qiagen; see Note 8) to the microfuge tube and incubate at 50°C for 10 min. Invert the tube several times during the incubation. The gel slice should be completely dissolved before moving to the next step.
5. Place a QIAquick spin column into a 2-mL collection tube. Add 0.75 mL of dissolved gel solution into the column. Centrifuge the column and collection tube unit in a microfuge at 15,000g and RT for 1 min.
6. Discard the flow-through and place the column back onto the same collection tube. Add the remaining gel solution onto the column and then centrifuge the column-collection tube unit at 15,000g and RT for 1 min.
7. Discard the flow-through, re-attach the collection tube, and add 0.75 mL of fresh QG buffer to the column. Centrifuge the column-collection tube unit at 15,000g and RT for 1 min.
8. Wash the column by adding 0.75 mL of PE buffer (Qiagen) and centrifuge the column and collection tube at 15,000g for 1 min. Discard the flow-through and spin the column-collection tube unit again at 15,000g for an additional minute, discarding the flow-through.
9. Place the column above a clean 1.7-mL microfuge tube. Add 50 μL of nucleotidefree H_2O or EB Buffer (Qiagen) to the column and centrifuge the column-collection tube unit at 15,000g for 1 min. Concentrate the purified DNA fragment by centrifuging the collection tube under vacuum for 20 min or until the final volume of the DNA sample is approx 5 μL.
10. To ligate the purified DNA fragment containing the host gene sequence into pB3-3, set up a ligation reaction by mixing 3 μL of the gel-purified DNA fragment (approx 200 ng), 1 μL of pB3-3 vector linearized with *HindIII* enzyme (approx 100 ng), 5 μL of 2X ligation buffer, and 1 μL of T4 DNA ligase (Promega) in a 0.7-mL microfuge tube.
11. Mix the reagents by flicking the tube several times. Centrifuge the tube briefly to collect the reaction mixture in the bottom of the tube and incubate overnight at 4°C.
12. Transform JM109 competent cells with the ligation product and isolate pB3-3 containing the plant gene sequence from the transformed cells as described anywhere else in this chapter.

Synthesis of BMV RNA Transcripts and Plant Inoculation

Before in vitro transcription, the three plasmids containing the BMV genome sequences should be linearized using *Spe* I (pF1-11) or *PshA* I (pF2-2 and pB3-3) restriction enzymes. RNA transcripts representing the three BMV genomic RNAs are prepared individually from the linearized plasmids in vitro and the products mixed before inoculating host plants.

RNA transcripts can be stored at –70°C before use. All reagents and materials used during the in vitro transcription reactions should be nuclease-free.

Synthesis of RNA Transcripts

1. Linearize pF1-11, pF2-2, and pB3-3 individually in 50-μL reactions containing 2 μg of

template DNA, 5 μL of 10X the restriction enzyme buffer, 0.5 μL of bovine serum albumin (100 mg/mL), and 1.5 μL of *Spe I* (pF1-11) or 1.5 μL of *PshA* I (pF2-2, pB3-3, and pB3-3 with insert) restriction enzymes. Incubate the reactions at 37°C for 1.5 h.

2. After incubation, add 50 μL of a phenol/chloroform/IAA solution to each microfuge tube. Mix the content thoroughly by vortexing the microfuge tubes for 20 s and then centrifuge at 15,000g and RT for 5 min.
3. Transfer the upper liquid phase (approx 50 μL) from each tube into a clean nuclease-free microfuge tube.
4. Add 50 μL of chloroform to each microfuge tube. Mix the contents thoroughly by vortexing the microfuge tubes for 15 s and then centrifuging the tubes at 15,000g for 5 min.
5. Transfer the upper liquid phase (approx 50 μL) from each tube into a clean nuclease-free microfuge tube. Add 5 μL of 3 M sodium acetate, pH 5.2, and 100 μL of ice-cold ethanol to each tube. Mix the content by flicking the tubes several times. Incubate the tubes at –20°C for 1 h or at –70°C for 20 min.
6. Centrifuge the microfuge tubes at 15,000g for 15 min and discard the supernatant.
7. Add 0.7 mL of ice-cold 70% ethanol to each tube. Flick the tubes several times and centrifuge them at 15,000g and RT for 5 min.
8. Discard the supernatant from each tube. Dry the pellets in a vacuum centrifuge for 5 min.
9. Resuspend each pellet in 15 μL of nuclease-free H2O. Visualize the efficiency of the linearization and estimate the concentration of each linearized template by electrophoresis through a 1% agarose gel containing 0.1 μg of ethidium bromide per milliliter of agarose solution. An aliquot of 1-kb DNA ladder containing a fragment of known concentration should be loaded on the gel to aid in determining the amount of product produced and the efficiency of the reaction. Transcripts from the individual linearized templates are synthesized by adding 10 μL of 2X NTP/CAP solution, 2 μL of 10X reaction buffer (both from mMessage mMachine kit, Ambion) and 6 μL of template DNA (approx 0.5 μg, representing BMV RNA 1, 2, or 3) into a nuclease-free microfuge tube.
10. The in vitro transcription reaction is initiated by adding 2 μL of T3 enzyme mix into each tube, mixing well by flicking the tubes several times and incubating tubes at 37°C for 1.5 h.
11. The product RNA transcripts can be used to infect a plant immediately, or stored at –70°C for later use.

Inoculation of Plant With Transcript

BMV RNA transcripts can be used to infect seedlings directly through mechanical inoculation. To achieve more successful infections, the viral transcripts should be inoculated first to *Nicotiana benthamiana* seedlings; this plant is easily infected and allows high titers of many viruses to accumulate in its leaves. Crude extracts from *N. benthamiana* leaves are then used to infect rice seedlings.

1. Seeds of *N. benthamiana* are sown in a soil mixture (Metromix 350; SunGro Horticulture,

Bellevue, WA) in small pots. At 3 wk after planting, individual seedlings are transplanted into 4.5-in. pots. Seeds of rice are planted directly into a soil in 4.5-in. pots. Rice is particularly intolerant of suboptimal soil conditions. Greenhouse temperatures for *N. benthamiana* and rice are 25/20°C (day/night). Seedlings are watered and fertilized as needed.

2. Mix in vitro transcripts representing BMV RNAs 1, 2, and RNA 3, harboring or not harboring a host-gene insert. Select two similar sized *N. benthamiana* seedlings at 5 d after transplanting. Dust two leaves of each plant with Carborundum (320 grit; Sigma) through cheesecloth layers and inoculate the leaves by gently rubbing 7.5 μL of mixed transcripts across the surface of each leaf. The inoculated plants are then moved to a growth chamber at 24/20°C (day/night). As controls, leaves of two *N. benthamiana* seedlings are inoculated with mixed transcripts representing BMV RNAs 1, 2, and 3, without the host-gene insert.
3. Monitor virus levels and stability of the plant target sequence in the virus in each inoculated *N. benthamiana* plant by immunocapture RT-PCR (IC) using primers specific for BMV RNA3 sequences as described elsewhere in this chapter.
4. Harvest leaf tissue from infected *N. benthamiana* plants, grind the tissue in 0.1 *M* phosphate buffer, pH 6.5 (1:10, w/v) at 4°C.
5. Dust leaves of 2-wk-old rice seedlings with Carborundum and inoculate them with the crude extracts prepared from the infected *N. benthamiana* leaves in step 4. The inoculated rice seedlings are grown in a greenhouse at 25/20°C (day/night) and monitored for virus infection and gene silencing through symptom observation, IC RT-PCR and semiquantitative RT-PCR described elsewhere in this chapter

Analysis of Virus Infection and Gene Silencing in Plant

The progress of infection by the BMV vector in the host can be monitored through IC RT-PCR. The stability of the plant-gene insert in the virus vector during systemic infection of the host can also be monitored using this technique.

Detection of Virus Through IC RT-PCR

This procedure involves the capture of BMV virions from crude plant extracts on walls of microfuge tubes precoated with an antibody specific for the BMV coat protein (CP). Reverse transcription reactions can then be performed in the same microfuge tube without the need to release viral RNA from the bound virions.

The resulting first strand cDNAs are then amplified via PCR using primers specific to the BMV RNA 3 sequence.

1. Dilute 1 μL of BMV CP antibody in 2.5 mL of coating buffer (coating buffer: 1.59 g Na_2CO_3 and 2.93 g $NaHCO_3$ to 1 L distilled H_2O. Adjust pH to 9.6 with HCl).
2. Add 30 μL of diluted antibody solution to each 0.6-mL microfuge tube (ISC Bioexpress) and incubate the tubes overnight at 4°C.
3. Wash each tube twice using 0.1 *M* phosphate buffer (PB), pH 7.0. After removing PB from the tubes, store them at −20°C for later use.

Table 12.3: Chronology of Key Eents in the Production and Commercialization of PRS-Resistant, Transgenic Papaa in Haaii

Year	Event
1940s	The papaya industry was started on Oahu island
1945	PRSV disease was discovered
1950s	Large production areas on Oahu were eliminated owing to PRSV
1950s	Papaya industry moved to the Puna area of Hawaii island mid-1970s Puna becomes the largest producer (95%) of Hawaii's papaya
1978	Work began on control methods for PRSV
1983	Mild mutant of PRSV HA (HA 5–1) for cross protection was isolated
1985	Concept of pathogen-derived resistance (PDR) codified
1985	In vitro translation of PRSV RNA was achieved providing basic information on the coding properties of the viral genome
1986	Work toward utilizing the PDR concept was begun by cloning the PRSV HA 5–1 *CP* gene 1987 The mild strain PRSV was used for crossprotection to manage PRSV under field conditions 1987 Tissue culture conditions to enable papaya transformation and regeneration was developed 1990 PRSV-resistant papaya R_0 line 55–1 hemizygous for the *CP* transgene is created by biolistictransformation
1991	Transgenic tobacco containing and expressing the functional PRSV *CP* transgene were generated
1992	The entire genome sequence of PRSV HA was determined
1992	Hawaii papaya industry faced serious disaster due to PRSV in Puna and rouging was begun
1992	Greenhouse evaluation of a R1 line hemizygous for the *CP* transgene of 55–1 was considered
1992	First field trial of 55–1 transgenic papaya was conducted in Waimanalo on Oahu island. During this time, cultivars Rainbow and SunUp, hemizygous and homozygous, respectively for the *CP* transgene found in 55–1 were developed.
1994	Hawaii's Department of Agriculture declared PRSV uncontrollable in Pahoa area of Puna and rouging was stopped in this area
1995	Field trial of SunUp and Rainbow began in Puna on Hawaii island
1996	Transgenic line 55–1 and its derivatives were deregulated by APHIS
1997	Kalapana area, the last area of Puna to be affected was severely infected and rouging was stopped here

(Table Contd.)

Year	Event
1997	Exemption from EPA was granted
1997	FDA approval was granted for the transgenic lines
1998	Bulk seed production of SunUp and Rainbow was completed
1998	License agreements were obtained from all parties allowing the commercial cultivation of transgenic papaya and its derivatives in Hawaii only
1998	Transgenic seed were distributed free to qualified growers
1998	100% of the Puna acreage was nontransgenic Kapoho
1999	90% of the farmers obtained the transgenic papaya seed and 76% of them planted them in the field
2000	Transgenic papaya made up 50% of the Puna acreage
2002	New PRSV-resistant cultivars are developed from the original 55–1 line

4. Collect two leaf discs by pressing leaf with the lid of a microfuge tube (approx 100 mg total) from each plant and grind them inside the microfuge tube in 100 μL of 0.1 M PB with disposable plastic pestles at RT.
5. Add 30 μL of crude leaf extract from each sample to a 0.6-mL tube pre-coated with the CP antibody and incubate the tube overnight at 4°C. 6. Wash each tube three times with 0.1 M PB.
7. Add 5 μL of solution containing a primer complementary to the 3' end of the BMV RNA 3 sequence (0.5 μL of 10 mM reverse primer in 4.5 μL of DEPC H_2O) to each tube and incubate the tubes at 70°C for 10 min followed by 2 min on ice.
8. Add 5.0 μL of premixed RT reagents (2 μL of 5X first-strand buffer, 1 μL of 0.1 M DTT, 0.5 μL of 10 mM dNTP Mix, 0.5 μL of RNasin, 0.5 μL of Super Script RT, and 0.5 μL of nuclease-free H2O) to each tube and incubate at 42°C for 1 h.
9. After 1 h, add 1 to 2 μL of RT reaction mix to a PCR tube containing 18.5 μL of mixed PCR reagents (2 μL of 10X PCR buffer, 2 μL of 25 mM *MgCl,* 0.5 gL of 10 μM forward primer (specific for the BMV CP gene sequence), 0.5 μL of 10 μM reverse primer, 0.4 μL of 10 mM dNTPs, 0.1 μL of Taq polymerase, 14.5 μL of nuclease-free H_2O).
10. Perform 30 cycles of PCR under predetermined conditions.
11. Analyze PCR products on a 1% agarose gel containing 0.1 μg of ethidium bromide per milliliter of agarose solution. Some quantitation of virus accumulation can be achieved by decreasing the number of PCR cycles so the system is not saturated. Products from BMV vectors maintaining the host-gene insert should yield a higher-molecular-weight fragment than that observed from the BMV vector with no host insert.

Analysis of Target-Gene Silencing Through Semiquantitatie RT-PCR

The level of plant-target-transcript silencing can be monitored through semiquantitative RT-PCR. The procedure described in this section uses primers specific for the target gene, complementary or identical to sequences outside the region of the gene expressed within the virus vector, and the gene encoding elongation factor-1a, as an internal control (other host mRNAs may serve as internal controls; the best are those whose levels do not fluctuate during virus infection).

Relative transcript levels for the gene of interest between various treatments are estimated by comparing the intensity of PCR product gel bands obtained after multiple PCR cycle numbers and normalizing the intensities according to estimates of the substrate RNA levels based on results from the internal control.

Isolation of TOTL RNA from LEF Tissues

1. Harvest leaf tissue from plants infected with the BMV vector containing or not containing the plant gene insert at 2 to 3 wk after inoculation. The harvested tissue can be used for RNA isolation immediately or stored at –70°C for future use.
2. Take 0.1 g of tissue from each sample, add liquid nitrogen, and grind using a mortar and pestle. Add 1 mL of TRIZOL reagent to each mortar while the tissue is still frozen and continue grinding for 1 min. Transfer the crude sap into cold 1.7-mL microfuge tubes and incubate the tubes at RT for 5 min.
3. Add 0.2 mL of chloroform to each microfuge tube. Invert tubes several times to mix the contents and incubate at RT for 5 min.
4. Centrifuge microfuge tubes at 15,000g for 15 min at 4°C.
5. Transfer the upper aqueous phase containing RNA to individual nuclease-free microfuge tubes. Add 0.5 mL of isopropanol into the aqueous phase in each tube. Mix thoroughly by inverting tubes several times. Incubate the tubes at RT for 10 min.

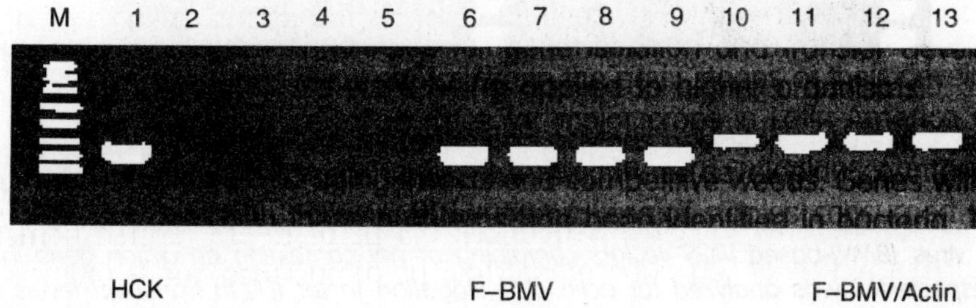

Figure 12.3: Presence of virus and plant-gene insert in inoculated plants determined by immunocapture reverse-transcription polymerase chain reaction. Brome mosaic virus *(BMV)-based virus-induced gene silencing (VIGS) vectors containing or not containing an action gene insert (F-BMV/Actin and F-BMV, respectively) were inoculated to rice plants. Four weeks after inoculation, BMV RNA3-specific primers flanking the actin insert site were used to determine the presence of BMV and actin insert. Templates for polymerase chain reaction were lane 1, purified BMV plasmid vector (pB3-3); lanes 2 through 5, extract for uninoculated plants (HCK); lanes 6 through 9, extract from plants inoculated with vector (F-BMV); lanes 10 to 13, extract from plants inoculated with vector carrying the actin insert (F-BMV/Actin). Each lane represents the analysis of a single plant. M, marker lane.*

6. Centrifuge the microfuge tubes at 15,000g for 10 min at 4°C.
7. Discard the supernatant. Add 0.7 mL of 75% ethanol to each microfuge tube. Flick the tubes several times and then centrifuge them at 15,000g for 5 min at 4°C.
8. Discard the supernatant. Dry the RNA pellets by vacuum centrifugation for 5 min at RT.
9. Dissolve each pellet in 30 µL of nuclease-free water.
10. Determine the RNA concentration for each sample by diluting 1 µL of isolated RNA in 500 µL of nuclease-free H_2O and measuring its absorbance (A) value at 260 and 280 nm in a UV spectrophotometer. Acceptable A260/A280 ratios for RNA are between 1.6 and 1.8. An A value of 1 at A_{260} and measured in a cuvet with a 1-cm path length approximates 40 µg/mL of RNA.
11. Add 1 µL of RNase-free DNase I to each RNA sample and incubate the samples at 37°C for 15 min to remove contaminating DNA.

Reerse Transcription

1. Add 10.5 µL of total RNA (2 µg), 1 µL of 10 *µM* Oligo(dT) primer, and I µL of dNTP Mix (10 mM each) into a nuclease-free microfuge tube.
2. Incubate the tube at 70°C for 5 min and then on ice for 2 min.
3. Add 4 µL of 5X first-strand buffer, 2 µL of 0.1 M DTT, and 0.5 µL of SUPERase In™ to the tubes. Mix the content thoroughly by flicking the microfuge tube several times, centrifuge to collect solution in base of tube and incubate the tube at 37°C for 2 min.
4. Add 1 µL of SuperScript™ II Reverse transcriptase to each tube and flick the tube several times to mix the content. Centrifuge to collect solution in base of tube.

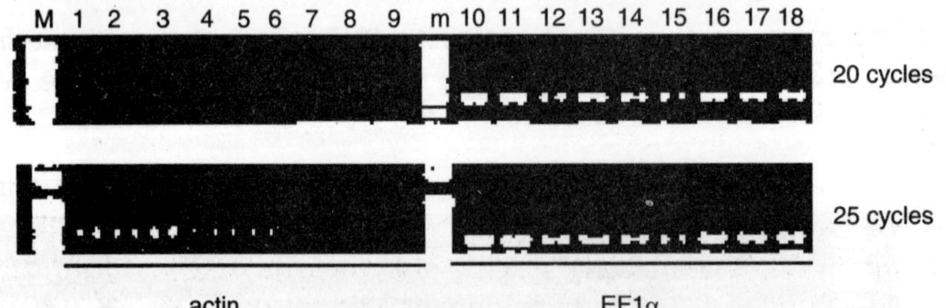

Figure 12.4: Actin transcript downregulation during virus-induced gene silencing (VIGS) determined by semiquantitative reverse-transcription polymerase chain reaction. RNA extracted from rice plants inoculated with Brome mosaic virus (BMV)-based VIGS vectors containing or not containing an action gene insert (F-BMV/Actin and F-BMV, respectively) was analyzed for actin and elogation factor (EF)-la transcript levels to determine the effectiveness of VIGS on actin messenger RNA expression. Primers specific for endogenous actin (550 bp; lanes 1-9) and EF-la (400 bp; lanes 10-18) gene were used to assess the host-encoded transcript levels during infection. Transcript levels were assessed after 20 and 25 cycles of polymerase chain reaction. To avoid signal from actin sequence expressed from F-BMV/Actin, the actin primers were specific for regions of the transcript flanking the actin sequence in the virus vector. Templates for reverse-transcription polymerase chain reaction were lanes 1-3 and 10-12, RNA from healthy plants; lanes 4-6 and 13-15, RNA from plants inoculated with F-BMV; lanes 7-9 and 16-18, RNA from plants inoculated with F-BMV/Actin. Each lane represents the analysis of a single plant. M, marker lane.

5. Incubate the microfuge tube at 37°C for 1 h. Inactivate the reaction by heating the tube at 70°C for 15 min. The resulting cDNA is the template for amplification during the subsequent PCR.

Polerse Chin Rection

1. For each sample, mix 4 µL of cDNA obtained from elsewhere in this chapter above with 53.6 µL of nuclease-free water, 8 µL of 10X PCR buffer, 8 µL of 25 mM MgCl2, 2 µL of 10 mM dNTP mix, 2 µL of forward primer (10 *µM),* 2 µL of reverse primer (10 *µM)* and 0.4 µL of *Taq* DNA polymerase (5 U/µL). Aliquot the 80 µL of reaction mixture equally into 4 PCR tubes.
2. Perform PCR under predetermined conditions. Remove one tube from the thermocycler at 20, 25, 30, and 35 cycles of reaction, respectively.
3. Visualize PCR products by electrophoresis in a 1.5% agarose gel containing 0.1 µg of ethidium bromide per milliliter of agarose solution. Determine the silencing result by comparing the band intensity of PCR products obtained after various PCR cycle numbers. RNA samples from plants infected with BMV not expressing a plant gene sequence should be included in the experiment as a control. RNA transcript levels from host genes that are minimally affected by virus infection (e.g., elongation factor-1a, ubiquitin, or actin) should be included in the same experiment as controls to normalize results based on template levels.

NOTES

1. Ampicillin can be first dissolved in sterilized H_2O (100 mg/mL H_2O) and stored at –20°C (stable for at least 2 mo). Addition of ampicillin to LB medium and preparation of LB medium plates is best done in a sterile laminar flow hood to avoid contamination. LB-agar medium made previously and stored at 4°C can be melted in a microwave set on "defrost" before being poured into Petri dishes. The solidified poured plates can be placed inside plastic bags and stored at 4°C for approx 1 mo.
2. Polyclonal antibody against BMV coat protein was prepared by injecting a rabbit with purified BMV virion. The antibody was mixed (1:1, v/v) with 50% sterilized glycerol and stored at –20°C. The antibody solution also contains 0.02% sodium azide (a chemical harmful to humans, but present as an antibiotic). Wear disposable gloves while working with solutions containing sodium azide.
3. A DNA fragment of 150 to 250 nucleotides should be amplified from a gene of interest through RT-PCR. Both forward and reverse primers should be designed based on the sequence of the gene. It is advantageous to compare sequences to be used for primers with other sequences from plants, both from the target gene and other plant sequences, to verify that they are unique (i.e., to avoid off-site targeting). Each primer should contain a HindIII restriction site at its 5' end so that the fragment can later be released from the T-Easy vector and inserted into the *HindIII* site within the BMV vector.
4. PB buffer contains chemicals harmful to human. Always wear a laboratory coat, disposable gloves, and eye protection while working with this solution.

Table 12.4: CP Gene Nucleic Acid Sequences of PRS-P Isolates From Various Geographic Regions.

Geographical origin	*GenBank accession no.*
Australia	U14736 U14737, U14738 U14740
Bangladesh	AY423557
Brazil	AF344640
	AF344647
	AF344647, AF344645, AF344646, AF344641, AF344639, AF344640, AF344650,
	AF344642, AF344644, AF344643
China	AF243496 X96538
India	AF305545
	AF063220
	AY238880, AY491011, AF238883, AY458617, AY458619, AY238884, AF120270,
	AY458618, AY238881, AY458620, AY238882, AY238885
Indonesia	AF374864
Japan	AB044339
Malaysia	AB044342
Mexico	AJ012649
	AJ012650
	AJ012099
	AF309968,
	AF319468, AF319493, AF319499, AF319502
Philippines	AF506902,
	AY587583
Sri Lanka	U14741
Taiwan	AB044341
	X78557
	X97251
Thailand	AF506898–900, AF506901, AF506902, AF506862, AF506889,
	U14743
	AB044340
USA-Florida	AF196839
USA-HA 5–1	D00595
USA HA	X67673
USA-Puerto Rico	AF196838
Vietnam	AF506862, AF506889
	U14742

5. In our experiments, we use 25 μL of competent cell suspension per transformation. The remaining competent cell suspension should be returned immediately to the -80°C freezer for future use. Repeated freeze-thawing decreases the competency of these cells for transformation, so it would be wise to aliquot cells into small volumes in separate tubes. The used pipet tips and Falcon tubes should be placed inside a biohazard bag for autoclaving. Wear disposable gloves while performing the transformation.
6. SDS is harmful to humans. Wear disposable gloves and eye protection while weighing out SDS and isolating plasmid DNA from the overnight cell culture. The used culture medium should be killed with Clorox. All used Falcon and microfuge tubes and pipet tips should be placed inside a biohazard bag for autoclaving.
7. Ethidium bromide may cause hereditable damage to humans. Always wear gloves, eye protection, and laboratory coat when running gels. TBE buffer and agarose gels containing ethidium bromide should be collected in waste containers and disposed of following proper procedures.
8. QG buffer contains chemicals harmful to humans. Always wear a laboratory coat, disposable gloves, and eye protection while working with this solution. Residual ethanol from PE buffer can affect later digestions and ligations. It can be removed completely from the column by the second centrifugation (for 1 min). When using water to elute DNA, make sure that the water pH value is between 7.0 and 8.5.
9. Phenol and chloroform are extremely toxic. Always wear a laboratory coat, disposable gloves, and eye protection while working with these chemicals.
10. TRIzol is toxic and in contact with skin. Always wear a laboratory coat, disposable gloves, and eye protection while working with it.

DEVELOPMENT OF GENETICALLY ENGINEERED RESISTANT PAPAYA FOR PAPAYA RINGSPOT VIRUS IN A TIMELY MANNER

The papaya industry in Hawaii started in the 1940s, and the *papaya ringspot virus* (PRSV) was discovered in 1945. By the 1950s, production on Oahu was affected, and the industry subsequently moved to Hawaii island into the area of Puna, which had no commercial production but was free of PRSV. However, by the 1970s PRSV was in the town of Hilo, approx 19 miles away from the papaya-growing area of Puna.

Because it was very probable that PRSV would eventually enter the Puna area, research was started by Gonsalves and coworkers in 1978 to develop control strategies for PRSV in Hawaii. Crossprotection was the first approach to be tried, using a mild nitrous acid mutant, designated HA.

The crossprotection strategy using the mild mutant HA was used to some extent on Oahu island and in Taiwan. However, protection was not sufficient to provide long-lasting economic benefits because of the effort involved in producing mild strains and inoculating plants and the fact that the mild strain produced significant symptoms on some popular cultivars, especially "Sunrise".

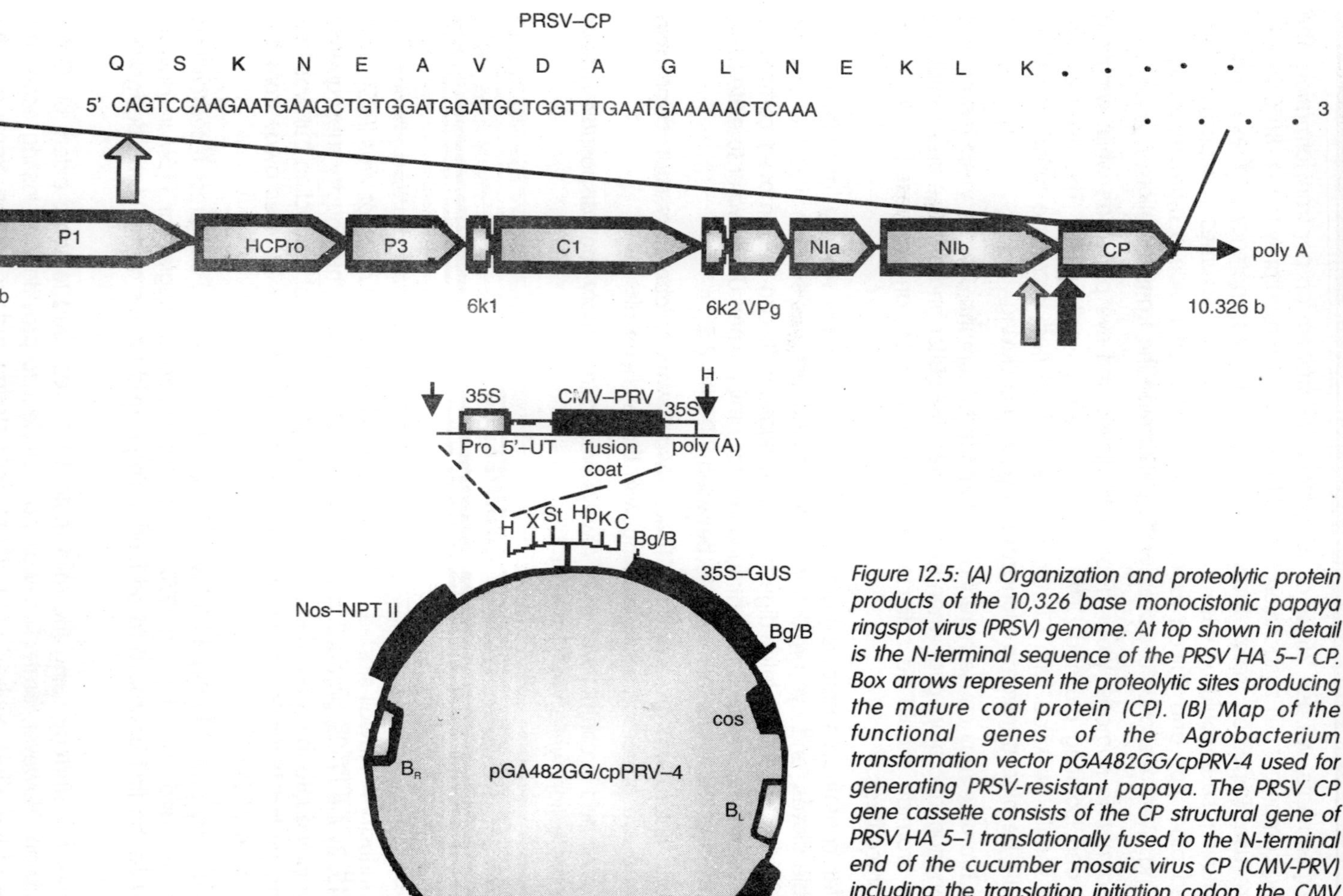

Figure 12.5: (A) Organization and proteolytic protein products of the 10,326 base monocistonic papaya ringspot virus (PRSV) genome. At top shown in detail is the N-terminal sequence of the PRSV HA 5–1 CP. Box arrows represent the proteolytic sites producing the mature coat protein (CP). (B) Map of the functional genes of the Agrobacterium transformation vector pGA482GG/cpPRV-4 used for generating PRSV-resistant papaya. The PRSV CP gene cassette consists of the CP structural gene of PRSV HA 5–1 translationally fused to the N-terminal end of the cucumber mosaic virus CP (CMV-PRV) including the translation initiation codon, the CMV 5'untranslated sequence (5'-UT) and the cauliflower mosaic virus 35S promoter (35S). The CMV-PRV gene cassette is flanked by selectable and visible marker genes, nptII and uidA (Gus), respectively. B_R and BL are the left and right borders of the transformation vector T-DNA sequence.

This approach was not widely used for longer because of its limitations. In the mid-1980s, an exciting, but yet unproven alternative approach was used. The report by Powell-Abel from Roger Beachy's group that transgenic tobacco expressing the coat protein *(CP)* gene of *tobacco mosaic virus* showed significant delay in disease symptoms caused by *tobacco mosaic virus,* spurred us to look towards this approach.

The general approach was coined "parasitederived resistance" (now called pathogen-derived resistance or PDR) by Sanford and Johnston, which offered a new approach for controlling PRSV. PDR is a phenomenon whereby transgenic plants containing genes or sequences of a parasite (in this case, the CP gene of PRSV) are protected against detrimental effects of the same or related pathogens.

In 1986, Gonsalves and coworkers began using the PDR concept by cloning the *CP* gene of the PRSV mild strain HA 5–1 and, finally, the resistant transgenic papaya (Rainbow and SunUp) was released for commercial cultivation.

PRSV-resistant transgenic papaya in Hawaii became the first commercialized transgenic fruit crop worldwide. The technical details are given later in this chapter. The transgenic work for Hawaii started earlier in anticipation that the PRSV would one day attack the Puna area.

In 1992, Hawaii's papaya industry faced a potential economic disaster when PRSV was discovered in the Puna district of Hawaii island, where 95% of the state's papaya was grown. By 1995, PRSV was widespread in Puna, and the industry was in a crisis situation.

Fortunately, our research had resulted in the development of a transgenic papaya that was resistant to PRSV at the right time. In fact, an initial field trial of the transgenic papaya was established on Oahu island at about the same time that PRSV was discovered in Puna.

The sequence of research events and more details for controlling the virus disease and transgenic papaya developments for Hawaii are well documented in reviews written earlier by Gonsalves and coworkers.

This chapter will mainly focus on the key technical aspects of methodology for genetic engineering the resistance in papaya against PRSV and development of suitable commercial transgenic resistant papaya in a timely manner.

PROPERTIES OF PAPAYA AND PRSV

Papaya *(Carica papaya L.)* is an important fruit crop and is widely grown in countries of the tropics and subtropics. It is a member of the Caricaceae family and believed to be originally from Southern Mexico and Northern Central America.

Papaya is a large herbaceous, dicotyledonous plant (up to 3–8 m height), with usually a single erect stem and a crown of alternate large palmate-lobed leaves. The inflorescences are borne in the axils of leaves. The plants are polygamous, with male, female, and hermaphrodite flowers.

Wild plants frequently are dioecious, with female or male flowers. Domesticated plants show different sexual types, including hermaphrodite flowers with different masculinity grades, as described by Storey in 1976. Papaya fruit size ranges from 255 g to 5 to 6 kg, and their colour ranges from pale to bright yellow-orange to red. The papaya fruit is most commonly consumed fresh but it is also processed for making fruit salad, juice, jam, jelly, pie, or ice cream flavoring. Unripe fruits can be eaten raw in salad, cooked in syrup and eaten as a dessert, and the boiled leaves can be used as a vegetable.

The fruit has digestive properties because of the presence of papain, and it also has great nutritional value, with high contents of vitamin A, vitamin C, calcium, potassium, and iron. Papaya trees can easily be cultivated in home gardens from seeds.

The tree produces fruit for consumption year round, usually starting at the age of 9 mo. Commercially, when trees are grown at a density of 1500 to 2500 per hectare, annual production can range from 125,000 to 300,000 lbs per hectare.

Fruits are harvested for 1 to 2 yr, after which the trees usually are too tall for efficient harvesting. The Food and Agriculture Organization (FAO) estimated that approximately six million metric tons of fruit were harvested in 2002, almost double of the 1980 harvest.

Brazil (25.2%), Nigeria (12.6%), India (11.8%), Mexico (11.6%), and Indonesia (8.6%) are the largest producers of papaya. Hawaii is the largest pr oducer of papaya in the United States. A major limiting factor for papaya cultivation in all geographic areas is a disease caused by PRSV.

PRSV infection is characterized by production of ring spot symptoms that develop on papaya fruits of infected trees in addition to a range of other symptoms, such as mosaic and chlorosis of the leaf lamina, water-soaked oily streaks on the petiole and upper part of the trunk, a distortion of young leaves that can resemble mite damage, and stunting. PRSV infection may cause lack or severe reduction of fruit production, and fruit that are produced are of poor quality and low sugar concentration.

In nature, PRSV is transmitted nonpersistently by numerous species of aphids to a limited host range of cucurbits and papaya. PRSV also produces local lesions on *Chenopodium quinoa* and *Chenopodium amaranticolour.*

The PRSV virions are nonenveloped, flexuous-rod in shape and measure 760 to 800 × 12 nm. The virus is grouped into two subtypes, PRSV-type P and -type W based on their infectivity. Type P infects cucurbits and papaya, whereas type W infects cucurbits but not papaya. Both P and W type viruses are serologically closely related.

The genome of PRSV consists of single-stranded RNA of 10,326 nucleotides with positive polarity and has the typical array of genes found in potyviruses. The genome is monocistronic and is expressed via a large polypeptide that is subsequently cleaved to yield all functional proteins.

The 381-kDa polyprotein is processed into eight to nine final products via three virus encoded proteinases (P1, HC-Pro, and NIa). Like other potyviruses, the proposed genetic organization of PRSV RNA is VPg-5' leader-P1 (63K)-HC Pro-P3 (46K)-CI-P5 (6K)-NIa-NIb-CP-3' noncoding regionpoly(A) tract. There are two possible cleavage sites, 20 amino acids apart, for the N terminus of the CP protein.

These two sites may be functional; the upstream site for producing a functional NIb protein (the viral replicase) and the other for producing the CP present in aphid-transmissible virions. The complete nucleotide sequence of the PRSV genome has been reported for the following geographical isolates: Mexico (Genebank AY231130), Hawaii (Genebank NC001785), Thailand strains P (Genebank AY162218) and W (Genebank AY010722), and Taiwan strains P (Genebank X97251) and W (Genebank AY027810). However, CP sequence of numerous strains has been analyzed from various laboratories.

PAPAYA RINGSPOT AND PAPAYA IN HAAII

The PRSV disease was discovered in 1945 on the island of Oahu, where a papaya industry of approx 500 acres was located. By the 1950s, large production areas on Oahu were eliminated, and the industry subsequently moved to Hawaii island into the area of Puna, which had no commercial production earlier.

Acreage of papaya increased to 650 by 1960 and to 2250 in 1990. In contrast, the acreage on Oahu fell to less than 50 by 1990. By the mid-1970s, Puna became the largest producer of Hawaii's papaya, contributing to *95%* of the total papaya production of the state.Remarkably, Puna remained free of PRSV for more than 30 yr.

Despite the presence of PRSV in Hilo and Keaau, communities only 19 miles away, Puna remained free of PRSV. The effective physical barrier of the lava rock terrain of Puna, together with the diligence by the Hawaii Department of Agriculture in surveying and rouging infected trees in the Hilo and Keaau areas, kept PRSV from spreading. However, it was very probable that PRSV would someday be found in Puna.

DEVELOPMENT OF TRANSGENIC PAPAYA FOR HAAII

The transgenic work for Hawaii started in 1986 with a team of scientists from different organizations, including Dennis Gonsalves, Richard Manshardt, Maureen Fitch, and Jerry Slightom. Steve Ferreira joined the team in the 1990s.

Engineering of the PRSV Transgene

The PRSV HA 5–1 strain from Hawaii was used as the source of the CP gene for the initial PDR construct because the goal was to create papaya resistant to Hawaiian PRSV strains. PRSV HA 5–1 is a mild virus form derived from the virulent Hawaiian strain PRSV HA by nitrous acid mutagenesis of extracts of infected squash. PRSV HA 5–1 was initially produced for crossprotection studies.

The PRSV HA 5–1 CP gene sequence originally was deciphered from a library of complementary DNA derived from purified virus particle RNA and from a partial peptide sequence of a subfragment of the purified CP. Today, CP genes from various geographic isolates can be directly amplified from total RNA of infected papaya tissue using polymerase chain reaction (PCR) primers to conserved regions flanking the PRSV CP gene thereby bypassing the need for viral particle purification for each new isolate.

The original construct for the PRSV HA5–1 CP gene was designed with concept that protein expression was required for PDR, since that was the prevailing thought at the time. Because the PRSV CP is produced from the extreme 3'-end of the polyprotein gene by posttranslational protease cleavage, there are no native translation signals specific for the CP sequence. Therefore to design a construct for protein expression of the PRSV CP transgene, a chimeric gene was made using the translation signals found in the leader sequence (5' untranslated RNA translational enhancer and initial 16 amino acid coding sequences) of the cucumber mosaic virus CP gene fused in frame to the structural sequence of the PRSV CP including the Q/S protein cleavage site and 51 nucleotides of the noncoding region (nucleotides 9257 to 10168).

This was accomplished by cloning the PRSV CP structural sequence from plasmid pPRV117 in between the *Cauliflower mosaic virus* (CaMV) 35S double enhancer promoter–translational leader sequence and CaMV 35S terminator of a cucumber mosaic virus expression cassette.

This PRSV *CP* expression cassette was finally cloned into pGA482GG, a modified version of the *Agrobacterium* transformation pGA482 that contained the nptII (neomycin phosphotransferase II) gene behind a nopaline synthase promoter and *a uidA* ((-glucuronidase [GUS]) gene behind a CaMV 35S promoter, used for kanamycin selection and colourimetric screening of transformants, respectively. Although we engineered the PRSV *CP* expression cassette in an *Agrobacterium* transformation vector, we used the plasmid only for biolistic transformation.

Expression of the CP from this vector was verified by the use of enzymelinked immunosorbent assay (ELISA) on the leaves of transformed papaya and tobacco. Other groups have shown that different leaders can promote the translation and accumulation of the PRSV *CP* in plants.

Since our original study, breakthrough research by Smith and colleagues with the potyvirus systems *tobacco etch virus* and *potato virus Y* showed that resistance with potyviruses was mediated by RNA, via the mechanism of posttranscriptional gene silencing (PTGS). Indeed, reports on CP-based resistance via protein expression are largely limited to the case of TMV. The authors and others also have reported evidence that PDR in transgenic papaya is mediated by RNA-based mechanisms.

Transformation of Papaa: The Original Approach

An essential element to obtaining the first transgenic papaya was development of tissue culture conditions and identification of a source of papaya tissue that could be efficiently procured, transformed, and regenerated into plants. The main tissue culture parameters were the concentrations of the synthetic auxin, 2,4-dichlorophenoxyacetic acid and sucrose in the induction medium used to proliferate zygotic embryos and somatic embryo cells and the choice of cultivar, which responded differently to these components.

The initial study tested embryogenic zygotic embryos, embryogenic calli, and somatic embryos derived from hypocotyls and zygotic embryos. In that study, it was established that 2,4-dichlorophenoxyacetic acid-treated zygotic embryos derived from immature seeds of 90- to 120-d-old green fruits had the highest transformation capacity (1.42% of bombarded embryos) after particle bombardment and antibiotic selection compared to embryogenic callus.

The biolistics approach was possible, in part, because the group had ready access to the newly invented gene gun at Cornell University and the help of John Sanford, a co-inventor of the technology. Biolistic transformation involved annealing the purified PRSV CP plasmid construct to tungsten particles followed by aseptic bombardment into papaya tissue.

After bombardment, the tissue was kept on "induction medium" for a total of 4 to 5 wk in the dark without antibiotic selection. The cells were then placed on "maturation medium" with antibiotic selection (75 mg/L kanamycin) for 4 wk in the light and then in maturation medium with higher selection (150 mg/L kanamycin) for 8 wk in the light, during which resistant embryos were able to proliferate. Development of resistant, green plantlets from embryos occurred on a "germination medium" with 150 mg/L kanamycin for 2 to 3 mo. Clones of resistant plantlets were then produced by microprogation and rooted in rooting media.

Leaves of kanamycin-resistant clones were tested for expression of GUS activity, which cause the leaves to turn blue in the presence of the substrate X-Gluc, and by PCR amplification and genomic DNA blot analysis to test for presence of the nptII and PRSV *CP* gene. DNA blot analysis also was a useful means to analyze the nature of the integration event (such as arrangment or copy number).

Because integration by the biolistic process is a random event, the relevant PRSV and marker genes of the transformation vector did not always cointegrate; thus, kanamycin resistance was not always correlated with GUS activity or presence of the CP gene. Two transformed lines that were positive for the CP gene were further analyzed by RNA blots to determine expression of the transgene. One line showed strong accumulation of the 1.35-kb RNA species predicted for the PRSV CP gene transcript.

Transformation and Tissue Culture of Papaa, Improements, and Alternate Protocols

Since the original successful transformation and regeneration report, our group reported a modified procedure for the production of somatic embryos that greatly increased transformation efficiency as well as a detailed, updated protocol for biolistic transformation. Somatic embryos were produced from zygotic embryos from seeds as in the initial procedure but were bombarded at a step under which active cell proliferation was taking place, 3 d after transfer to fresh induction media.

A second difference was that the antibiotic selection period was performed only 7 d after bombardment, but ceased earlier, with the latter induction and maturation steps performed without antibiotic selection. This altered selection scheme allowed for effective amplification of transformed embryos which proliferated vigorously at the later stages in media without antibiotics. Protocols for efficient production of somatic embryos from hypocotyls and subsequent regeneration were developed and found to be efficient for papaya transformation.

The advantage of using hypocoptylderived somatic embryos is that it is easier to procure than the zygotic embryos which have to be individually excised from seeds of immature fruit of the correct age. It appears that antibiotic selection can reduce the number of transgenic plants that are recovered. Several studies have addressed this issue by limiting the period under antibiotic selection and allowing regeneration of plants to occur without selection as described previously, judicious use of alternative antibiotics as in the case of papaya root explant selection and regeneration, or the elimination of antibiotic selection entirely by use of green fluorescent protein as a visual marker.

Selection systems for papaya transformation using alternative antibiotics such as hygromycin have also been successfully tested. Subsequent to the initial successful transformation of papaya by particle bombardment, protocols for transformation of papaya by *Agrobacterium* were established and improved, providing an alternative means for production of transgenic papaya from hypocotyls, zygotic embryos, or petioles. One potential advantage of Agrobacterium-mediated over biolistic-mediated transformation is that, for the most part, the structure of the integrated transgene DNA is predictable, consisting of the marker gene and gene of interest bordered by the *Agrobacterium* binary vector transfer DNA.

A recent study reported high efficiency of cotransformation by the biolistic method of two plasmids into papaya, one carrying a marker gene and the other carrying grape stilbene synthase gene involved in *Phytophthora* resistance (>50%). This new twist to the biolistic transformation approach of papaya might theoretically allow independent insertion of the marker gene and gene of interest and allow segregation and isolation of a marker-free plant.

In addition to improvements in tissue culture techniques and approaches for transformation, application of recent refinements in papaya clonal propagation systems should play a large role in the high-throughput development, screening, maintenance, and distribution of new, pathogen-resistant elite papaya cultivars. For example, efficient micropropagation protocols can allow the propagation of hermaphrodite clones of superior uniformity and elimination of the labor involved in planting and thinning of trees of the undesirable sex.

Testing of Transgenic Papaa for Viral Resistance in the Greenhouse

In the study leading to the identification of the first pathogen-derived resistant papaya line, all transgenic papaya lines that were positive for the CP gene were screened for resistance to the virulent Hawaiian isolate, PRSV HA from which the attenuated strain PRSV HA 5–1 was derived. Six transformants were of the Sunset cultivar and three of the Kapoho cultivar. Four contained the PRSV CP gene (two Sunset- and two Kapoho-derived lines), whereas the other 5, along with 35 nontransformed plants, served as controls.

Three to fifteen replications were performed on clones of test lines that were produced by micropropagation. Manual inoculation was performed by dusting the four or five youngest fully expanded leaves of test plants with carborundum followed by rubbing with potassium phosphate buffer extracts of *Cucumis metuliferous* leaves infected with PRSV HA. Evaluation of symptoms was performed after 21 d. Of the four transgenic lines tested, one Kapoho line was completely susceptible, one Kapoho and one Sunset line had intermediate (25–33% of the total test plants became infected) levels of resistance, and one Sunset line had complete immunity.

Of the lines with intermediate resistance, there was a delay of 3 to 17 d before the onset of symptoms. Infection was to some extent affected by the age of the plant at inoculation, with earlier inoculations more likely to cause infection. Line 55–1 was symptomless throughout. Extracts from symptomless leaves of 55–1 as well as 19– 1 and 60–3, were tested by a virus recovery assay (in this case, the ability to infect the local lesion host *C. quinoa)* and found to be negative.

Although 39–1 was not resistant, it still showed delayed symptom development compared with nontransgenic controls. Although line 55–1 was virtually immune to the virulent HA strain of PRSV in the initial study, crossprotection studies have shown that the attenuated HA 5–1 did not effectively inhibit infection by Thailand or Taiwan isolates.

Thus, line 55–1 was further tested for resistance to other strains of PRSV in a subsequent greenhouse study by Tennant et al.. To produce the numerous transgenic test plants for this study, the primary transformant (R_0) 55–1 plant, which was female, was crossed with pollen from Sunrise (a sibling line of Sunset). The R_1 transgenic progeny from this cross were identified by the ELISA assay for presence of the nptII gene.

The transgene segregated at a ratio of 1:1 confirming a single transgene insertion in the original 55–1 plant. The R1 progeny hemizygous for the transgene were first rigorously tested by inoculation with the Hawaii PRSV HA isolate because this was to be one of the actual challenges in the papaya fields in Hawaii. Three main approaches to study infectivity of the severe isolate HA were followed:

1. One to two mechanical inoculations.
2. Multiple mechanical inoculations.

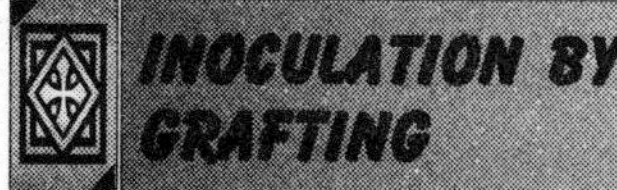

INOCULATION BY GRAFTING

For the first approach, the three youngest fully expanded leaves were inoculated with *C. metuliferous* extracts of different dilutions by the carborundum dusting method. Ten transgenic and 10 nontransgenic plants were tested for each dilution. Symptoms were monitored for 6 wk and plants were reinoculated if no symptoms were apparent.

Three weeks after the second inoculation, leaf extracts from these plants were tested for infectious virus by the virus recovery assay. For the multiple inoculation assay, new growth on papaya plants were inoculated every 2 to 4 wk for 10 mo with PRSV infected *C. metuliferus* extracts. For inoculation by grafting, 10 transgenic seedlings were grafted to non-transgenic, HA 5–1 infected trunks. The result of this study showed that there was no replication or movement of the virus in the transgenic plants.

The very vigorous resistance of the transgenic papaya to the endemic Hawaiian PRSV isolate added impetus for the initiation of experimental field trials in Hawaii. The R1 transgenic papaya also were tested for resistance to other geographical isolates of PRSV. Eleven PRSV isolates were studied: two from Mexico, two from Florida (F, G), one each from the Bahamas, Australia, Brazil, China, Okinawa, Ecuador, Guam, Thailand, Jamaica, and four from Hawaii (HA 5–1, HA, HA-Oahu, and HA-Panaewa).

All were serologically indistinguishable from HA 5–1 using antibodies to the HA 5–1. The plants to be inoculated were 5- to 8-wk-old, 6- to 15-cm high, with 6 to 10 leaves. The transgenic R1 seedlings (10 per virus isolate) were inoculated with the various PRSV strains from diluted extracts of PRSV infected *C. metuliferus* leaves.

All plants were observed daily for 6 wk, and symptomless plants reinoculated and tested by the virus recovery assay. Before the creation of the transgenic papaya, crossprotection was the only other direct means to control the severity of PRSV infection in papaya. Therefore, crossprotection experiments were performed so that resistance by transgenic plants could be compared with a tested approach.

Test plants for crossprotection experiments were performed by inoculation of the attenuated HA 5–1 from *C. metuliferous* extracts to nontransgenic papaya. Confirmation of infection was determined by ELISA using antibody against HA 5–1 because plants infected with this virus strain showed little or no overt symptoms. The infected plants were then challenged with *C. metuliferus* extracts of the various geographical isolates exactly as was done for the R1 transgenic line.

Viruses from Guam, Brazil, Thailand, Ecuador, and Okinawa induced severe symptoms on all transgenic plants, although these were not as severe as that of nontransgenic plants. Australia, China, and Jamaica had an attenuated phenotype on all transgenic test plants but, as in the case of the severe isolates, there was a 7- to 10-d delay compared with the nontransgenic control plants at the monitoring days of 10, 21, and 42 d after inoculation.

Complete immunity was found for HA and HA-Panaewa, but severe symptoms were found in 6% of the plants inoculated with the Hawaii strain HA-Oahu. The virus strains from Bahamas, Mexico, and Florida showed a fraction of plants with severe phenotype, whereas others were free of symptoms. Interestingly, the plants without symptoms remained this way after reinoculation.

The results of the crossprotection studies were similar to the transgenic studies. However, a major difference was that severe symptoms could be obtained in virus recovery assays from HA or HA-Panaewa symptomless, crossprotected leaves but not from symptomless leaves from transgenic plants. Thus, symptomless leaves of crossprotected plants harbored infective, virulent virus whereas those of transgenic plants did not.

Initial Field Testing

Although results on the R_1 generation of 55–1 hemizygous (CP/+) for the transgene were indicating robust resistance to Hawaii PRSV isolates HA (from Oahu island) and HA-Panaewa (from Hawaii island) field trials in Hawaii were already initiated in attempt to push the progress of the original, promising R0 transformant 55–1.

Timing was critical because the virus was first identified in the Pahoa area of Puna district on the island of Hawaii, the heart of the state's papaya growing industry, in May 1992. By October 1994, the state of Hawaii Department of Agriculture declared the Pahoa area uncontrollable and stopped marking trees for rouging, that is, the activity of cutting down infected trees to curb the buildup of virus inoculum.

The purpose of the first field trial initiated in 1992 was to determine whether the resistance first exhibited by the transgenic plants under greenhouse conditions would hold up under mechanical and aphid transmitted inoculations under field conditions. Although CP-mediated protection in other crops had been evaluated in the field, this field trial was to be a first for a perennial crop, and therefore it was a crucial test for robustness and durability of resistance over a long period.

The field trial was conducted at the University of Hawaii experimental station in Waimanalo, situated at 15 m above sea level on the island of Oahu. The plants for this study were produced clonally by microprogation to insure genetic uniformity. A total of three plant types were tested, including the test plant R0 line 55–1, which was hemizygous for the transgene, and two controls, a transgenic line carrying no PRSV CP gene (Sunset line 62–1), and a nontransgenic parental line Sunset.

Inoculation was performed by either of two means, mechanical or vector (aphid transmitted), for each of the three sample types for a total of six experimental treatments. Ten replicates of each treatment set was performed, for a total of 60 plants, arranged in 10 rows, 3 m between rows and 2 m between plants in a row, and randomized with respect to planting and type of inoculation (mechanical vs aphid transmitted).

The source of inoculum was from an infected papaya found at the University of Hawaii Horticulture facility. Seedlings were inoculated at approx 4 mo with symptomless plants reinoculated after approx 3 wk. Thereafter, symptoms were monitored every 2 wk during the first 4 mo and at various intervals for 2 yr. Plants were evaluated based on leaf symptoms, the girth of the trunk at 45 cm above ground level and for the presence of PRSV by double antibody sandwich-ELISA.

Transgenic plants exhibited normal leaves, had trunk girths averaging 14.6 cm at 18 mo vs 9.3 from sensitive, nontransgenic plants and were negative for PRSV by the double antibody sandwich-ELISA. Manual inoculations resulted in a quick manifestation of infection compared with aphid transmitted infection, but the resulting phenotypes were the same.

By the end of 5 mo, all nontransgenic plants were severely infected and near the end of the first year, die back of nontransgenic plants was observed because of weakening and fungal root infections. At the end of 2 yr, in 1994, a complete loss of these plants occurred.

Development of Cultiars "SunUp and Rainbo"

Given the imposing PRSV presence on the island of Hawaii and the performance of line 55–1 and its R_1 derivatives in field trials and in the greenhouse, respectively, we made an active attempt to use this new germplasm for the direct development of commercial cultivars during the period of this first field trial.

From the greenhouse study on R1 lines, it was known that the original line 55–1 contained a single copy of the transgene *(cp/+)* in the background of Sunset, a red-fleshed cultivar and showed good resistance to Hawaii isolates of PRSV but not those of other geographic areas. Subsequently, the SunUp variety, which is homozygous (CP/CP) for the transgene but is otherwise identical to Sunset, was created as the R_3 generation of the original transformant 55–1.

This germplasm held the hope for the development of new resistant varieties because crosses with any other nontransgenic variety would yield 100% progeny that would be hemizygous for the *CP* gene *(CP/+)*. In Hawaii, the yellowfleshed Kapoho variety is by far the more popular among farmers and consumers and has a pyriform shape and medium size, which are desirable commercial characteristics for packing and shipping.

Thus, in attempt to combine the PRSV resistance and Kapoho characteristics, Rainbow, a F1 hybrid between SunUp and Kapoho that was yellow-fleshed and hemizygous for the transgene *(cp/+)* was created. The resulting Rainbow cultivar bore pear-shaped fruit with yellow-orange flesh as anticipated and together with SunUp was ready to be tested in field trials to begin in 1995.

1995 Field Test in Puna

By late 1994 an application for a field trial was submitted to Animal and Plant Health Inspection Service (APHIS). The field trial was allowed on the stipulations that the test field was sufficiently isolated from nontransgenic commercial orchards to minimize the chance of transgenic pollen escaping to those fields, (2) fruits of abandoned trees left in the area had to be monitored for the possible introgression of the transgene, and (3) all fruits had to be buried on site.

The field trial began with the planting of 3-mo-old seedlings on a portion of an actual farmer's field in Puna in October, 1995 under a permit from APHIS. The test field was at least 0.4 km from surrounding commercial fields.

In contrast to the initial trial using line 55–1, this trial involved the transgenic varieties SunUp and Rainbow, and nontransgenic Sunrise, grown by the farmer, and involved mechanical inoculation only at a latter phase, after aphid transmitted virus inoculation was well underway. Other test plants included Sunset, a nontransgenic version of Rainbow (Sunset × Kapoho), and line 63–1, which was a transgenic Sunset containing the CP gene of PRSV HA 5–1 and had been obtained during the initial transformation experiments that created line 55–1.

Plants were spaced 1.7 m apart within one row, 8 plants to a row with rows 3.4 m apart. Each two row set was replicated four times in a randomized (complete block) design. This block of transgenic plants was surrounded on all sides by two rows of nontransgenic Sunrise plants.

A similar field situated adjacent to the test plot was set up to simulate a commercial type planting. Altogether, the trial was designed to obtain data relating to how these lines, especially Rainbow, would perform under conditions closely simulating that of commercial production.

Like other field tests and unlike greenhouse conditions, many other factors mitigating the health and plant productivity would come into play, such as natural water fluctuations, pest challenges by broad mites and leaf edge roller mites, and root rot caused by *Phytophthora palmivora*. In the field trial, the virus source was a PRSVinfected orchard 24 m upwind of the test plot and the susceptible Sunrise bordering the block of transgenic plants.

Mechanical inoculation was not initiated until 2 mo after the initial infection was observed, and then only on every fifth non-infected plant in the Sunrise border rows. PRSV infection first occurred at 3.5 mo after transplanting. PRSV infection symptoms included water-soaked streaking on leaf petioles, chlorotic mosaic and veinclearing on leaves, leaf distortion and shoestringing of leaves, and ringspots on the fruit.

One year after planting, 8 mo after the initial infection, and 3 mo after the initial harvest, all Sunrise plants were infected. None of the SunUp or Rainbow plants were infected during the 2.5-yr period of the trial in the replicated field.

However, three Rainbow plants of approx 5000 (in a field not yet culled of female plants) in the block of Rainbow plants tested for commercial production were infected 4 mo after transplanting. PRSV also was observed at a rate of 1.3% on papaya more than 20 mo old; however, symptoms occurred only on some small lateral shoots and not on the main growing shoot.

In both types of infection, recovery assays were performed and confirmed the presence of infective PRSV. Fruit data were taken at bimonthly intervals starting at 15 mo after an initial harvest period of 5 mo. Rainbow yields were 126 tons of marketable fruit per hectare per year compared with the average 35.2 tons per hectare per year during the 5-yr period before the discovery of PRSV in Puna (1998–1992). Average fruit weight was 635 to 771 g, and the refractory solids were higher than the 11% minimum of grade A fruit.

Nontransgenic Sunrise, on the other hand, was commercially unacceptable 5 mo into the harvest period. The field trial in Puna was timely as PRSV was ravaging the papaya industry of Hawaii during this period, with the discontinuation of rouging in September 1997 in Kalapana, the last area of Puna district to be infected. In fact, it took only 5 yr from first detection to total devastation of the entire papaya growing area of Puna. Subsequently, a plan was set to move papaya plantings to areas uninfected by PRSV, eradication of papaya and cucurbits in the Puna area, and a 1-yr moratorium on the planting of papaya in the Puna area.

Greenhouse Analsis of Rainbow and SunUp

The field trial data indicated that the new cultivars SunUp and Rainbow would be viable for direct commercial applications in Hawaii. During this period, a rigorous study also was conducted on the resistance properties of these new lines to other geographic isolates of PRSV under closely monitored greenhouse conditions of the Cornell facilities in Geneva.

In this study, resistance of SunUp and Rainbow were evaluated against six PRSV isolates (three from Hawaii, OA, KE, and KA and three from Brazil, Thailand, and Jamaica) by mechanical inoculation. For the cultivar Rainbow, resistance was only observed against the homologous HA isolate.

However, when challenged by the other Hawaiian PRSV isolates, there was a delay in the development of symptoms and a period of 2 to 28 d, during which new leaves did not display symptoms, followed by their reappearance, but in a less severe form.

The symptoms were severe, with no recovery from isolates from outside Hawaii, although a lag in symptom appearance was observed. The resistance exhibited by Rainbow mirrored that of the R0 line 55–1, both hemizygous for the transgene, indicating that the hybrid combination did exert obvious influence on transgene function.

Table 12.5: Age-Dependent Resistance Properties of Rainbo, SunUp and Sunrise Papaa to PRSV Isolates From Haaii and Other Geographic Sources.

			Resistance (%) to PRSV isolates							
	Age	*Height*		*Hawaii*			*Outside Hawaii*			
Cultivar	**(wk)**	**(cm)**	**HA**	**OA**	**KA**	**KE**	**JA**	**BR**	**TH**
Rainbow	6–15	6–9	86	14	0	0	0	0	0
SunUp	3–16	3–20	100	100	100	100	100	100	0
Sunrise	6–15	6–9	0	0	0	0	0	0	0
Rainbow	14	6–20	100	62	33	0			
	17	34–60	100	100	100	100			
SunUp	16	15							0
	23	17–59							75
	29	70–117							100
Sunrise	14–29	24–69	0	0	0	0			0

In contrast to Rainbow, SunUp was resistant to all PRSV isolates except the Thailand isolate. For the Thailand isolate, there was a long delay of 4 to 6 wk before symptoms manifested. The difference in susceptibility between Rainbow and SunUp1 suggested that transgene dosage could play a strong role in determining viral resistance.

In attempt to understand the basis of the dosage effect, transcription rates of the CP transgenes were determined and found to be two times higher in SunUp compared with Rainbow as expected. Paradoxically, transgene transcripts accumulated to a lower steady state level in SunUp as compared with Rainbow.

Taken together with the observation that SunUp is more resistant than Rainbow, the evidence suggested that resistance is mediated by a mechanism related to PTGS and that the increased copy number found in SunUp influenced the efficiency of this silencing.

A comparison of the nucleotide sequences of the CP genes of the various PRSV isolates tested indicated strongest homology between the HA 5–1 sequence and Hawaii isolates and the least overall homology to the Thailand CP gene (89.5%), particularly in the N terminus (83.7%), and a 3'-noncoding region of 35 nt (89.4%).

Thus, in this study there was good correlation between the level of CP sequence homology and resistance; the closer the homology between the CP transgene and the challenging virus, the better the resistance.

A system for producing infectious PRSV RNA transcripts in vitro allowed for the testing of chimeric contructs consisting of the HA virus genome with all or portions of its CP gene replaced with that of an infective isolate, in this case PRSV YK from Taiwan. Like the CP gene from the Thai isolate, YK has a relatively low homology (89.8%) with the CP gene of HA.

The PRSV YK *CP* sequence did in fact confer infectivity to the chimeric HA virus on Rainbow and to a lesser extent SunUp with some segments of the YK CP gene appearing to have more influence on infectivity than others.

This result indicates that the cognate CP sequence of the infecting virus is in fact an important determinant for CP transgene-based resistance but that the position of homology might have also play a role. Interestingly, the chimeric virus containing the entire CP gene and 3'-untranslated region of PRSV YK caused attenuated symptoms compared to transcripts produced by constructs expressing on YK sequences.

These results indicate importantly that viral sequences other than the CP gene can also influence infectivity on a transgenic plant expressing the CP gene. Recent data strengthens this observation.

Previous results indicated that a R_1 line hemizygous for the CP transgene showed better resistance when inoculated at later stages of growth. In this study, Rainbow and SunUp of different ages and sizes (14 wk/13 cm, 17 wk/Rainbow 46 cm) and (16 wk/15 cm, 23 wk/38 cm, and 29 wk/93 cm) were inoculated with Hawaii and Thai isolates and evaluated after 56 d.

The older and larger Rainbow plants were completely resistant to all Hawaii isolates, whereas the larger SunUp plants were completely resistant to even the Thai isolate. This result concurred with the resistance properties of R1 plants in the earlier study. The drastic contrast in performance of the Rainbow plants with regards to age graphically demonstrates that subtle factors influence

Table 12.6 : Summarized Reactions and CP Nucleotide Sequence Homologies of PRSV Isolates to PRSV HA 5–1 Inoculated to Rainbow and SunUp Papaa

	% Homology to transgeneCP					Reaction to isolates	
PRSV isolates	*N*	*core*	*C*	*3'-ncr*	*overall*	*Rainbow*	*SunUp*
Hawaii-HA	99.3	99.8	100	100	99.8	R	R
Hawaii-OA	97.3	98.0	100	95.7	97.9	sR	R
Hawaii-KA	95.3	97.1	98.3	93.6	96.7	sR	R
Hawaii-KE	95.3	97.1	98.3	93.6	96.7	sR	R
Jamaica-JA	89.3	95.0	91.5	69.6	92.5	S	R
Brazil-BR	84.4	93.9	98.3	73.3	91.6	S	R
Thailand-TH	83.7	90.7	91.5	89.4	89.5	S	sR

viral resistance. Factors such as these may have contributed to the durable resistance of Rainbow observed under actual field conditions and reemphasizes the importance of the field trials.

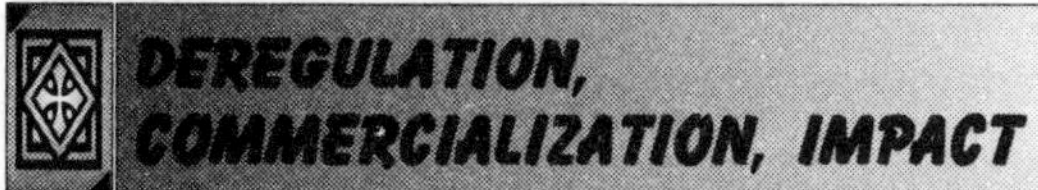

DEREGULATION, COMMERCIALIZATION, IMPACT

In 1995, efforts to deregulate the transgenic papaya were proactively initiated by the research team due to the ongoing devastation of the industry by the killer PRSV and an interest of the papaya industry in getting the transgenic line commercialized. The following subheading is taken largely from parts of Gonsalves.

Deregulation and Commercialization

Three major governmental agencies namely APHIS, the Environmental Protection Agency (EPA), and the Food and Drug Administration (FDA), deregulate the transgenic plants and their products in the United States. APHIS was largely concerned with the potential risk of transgenic papaya on the environment. Two main risks were of heteroencapsidation of the incoming virus with CP produced by the transgenic papaya and of recombination of the transgene with incoming viruses.

A third concern was that escape of the transgenic genes to wild relatives might make the relatives more weedy or even make papaya more weedy because of resistance to PRSV. However, this was of no consequence because there are no papaya relatives in the wild in Hawaii, nor is papaya considered a weed there, even in areas were there is no PRSV. In November 1996, transgenic line 55–1 and their derivatives were deregulated by APHIS.

According to the EPA, the CP transgenes are a pesticide because they confer resistance to plant viruses. A pesticide is subjected to tolerance levels in the plant. In the permit application, we petitioned for an exemption from tolerance levels of the CP produced by the transgenic plant.

We contended that the pesticide (the CP gene) was already present in many fruits consumed by the public because much of the papaya eaten in the tropics is from PRSV-infected plants. In fact, we had earlier used cross protection (the deliberate infection of papaya with a mild strain of PRSV) to control PRSV. Fruit from these trees was sold to consumers. Furthermore, there is no evidence to date that the CP of PRSV or other plant viruses is allergenic or detrimental to human health in any way.

Finally, measured amounts of CP in transgenic plants were much lower than those of infected plants. An exemption from tolerance to lines 55–1 and 63–1 was granted in August 1997. The FDA is concerned with food safety of transgenic products.

This agency follows a consultative process whereby the investigators submit an application with data and statements corroborating that the product is not harmful to human health. Several aspects of the transgenic papaya were considered: the concentration range of some important vitamins, including vitamin C; the presence of *uidA* and *nptII* genes; and whether transgenic papaya had abnormally high concentrations of benzyl isothiocyanate.

This latter compound has been reported in papaya. Approval by the FDA was granted in September 1997. In the United States, a transgenic product cannot legally be commercialized unless it is fully deregulated and until licenses are obtained for the use of the intellectual property rights for processes or components that are part of the product or that have been used to develop the product.

The processes in question were the gene gun and PDR, in particular, CP-mediated protection. The components were translational enhancement leader sequences and genes *(nptII, uidA,* and CP). This crucial hurdle involved legal and financial considerations beyond our means and expertise.

These tasks were taken up by the industry's papaya administrative committee and its legal counsel, Michael Goldman. The license agreements were obtained from all parties in April 1998, allowing the commercial cultivation of the papaya or its derivatives in Hawaii only.

On May 1, 1998, seeds were distributed free to growers who qualified by watching an educational video and signing an agreement that restricted growing of transgenic papaya only in Hawaii. Fruits can be sold outside Hawaii, provided that the importing state or country allows the importation and sale of transgenic papaya.

Impact of Transgenic Papaya

The development, commercialization, and rapid adoption of transgenic papaya cultivars, that is, Rainbow, SunUp, resistant to PRSV has significant socioeconomic impact on Hawaii. The transgenic papaya in Hawaii was adopted very well by the papaya growers and consumers. Adoption by farmers of the transgenic papaya was very high.

Adoption was defined as to whether the farmer had planted the seeds, and not based on whether the farmer had signed up and obtained seeds. By September 1999, 90% of the farmers had obtained transgenic seeds and 76% of them had planted (adopted) the seeds. Personal interviews were conducted with 93 of the 171 farmers who had registered to obtain transgenic papaya seeds in 1998.

To the question on why they wanted to plant Rainbow, 96% said that it was because of the ability of Rainbow to resist PRSV. A detail about adoption on transgenic papaya in Hawaii can be obtained in an American Phytopathological Society feature story, "Transgenic virus-resistant papaya: the Hawaiian 'Rainbow' was rapidly adopted by farmers and is of major importance in Hawaii today".

The release of the transgenic papaya resulted in an increase of papaya production in Hawaii and Puna. In 1992, Puna produced 53 million of the state's 55 million pounds of fresh papaya. The production remained high for 2 yr following the discovery of PRSV in Puna as the result of massive efforts to control the spread of the virus. However, by 1995 papaya production in Puna had decreased to 39 million pounds and was down to 26 million pounds in 1998 when transgenic seeds of cultivars were released to farmers.

Production of papaya in Puna increased starting in 2000 and peaked at 40 million pounds in 2001 with 35 million pounds being produced in 2002. The effect of PRSV on papaya production in Puna also can be seen by the decrease in the total percentage of Hawaii's fresh papaya production that was produced in Puna.

In 1992, Puna accounted for 95% of the total production, but this figure subsequently dropped to 65% in 1999 followed by a recovery to 84% in 2002. The impact of the transgenic papaya in increasing papaya production in Puna is also seen by analyzing the relative bearing acres of

Table 12.7 : Fresh Papaa Productiona in the State of Haaii and in the Puna District From 1992–2002

	Fresh papaya utilization in Hawaii		
		Total	Puna
Year	(x1000 lbs)	(x1000 lbs)	%
(Virus in Puna) 95	1992	55,800	53,010
1993	58,200	55,290	95
1994	56,200	55,525	99
1995	41,900	39,215	94
1996	37,800	34,195	90
1997	35,700	27,810	78
(Transgenic seeds released)			
Year	(x1000 lbs)	(x1000 lbs)	%
1998	35,600	26,750	75
1999	39,400	25,610	65
2000	50,250	33,950	68
2001	52,000	40,290	77
2002	42,700	35,880	84

Table 12.8 : Bearing Acres in Puna of Nontransgenic Kapoho and Transgenic Rainbow and the Relationship to Production (1000 lbs) of Fresh Fruit Useda

Year	Bearingacres	%Kapoho	%Rainbow	Production
1998	1640	100	0	26,250
2000	1190	32	50	33,950
2001	1675	39	41	40,290
2002	1385	49	37	35,880

Rainbow and the nontransgenic Kapoho. In 1998, Puna production was 26 million pounds from 1640 acres of bearing Kapoho, because Rainbow had not yet produced mature fruit. In 2000, Puna production had increased to 34 million pounds from 1190 bearing acres, with Kapoho comprising 32% and Rainbow comprising 50% of the acres.

In 2001, 40 million pounds were produced from 1675 bearing acres with Kapoho and Rainbow accounting for 39 and 41% of the acreage, respectively. In 2002, the bearing acreage dropped and the amount of Kapoho rose to 49%, whereas Rainbow remained steady at 37%. Production dropped from 40 million pounds in 2001 to 36 million pounds in 2002.

These data suggest that Rainbow accounts for at least half of the fresh fruit production in Puna. Furthermore, production of similar amounts of papaya can be obtained with less acreage. This latter observation is attributed to the higher level of production of Rainbow compared to nontransgenic Kapoho.

One of the major contributions that the transgenic papaya has made to the papaya industry is that of helping in the economical production of nontransgenic papaya. This has occurred in several ways.

First, the initial largescale planting of transgenic papaya in established farms along with the elimination of abandoned virus-infected fields drastically reduced the amount of available virus inoculum. The reduction in virus inoculum allowed for strategic planting of nontransgenic papaya in areas that were free of infected plants and were not surrounded by areas of infected plants, such as had been present in 1992.

Although definitive experiments have not been conducted, it seems that transgenic papaya can provide a buffer zone to protect nontransgenic papaya that are planted within the confines of the buffer. The reasoning is that viruliferous aphids will feed on transgenic plants and thus be purged of virus before traveling to the nontransgenic plantings within the buffer.

This approach also has the advantage that it allows the grower to produce transgenic and nontransgenic papaya in relatively close proximity. Timely elimination of infected trees would need to be practiced to delay large-scale infection of the nontransgenic plants.

Diersified Cultivation of Papaya in Hawaii

The release of PRSV-resistant papaya provided options for papaya growers. Before the release of transgenic papaya, growers on Oahu, for example, farmed only small plots of papaya because of the effect of PRSV on production. Growers on Oahu now enjoy a niche market, growing Rainbow papaya for residents in Honolulu and other urban areas of the island.

Besides SunUp and Rainbow, new varieties have since been created starting with Rainbow F_2 plants. Rainbow F_2 female plants were chosen to initiate the backcrossing program since they were 50% homozygous for all characteristics.

Multigenerational backcrosses have been made to Kapoho and the variety Kamiya in attempt to transfer the PRSV-resistance to these varieties. Kamiya bears fruit with smooth waxy skin, and deep orange flesh colour which is popular on the island of Oahu.

Backcrossing with Kapoho was performed in attempt to obtain a PRSV resistant line with firm-fleshed fruit qualities of this parent important for shipping and handling. Other important agronomic traits are that Kapoho is adapted to the shallow, lava fields, whereas Phytophthera-tolerant Kamiya is adapted to deep soils.

During the multiyear backcross program, the use of embryo rescue, harvesting seed for tissue culture germination from 2-mo-old, immature fruit rather than waiting for seed from mature, 5-mo-old fruit reduced the breeding program from 4 to 3 yr.

Micropropagation and rooting of cuttings were used to produce clones of random and selected progeny which allowed the testing of genetically identical lines in field trials at different geographical locations. In addition, PCR was used to identify lines segregating for the transgene as well as to determine the sex to allow direct targeting of relevant lines.

The resulting Kapoho backcross 1 (Rainbow F2 X Kapoho) and Kamiya F1 hybrid (Rainbow F2 X Kamiya) were named Poamoho Gold and Laie Gold, respectively. Micropropagation of these papaya varieties, particularly Laie Gold ensure the production of only hermaphrodite plants demanded by the market, earlier and lower bearing trees with initially higher yields, and provides selected, superior clones that could result in improved quality and yield.

It is important to note that because these PRSV-resistant cultivars derived from the original, deregulated transgene line 55–1, they also could be directly used for commercial application, circumventing the time-consuming and complicated process of approval required if a novel transformation event was involved and allowing the development of cultivars to meet niche market needs on Oahu island and possibly other islands.

Since the introduction of transgenic papaya, the number of cultivars available to papaya growers in Hawaii has thus actually increased. As noted earlier, Kapoho accounted for 95% of Hawaii's papaya market in 1992. Now Rainbow and Kapoho are dominant and the transgenic SunUp, Laie Gold, and the nontransgenic Sunrise make up a small but significant part of the cultivars grown in Hawaii.

IMPORTANT FACTORS INFLUENCING THE TIMELY DEVELOPMENT OF TRANSGENIC PAPAYA IN HAWAII

Given the speed and severity of the PRSV epidemic on Hawaii island which began with the discovery of PRSV in Puna in 1992 and reached the level of disaster within a span of 5 yr in 1997, quick action was necessary to bring about the actual commercialization of the transgenic papaya and bring about the repair of the papaya industry.

Key factors contributing to this timely development are outlined below:

1. R_0 plants were directly tested in the field before full characterization. The first report of the transgenic papaya together with initial characterization for resistance was published in 1992, the year the first field trial on the R0 line 55–1 was initiated. At this time, it was not fully characterized genetically, and its resistance characteristics were not fully known, particularly to other geographic isolates of PRSV. However, field data were essential in assessing the potential of using transgenic papaya for applied, commercial applications.
2. The first field trial was used to test resistance under field conditions as well as a venue to develop practical cultivars. Two potential cultivars, Rainbow and SunUp, hemizygous and homozygous, respectively, for the PRSV *CP* transgene were developed while resistance properties were still being performed on the original R0 plant line. These cultivars showed promising horticultural characteristics and were ready for field testing by 1995. By 1994, data from this first field trial of R0 plants was near completion with results indicating that line 55–1 showed complete resistance in the field to a PRSV isolate found in Hawaii.
3. Resistance of new transgenic plants to other geographical isolates was rigorously tested at an early stage. Resistance of an R1 generation of transgenic papaya to various geographical PRSV isolates in addition to those from Hawaii was rigorously tested, concurrent with the initial field trial and development of new transgenic cultivars, and the results were reported in 1994 at the end of the first field trial. These results reconfirmed the vigorous resistance of the transgenic papaya to local Hawaiian PRSV isolates while providing an early warning on the limitations of their use against other geographic isolates and in other areas. During this period, the R0 line 55–1 was documented to contain a single transgene.
4. The second field trial in Puna was set up in a timely manner, using actual cultivars of interest and at an actual site of production. The large field trial in Puna was begun in 1995, using the newly developed transgenic-derived cultivars, SunUp and Rainbow only 3 yr after initiation of the first field trial on the R0 line. Although greenhouse trials on these particular cultivars were ongoing and not completed at this time, the previous R1 line data and preliminary data were sufficiently positive to warrant a direct field test under commercial conditions.
5. Deregulation so crucial to commercialization and release was begun well in advance, during the field trial in Puna. Application for deregulation of the transgenic papaya including 55–1 and a second transgenic line derivative 63–1 was begun before the completion of this field trial.
6. Seed production was initiated before papaya was commercialized. Cultivars SunUp and Rainbow were created by the end of 1994 for the beginning of the field trial in Puna. By the end of the final field trial and before the release of the transgenic cultivars in 1998, enough seeds of the cultivar Rainbow to plant 1000 acres were already bulked up on the island of Kauai, ready for distribution to farmers who by this time had been forced to abandon their papaya fields owing to the devastation caused by PRSV.

7. Commitment of researchers and acceptance of the consumers. A major factor in the success of the transgenic papaya in Hawaii was the combination of researchers each which contributed a unique aspect to the project. Personnel at the University of Hawaii monitored the virus epidemic at the forefront and managed field trials and developed the papaya tissue culture, transformation, and regeneration protocols. Jerry Slightom of the UpJohn company and John Sanford at Cornell University were key members contributing to expertise in biotechnology and such tools as vectors and development of the gene gun and PDR, respectively. Dr. Dennis Gonsalves of Cornell University provided an intimate knowledge and experience in the field of PRSV and PDR. Finally, the transgenic papaya was enthusiastically embraced by both farmers and consumers alike, without whose support and acceptance the success of such an undertaking would not be possible.

DEVELOPMENT OF TRANSGENIC PAPAYA IN OTHER GEOGRAPHIC AREAS

Because PRSV is a worldwide problem on papaya, which is widely grown in the tropics, other countries have showed interest in developing the technology for their use. Thus, a program was set up by one of the authors (D.G.) to develop and transfer the technology to Brazil, Jamaica, Thailand, and Venezuela. This section summarizes the status of the program and the results obtained on transgenic papaya developed in other laboratories.

Brazil

The CP gene of PRSV isolate from the Southeast region of the state of Bahia was used to engineer the PRSV CP constructs and transform papaya via biolistic and *A. tumefaciens-mediated* transformation. The resulting R0 plants appeared to be resistant to the homologous virus as well as to the Hawaiian strain PRSV HA and also an isolate from Thailand.

Candidate resistant lines were sent to Brazil in 1999, where they were subsequently analyzed up to the third generation. Thirty-two transgenic papayas were tested in the field in Brasilia in a 900-m^2 plot where they showed good resistance. The main purpose of the initial tests was to produce seeds and then test the plants in producing areas in the states of Ceara, Espirito Santo, and Bahia.

Unfortunately, the program has come to a stand still because of the strict regulations imposed on genetically modified organisms (GMOs) by the home regulatory body, the Brazilian Technical Committee for Biosafety (CTNBio). Currently, a small field trial is being carried on at the experimental station at Cruz das Almas.

Recently, Brazil has passed legislation to allow further testing of transgenic plants and it is hoped that more intensive work can be resumed to identify and eventually commercialize the resistant line.

Jamaica

A virus isolate from the island of Cayman was used in the construct. Two versions of the transgene were made: one with a translatable CP gene and the other a nontranslatable CP gene.

Transgenic plants were obtained after bombardment into Sunrise (solo type) papaya somatic embryos.

Under greenhouse conditions with manual inoculation, high (78%) resistance was found for the translatable *CP* construct compared with only 10% for the nontranslated construct. However, even for the sensitive plants, a delayed recovery was observed in which an initial sensitive phenotype disappeared in subsequent new growth. R0 plants were transferred to Jamaica in 1998.

Initial field trial was conducted with R0 in 1998 and the resistance to field sources of PRSV of the homologous type in the field was similar to mechanical inoculation in the greenhouse with *80%* of the transgenic papaya carrying the translatable *CP* gene compared to 44% for the nontranslated construct.

Field trials conducted in 1999 on R1 plants showed a similar trend to parental lines with 58% resistance. Thus, it appeared that in the case of the transgenic papaya in Jamaica, greater resistance was correlated to translatability of the CP. The horticultural, nutritional and other components fell within the range documented for nontransgenic papaya. Results of rat feeding trials showed no adverse health affects attributed solely to transgenic papaya fruit.

Taken together, these data indicated that transgenic papaya was safe for consumption. As in other studies, one goal of the project in Jamaica was the development of local resistant cultivars by crossing in the resistance transgene into locally favoured cultivars.

Molecular traits deemed useful for such development is the identification of transgenic lines with simple genetics (i.e., a single transgene insert) that can be maintained by selfing and the availability of local cultivars with attributes that are attractive to the local market, which in the case of Jamaica included the large-fruited "Santa Cruz giant" and "Cedro varieties."

The target release date of the Jamaica transgenic cultivars for 2002 initially set up in 2000 has been delayed because of the lack of support for a third and final field trial. Thus the release of transgenic papaya in Jamaica is currently on hold.

A further complication is that Europe is a significant market for Jamaican papaya, and unless the papaya is deregulated by the European Union, transgenic papaya will not be able to be exported to Europe or elsewhere. At this moment, there is still a "GMO controversy" with the deregulation of transgenic products in Europe.

Conceivably, this could hold back the commercialization of transgenic papaya in Jamaica for fear that the transgenic papaya from Jamaica may somehow be mistakenly shipped to Europe before it is deregulated in Europe.

Thailand

Two popular cultivars, "Khakdum" and "Khaknun," were targeted for transformation using the nontranslatable CP gene of a PRSV isolate from Northeast Thailand. Several transgenic resistant lines of CP transgenic Khakdum and Khaknun were identified in greenhouse inoculation tests at Cornell University, New York.

A number of potential R0 lines were delivered to Thailand in July 1997. Work was immediately started by Dr. Nonglak and Ms. Prasartsee on the propagation, seed increase, and testing for resistance of the potential lines. Ms. Prasartsee has headed the work since 1997. By 1999, field trials of the R1 generation showed excellent results. By the year 2002, an R_3 line of Khaknun had been selected and showed excellent PRSV resistance and horticultural characteristics.

In comparative tests, the transgenic line showed that 97% of the progeny were resistant under intense disease pressure and yielded 63 kg of fruit per tree in the first year, whereas nontransgenic papaya yielded only 0.7 kg per year. Crosses between independent lines of Khakdum recently have been shown to also show good resistance under greenhouse and field conditions.

Concurrently, molecular characterization, biosafety experiments, analysis of transgenic fruit for food properties and food safety, and intellectual property rights were initiated using material that has been selected for eventual deregulation and commercialization.

Nearly all biosafety experiments that are mandated by the national committee on biosafety have been completed. Tests on food safety and other characteristics, such as vitamin, amino acids, soluble solids and other profiles, are being done and should be completed in the near future. However, recent controversial events relating to GMOs will very likely slow down the process of deregulating the transgenic papaya.

Venezuela

The CP gene of the two isolates were cloned in the plant transformation vector in sense/ translatable, sense/untranslatable and antisense forms. The plant transformation vectors with the cloned genes were sent to Venezuela where they were employed to transform a local papaya "Thailandia Roja" via *A. tumefaciens.* A few putative transgenic lines were recovered in 1997 and subsequently crossed.

The R_0, R_1, and R2 generations were molecularly characterized and tested for their resistance against PRSV strains from Venezuela, Hawaii, and Thailand. The resistance seemed to be RNA mediated, and R1 and R2 plants showed a promising level of resistance not only to local isolates but also to different geographic isolates of PRSV such as isolates from Thailand and Hawaii.

Two hermaphrodite plants showing high level of resistance from the R2 generation were identified and kept for further multiplication and testing. A small-field plot test was conducted at Lagunillas, Merida with a special permit from the Ministry of Health of Venezuela.

The plot was planted with R1 individuals previously selected in the greenhouse as PRSV-resistant. Later on, the performance of these plants was very good in the field under the local virus pressure. When the transgenic PRSV-resistant papaya were set to flower, unexpected problems started to emerge by the GMO activists and the field trial was set on fire by some of their members in December 2000. Currently, Dr. Fermin has seeds of resistant plants from R_2 generations, but their use in field trials is still in doubt.

Taian

Transgenic papaya resistant to PRSV were successfully developed by Dr. Shyi-Dong Yeh's team of the National Chung Hsing University following an approach similar to that used for developing the cultivars Rainbow and Sunup for Hawaii, i.e., by use of CP-mediated resistance.

In his lab the CP gene of a local PRSV isolate, YK was inserted as a transgene in the Taiwanese papaya cultivar, Tainung No. 2. Transgenic Tainung No. 2 papaya were obtained by Agrobacterium-mediated transformation rather than the biolistic method used for developing Rainbow and SunUp. Transgenic, resistant papaya lines carrying the CP gene of PRSV YK were generated and four transgenic lines resistant in greenhouse experiments were evaluated under field conditions for their resistance properties and fruit production from 1996 to 1999. Performance of the transgenic lines in the field trials was found to be similar to that of Rainbow and SunUp.

None of the transgenic lines showed severe symptoms of PRSV infection, whereas 100% of the nontransgenic plants were severely infected 3 to 5 mo after planting. The transgenic plants (20–30%) did exhibit mild symptoms in the first and second field trials but this did not reduce the yield or fruit quality of these plants.

The transgenic lines were not only protected from virus infection, but also produced 11–56% more marketable quality papaya compared with nontransgenic papaya. In another transformation experiment, Dr. Yeh's laboratory obtained 45 putative transgenic lines, which exhibited PRSV-resistance ranging from delay of symptom development to complete immunity after inoculation in the greenhouse.

Similar to Rainbow and Sunup, molecular analysis of nine selected lines that exhibited different levels of resistance revealed that the expression level of the transgene was negatively correlated with the degree of resistance, suggesting that the resistance is manifested by a RNA-mediated mechanism.

Segregation analysis showed that the transgene in immune line 18–0–9 had an inheritance of two dominant loci and the other four highly resistant lines had single dominant loci. Seven selected lines were tested further for resistance to three heterologous PRSV strains that originated in Hawaii, Thailand, and Mexico.

Six of the seven lines showed varying degrees of resistance to the heterologous strains, whereas one line, 19–0–1, was immune not only to the homologous YK strain but also to the three heterologous strains. In contrast to the transgenic papaya Rainbow, line 19–0–1 from Taiwan was reported to be immune to other heterologous isolates of papaya, whereas Rainbow and Sunup showed a very high degree of durable resistance to the homologous isolates of PRSV as other lines.So far none of the transgenic plants from Taiwan have been deregulated or commercialized.

Australia

PRSV was only described in Australia as late as 1991 and as yet has not been a major factor limiting production of papaya there. Unlike the spread of virus typical in other countries, strict quarantine measures in Southeast Queensland, where PRSV was first discovered has restricted its spread to the major production region of tropical North Queensland.

Nevertheless, given its history in other parts of the world, researchers in Australia have been active in the production of transgenic papaya that are resistant to Australian PRSV isolates. To this end, transformation and regeneration techniques for papaya were developed. Based on CP nucleotide sequence data comparisons of isolates from within and outside of Australia, domestic PRSV isolates have been shown to vary by only 2%.

The PRSV isolate used for transgenic studies was obtained from Southeast Queensland. The transgene was designed with a premature stop codon in the PRSV *CP* sequence, thus it was expected that a functional CP protein would not be expressed.

Transformation was facilitated by biolistic transformation of secondary somatic embryos of cultivars GD3–1–19 and ER6–4, local cultivars that are also used for production and breeding.Only two resistant lines were regenerated for each cultivar, each with multiple inserts, and interestingly each male plant.

According to RNA blot analysis, the plants with the best resistance exhibited the least detectible message, strongly suggestive of the involvement of an RNA silencing mechanism for viral resistance. Copy number appeared also to play a role in the level of resistance of R0 plants, as those with single copies were more susceptible.

This observation of gene copy number dependence is consistent with that found for RNA-mediated silencing and PRSV resistance of the original Hawaiian transgenic papayas.

Florida (United States)

Surveys of the PRSV CP gene sequences of Florida isolates indicate a closer relationship to PRSV sequences from Puerto Rico and Mexico compared to those isolated from more distant locations.

Thus, this study on transgenic papaya carrying the PRSV CP of a Florida isolate was done with the intention of producing of cultivars resistant to PRSV of the Caribbean region. The source of the CP sequence was the Florida PRSV H1K isolate.

Four different types of constructs, a sense, antisense, frame shift and stop codon mutation of the CP gene were made. The construct was transformed into immature zygotic embryos of the experimental cultivar F65, which is an ancestor of the PRSVtolerant cultivar Red Lady, by *Agrobacterium-mediated* transformation.

None of the resulting plants were immune when inoculated with the homologous PRSV at 10 wk, but moderate-to-highly resistant individuals were identified from among each construct type when inoculation was done at later stages.

Of interest, the lines derived from sense and antisense constructs were found to be infertile. The remaining stop codon and frame shift mutation constructs lines were fertilized with pollen from the local cultivars Red Lady and Experimental No. 15 and cultivars grown at the University of Puerto Rico including "Puerto Rico 6–65," "Tainung No. 5," "Solo 40," and Sunrise.

The Puerto Rico 6–65 and Tainung No.5 are PRSV-tolerant varieties, whereas Solo 40 and Sunrise are highly sensitive. Resistance of the R1 progeny seemed to be influenced to some extent by the particular cross, with progeny from the tolerant Tainung No. 5 and sensitive Sunrise more sensitive than that of other combinations.

Hong Kong (China)

Researchers in China reported the first case of using the PRSV replicase gene for the production of PRSV-resistant papaya. The replicase gene was cloned from pumpkin leaves infected with PRSV. For the papaya replicase construct, the 3'-end of the gene was deleted and additional codons were added to the 5'-end of the gene.

Transformants were obtained by *Agrobacterium*mediated transformation of embryogenic calli of the cultivar Tai-nong-2. The resulting transformants showed varying levels of resistance in response to mechanical inoculation, including apparent complete immunity in the greenhouse.

The mechanism of resistance, whether protein or RNA-mediated, is unknown. The greenhouse test seems to be encouraging, but the fate of these transgenic lines under natural field conditions will have to await results of field data.

NEW APPROACHES

PDR has been successful for controlling plant virus diseases by exploiting different genes coding for viral proteins, but the CP gene is, by far, the most widely used to engineer transgenic resistance. The transgenic papaya in Hawaii is a successful example for the utilization of PDR in controlling PRSV.

In that case, the CP was expressed in transgenic plants. It is now conclusive that transgenic resistance in papaya is RNA mediated through posttranscriptional gene silencing. In fact, the prevailing mechanism for transgenic resistance is via posttranscriptional gene silencing or PTGS.

As shown by our work in Hawaii, Rainbow, which is hemizygous for the CP gene, had a much more narrow base of resistance than SunUp, which is homozygous for the same CP gene. Thus, our early approaches for the developing countries have been to use the CP gene from PRSV strains of the target countries. In our quest to develop transgenic papaya with a broader range of resistance, we took advantage of our work on tospoviruses, which is described in the next section.

Segmented Gene Approach

Evidence that viral transgenes could only confer resistance against closely related viruses stimulated new approaches to create transgenic plants that could be simultaneously resistant to different viruses. The first approach involved the use of multiple *CP* or nucleoprotein *(N)* transgenes regulated by independent promoters and terminators.

Two main disadvantages precluded the extensive use of this approach: concerns about the introduction of increasingly higher amounts of foreign DNA into crop plants, and limitation on the number of CP genes that could be simultaneously transformed into the plant.

A second, more feasible approach toward multiple virus resistance was developed when it was realized that virtually any segment of DNA derived from the *N* gene of *tomato spotted wilt virus* (TSWV) conferred resistance to TSWV in transgenic tobacco.

These *N* gene fragments could be as short as 100 by if they were fused to a longer fragment of non-related DNA, such as the 720 by *GFP* gene. As a consequence of this finding, chimeric transgenes consisting of the CP genes of *turnip mosaic virus* fused to a 217-bp segment of the *N* gene of TSWV were designed and shown to confer resistance to both viruses in *Nicotiana benthamiana.*

The strategy has been extended to make transgenic *N. benthamiana* plants resistant to three different tospoviruses. Experiments are underway to test the strategy of pasting multiple 240-bp long fragments of CP genes from different geographical isolates in different combinations and orientations for their efficacy in PRSV resistance.

Resistance is predicted to be specific for all those isolates that share a high degree of similarity to any of the fragments in the chimeric gene. The downside of this approach is that even though segments of the chimeric gene are short, there is still a limitation on the number of PRSV segments which can be engineered together and consequently there is still a limitation to which PRSV isolates a given transgene construct with be effective against. Even so, this is the first rational approach to widen transgenic resistance without increasing the amount of foreign DNA delivered to the target plant. Papaya plants transformed with multiple PRSV CP genes fragments are still under evaluation, but the first virus challenges indicate that we have been able to widen resistance to PRSV using the transgenic system described here.

Synthetic Gene Approach

A new approach to widen the range of resistance involving the creation of transgenes with a single, short DNA segment (—250 bp) able to confer resistance to multiple viruses is currently being tested. This approach does not involve the use of native CP genes, chimeras or multiple transformations with selected sequences and would reduce the amount of foreign DNA inserted in the plant.

Instead of searching for a natural variant of the CP gene able to confer a broad-spectrum resistance, we created it by rational design. To test the feasibility of this approach, we used *N. benthamiana,* and the transgenes encoding the third fourth of the TSWV *N* gene which has been previously shown to confer resistance to TSWV as a model system.

On the basis of the nucleotide sequence of the *N* gene fragment, we designed and synthesized a novel sequence from oligo nucleotides that was highly similar (90% in this case) to the corresponding fragments of three distantly related tospoviruses, TSWV, *groundnut ringspot virus,* and *tomato chlorotic spot virus,* which are normally only approx *78%* similar at the nucleotide level.

The aim was to introduce nucleotide changes at specific locations that would create a synthetic gene equally distant to all sequences used for its creation. Importantly, the nucleotide changes were designed to create short stretches of 20 nt or more of total identity to the different genes at different points along the synthetic gene, which is a prerequisite for targeting RNA degradation by PTGS. This synthetic sequence allowed us to obtain transgenic *N. benthamiana* plants resistant to two different tospoviruses *groundnut ringspot virus* (24% resistance) and TSWV, and potentially to a third one (TCSV) that could not be used in our laboratory.

We have designed a PRSV-derived single sequence aimed at targeting different isolates of the virus via PTGS. The engineered sequence is equally distant at the nucleotide level to all PRSV isolate sequences that we compiled and used for its design. We targeted the variable (5') and conserved regions (3') of the PRSV CP.

The two synthetic genes were each independently cloned in tandems of three, and transcriptionally fused to a silencer gene which in this case, was half the N gene of TSWV. The two different constructs have already been cloned in a plant expression cassette and somatic papaya embryos have been transformed by the biolistic method.

Results derived from these experiments might lead to the design and creation of a synthetic gene in which we would be able to play not only with sequence similarity, but also with the secondary and tertiary structures of the protecting transgene. The long-term goal of these experiments is to create a short, synthetic sequence able to confer universal and durable resistance to PRSV.

13

Chapter

FUNGICIDES

Many early systemic fungicides targeted powdery mildews. Screening analogs of the organophosphorus insecticide, diazinon, during the early 1960s at the then ICI Agrochemicals' Jealott's Hill Research Station uncovered a series of 2-amino-4-hydroxypyrimidines with both protectant and systemic activity against several important powdery mildews of non-woody crop plants.

This led to the development of dimethirimol (Milcurb, ICI Plant Protection, Fernhurst, UK) as an effective soil drench for powdery mildew control in both protected and outdoor cucurbit crops.

The 2-ethylamino analog, ethrimol, soon followed and, because of its better activity than dimethirimol, especially against cereal mildews, it was widely adopted in the United Kingdom as a seed treatment formulation (Milstem, ICI Plant Protetion, Fernhurst, UK) against barley powdery mildew (*Blumeria = Erysiphe graminis f.sp. hordei*).

Controlling foliar diseases with a seed treatment was then novel, and success depended on the development of new machinery to accurately load the bulky fungicide dose onto the seed, and careful calibration of seed drills.

The sulphamate ester of ethirimol, bupirimate, does penetrate woody tissue, and its introduction extended control to include powdery mildews of orchard crops, soft fruit, and ornamentals. Despite problems of resistance (see below), this fungicide group is still available, although dimethirimol has very few uses, if any.

Ethirimol is now only available as a mixture partner with the sterol demethylase inhibitor (DMI) fungicide, flutriafol, and thiabendazole as the seed treatment Ferrax, whereas bupirimate is available both as a single product (Nimrod, Syngenta, Whittlesford, UK) and as a mixed spray formulation with triforine (Nimrod-T) for use on ornamental crops.

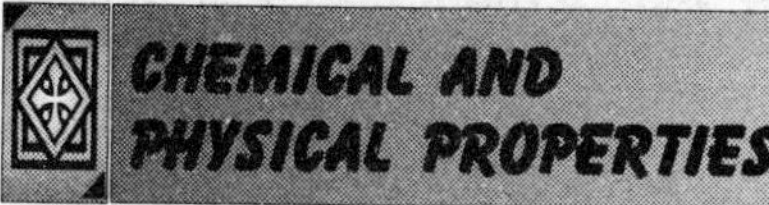

CHEMICAL AND PHYSICAL PROPERTIES

The main properties of this group are shown in Table 1. More detailed information about solubilities in different solvents, CAS registry number, IUPAC and chemical abstracts names, and patent numbers can be found in *The Pesticide Manual.* Both dimethirimol and ethirimol are colourless crystals, whereas bupirimate is a pale tan-coloured waxy solid. All are rapidly decomposed by ultraviolet light in aqueous solutions. Bupirimate has some useful vapor activity.

BIOLOGICAL PROPERTIES

2-Aminopyrimidines are effective only against powdery mildews on a wide range of field, orchard, ornamental, and protected crops. Activity is both protective and curative. When used as seed treatments or soil drenches, uptake depends on acidity and organic/ matter content, which should not exceed 20%. All are systemic with varying degrees of translaminar activity. Spray application rates are generally between 200 and 400 g a.i./ha.

Degredation in plants occurs rapidly through N-dealkylation, conjugation to form glycosides, and hydroxylation of the n-butyl group, which means that harvest intervals can be as little as 1 day in some crops, but more generally up to 2 weeks.

There have been few reports of phytotoxicity, although leaf spotting may occur on protected cucurbits during periods of low light intensities.

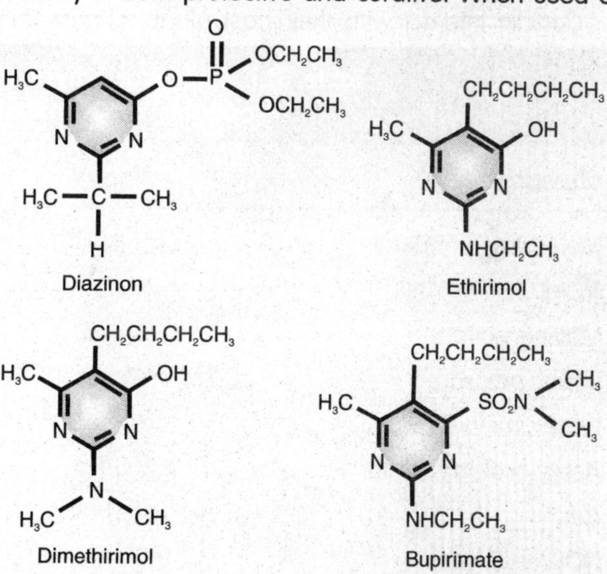

Figure 13.1: Chemical structure of diazinon and 2-aminopyrimidine fungicides.

Toxicology, Ecotoxicology, and Environmental Fate

Mammalian toxicities are low and few harmful effects have been reported, although bupirimate is a mild eye and skin irritant in rabbits. Both dimethirimol and ethirimol have very low toxicities to a wide range of birds; aquatic organisms, including rainbow trout, daphnia, and algae; beneficial arthropods; and earth worms. Bupirimate is considered potentially harmful to some aquatic organisms and consequently has an EC risk of R53.

These properties make 2-aminopyrimidines useful components in integrated pest management (IPM) systems where powdery mildews become a problem. Information on their environmental fate is limited. Half-life in sandy soils is generally around 4 to 6 months. Metabolism within plants is rapid, and following oral administration, they are eliminated from mammals principally through urine. After just 24 hours, 68% of the dose of bupirimate administered to mammals is eliminated in the urine.

Mode of Action

Powdery mildews penetrate their hosts directly through the leaf cuticle using a specialized structure called an appressorium. Although all stages of mildew development are affected by 2-aminopyrimidines, disrupting appressria formation is the most critical stage, and it is the reason why these fungicides are protectants.

Key biochemical events controlling appressoria formation occur early in the infection process, and for barley powdery mildew, conidia must be exposed to sufficient ethirimol within 8 hours of inoculation to stop appressoria from forming.

How conidia receive 2-aminopyrimidines is not clear, but the fungicides are exuded to some extent onto the leaf surface and, espcially bupirimate, redistribute as a vapor.

Conidia interact with their host plants almost immediately after they land on leaf surfaces,

Table 31.1: Chemical, Physical, and Toxicological Properties of 2-Aminopyrimidines.

Characteristics	*Ethirimol*	*Dimethirimol*	*Bupirimate*
Molecular formula	$C_{11}H_{19}N_3O$	$C_{11}H_{19}N_3O$	$C_{13}H_{24}N_4O_3S$
Molecular weight	209.3	209.3	316.4
Melting point in ° C	159—160	102	50-51
Vapor pressure	0.267 mPa at 25°C	1.46 mPa at 30°C	0.1 mPa at 25°C
Water solubility	150 mg/l at 20°C	1200 mg/l at 25°C	22 mg/l at 25°C
Acute oral toxicity for rats	6.3 g/Kg	2.35 g/Kg	<4.0 g/Kg
NOEL in 2-year			
feeding tests	—rats	200 mg/kg diet	300 mg/Kg diet
100 mg/Kg diet			
—dogs	30 mg/Kg daily	25 mg/Kg daily	15 mg/Kg daily
Toxicity class			
WHO (a.i.)	III	III	III
	III	III	III
EPA formulation	IV	III	III
EC risk	Xn; R21	Xn; R21	Xi; R43 N; R51; R53

and where a primary germ-tube exists, access is quickly gained to host metabolism. Early work to identify the target site recognized limitations imposed by obligate parasites on biochemical approaches, and it sought to reverse toxicity by adding various metabolites.

A common theme throughout these experiments was that adenine and adenosine were good reversal agents, whereas several purine analogs acted like 2-aminopyrimidines and inhibited appressoria formation. Ethirimol-resistant baley mildew strains generally showed some level of cross resistance to these analogs. Surveys of enzymes involved in purine salvage in conidial extracts from several powdery mildews, coupled with structure-activity studies using 2-aminopyrimidine analogs, identified adenosine deamnase (ADAase; EC 3.5.4.4) as a target.

ADAase catalyzes the almost irreversible conversion of adenosine to inosine (and deoxyadenosine to deoxyinosine) and is a key enzyme in many fungi for the salvage of adenosine nucleotides to guanosine nucleotides. 2-aminopyrimidines are noncompetitive inhibitors with *a* K_i of 2.32×10^{-5} M for ethirimol at pH 7.8 and deoxyadenosine as the substrate.

Why inhibition of ADAase affects mildew development is not clear, but the enzyme ensures an adequate supply of guanosine nucleotides for the demands of nucleic acid biosynthesis during rapid growth. The enzyme indirectly regulates AMP levels and, hence, the energy charge within a cell, which is a key factor controlling overall metabolic activity.

Selectivity is easily explained as plants lack ADAase and salvage purines at the nucleotide level instead. The reason for the extreme specificity toward powdery mildews is not known.

Resistance Issues

Cucurbit powdery mildew (*Sphaerotheca fuliginea*) seems an excellent predictor of fungicides likely to cause high resistance risk, and within 2 years of the introduction of dimethirimol into glasshouses in Northwest Europe, control failures of this disease were reported.

These were confirmed as being caused by resistance, and the product was withdrawn, although control failures have not been encountered in field grown cucurbits.

A more gradual decrease in sensitivity occurred in barley powdery mildew, but by 1973, ethirimol was no longer recommended for use on winter barley in an effort to limit carryover of resistant strains to the then more important spring sown crop.

Resistance factors of 100-fold were common. Although the resistance risk of 2-aminopyrimidines is significant, resistant strains are generally less fit than are wildtype ones, and it has been possible to manage use of these fungicides through use of appropriate mixtures.

Both laboratory and field studies linked the subsequent development of DMI resistance in barley powdery mildew to increased sensitivity to ethirimol, and this negative cross resistance was one factor in the introduction of the mixed seed treatment Ferrax.

The mechanism(s) of resistance is not understood, although it is not due to detoxification, nor does the target enzyme alter its kinetic properties.

The gradual decline in sensitivity coupled with genetic analysis suggests that several genetic factors are involved. One of these genes may have been identified, although high experimental variation associated with bioassays makes firm conclusions difficult.

FUTURE DIRECTIONS

Despite the lack of ADAase in plants, and its importance in fungal metabolism, as far as the author is aware, serious attempts have not been made to broaden activity associated with this target to other plant pathogens. This is partly explained by fear that new chemistries may interfere with human ADAase, and its absence or inhibition causes severe immunodeficiency, although this factor is exploited in transplant surgery. Over the years, several new chemistries active only against powdery mildews have been evaluated in the author's laboratory.

Some compounds were cross resitant with 2-aminopyrimidines, and others were not. A common feature has been strong inhibition of appressoria formation, and recent studies with one of these compounds, quinoxyfen, has exposed a possible involvement with the signaling pathways governing appressoria formation. 2-Aminopyrimidines, and especially mildew strains resistant to them, may play a useful part in defining these signaling pathways that offer novel opportunities for fungicide action.

FUNGICIDES

Origins

The damaging effects of plant diseases have been recognized since time immemorial. There are a number of references to the affliction of crops by blasts, mildews, and the like in the bible and in classical Latin writings, but the recognition that this damage was caused mostly by fungi came very much later. It was in the mid-19th century, through the publications of the Reverend M. J. Berkeley, Anton de Bary, the Tulasne brothers, and other pioneers of plant pathology, that certain fungi became known to be causal agents of crop disease.

However, effective methods for preventing or decreasing the damage caused by fungal diseases of crops began to emerge somewhat earlier than this. Disinfection of wheat seed to protect against bunt disease, which is now known to be caused by the fungus *Tilletia caries*, was advocated around 1750 by Jethro Tull. The seed was sprinkled with brine (aqueous solution of sodium chloride) and dried with lime.

The use of brine is said to have originated from the planting of seed salvaged from a shipwreck. The efficacy of this treatment was soon demonstrated by Mathieu Tillet in field experiments, which were probably the first fungicide trials ever done and were remarkably well conducted. Later, Benedict Prevost found that treatment of bunted seed with copper sulfate gave better control.

Copper preparations, especially powdered copper carbonate, were used for many years until they were superseded by organomercurials, first reported in 1913 by the company I.G. Farben-Industrie AG. Phenyl- and alkyl-mercury seed treatments proved even more effective than copper treatments against wheat bunt, controlled a broader range of seed-borne diseases of cereals, and were less injurious to the seed.

Being cheap as well as efficacious, their use became routine on wheat and barley in many countries and persisted in some of them up to the early 1990s. Regarding diseases of foliage and fruit, the first fungicide to become widely used was lime-sulfur. This originated in the work and writings of William Forsyth, gardener to King George the Third of Great Britain and Ireland.

Having done field trials in Hyde Park in London, he recorded in 1802 that a mixture of sulfur, lime, tobacco, and elder-buds, when applied to fruit trees, suppressed powdery mildew. Lime-sulfur and, later, wettable sulfur preparations became increasingly used in Europe and the U.S. during the 19th century. Their first application to grape vines, which became by far the largest crop to be treated with sulfur, was also made in England by a Mr. Tucker of Margate in Kent, after whom the grape powdery mildew (now *Uncinula necator*) was first named as *Oidium Tuckeri.*

Later, sulfur formulations found disease targets other than powdery mildews, for example apple scab (caused by *Venturia inaequalis*). Then came the well-known discovery in ca. 1885 of Bordeaux mixture by Alexis Millardet, originating from its application to grape vines in order to deter thieves. Many other fungicide formulations based on copper compounds, notably copper oxychloride and cuprous oxide, were to follow.

These gave control of downy mildew of grapes, caused by *Plasmopara viticola*, potato blight, caused *by Phytophthora infestans*, and other diseases not controlled by sulfur. Since these earlier pioneering efforts in disease control, which focused on a few crops in Europe, the enormity of the damage caused by fungi to crops of many kinds throughout the world has become clear.

Oerke, from a study of eight major crops, concluded in 1996 that crop disease globally causes a potential yield loss of 17.5%, and that this is reduced to 13.5% through control measures. The adoption of hygienic cultural practices, especially crop rotation, and the planting of disease-resistant varieties have decreased or prevented disease-induced losses in many crops, but fungicide application has been over many years, and still remains, the predominant means of crop disease management worldwide.

Fortunately, the number, diversity, effectiveness, and safety of agricultural fungicides have all increased greatly, especially over the last three or four decades.

Evolution of the Modern Fungicide Armory

The first-invented fungicides, based on sulfur, copper, and mercury, served the farmer well. However, they were not fully effective against certain diseases, especially after infection had occurred, and tended to injure crops so that, in the absence of disease, they sometimes reduced yields or crop quality.

There was much scope for other fungicide treatments with improved properties. In 1934, the discovery of the dithiocarbamate fungicides was reported by Tisdale and Williams, working at E. I. du Pont de Nemours and Co., and independently by Martin in England. Over the next 10 years, major fungicides such as thiram, zineb, and maneb emerged within this family.

The modern relative, mancozeb, is used very widely today in many crops. The dithiocarbamates proved as effective, in some situations more so, and less injurious to crop plants, compared with copper, sulfur, and mercury.

However, like these earlier materials, they are surface fungicides. They penetrate little into plants, and so cannot affect established infections, which generally extend well below the plant surface.

Moreover, treatments to foliage and fruit have to be repeated at frequent intervals in order to replace losses through weathering and to protect new plant growth. Over the next 30 years, several other types of surface fungicide were introduced.

The most important are proably the phthalimides, exemplified by captan, discovered by the Standard Oil Development Company together with Rutgers University, which was followed by captafol and folpet. Later chlorothalonil, the only member of its class and still widely used in cereals and other crops, was developed by the Diamond Alkali Company.

Systemic fungicides, which would penetrate and move in the treated plants and, hence, might eradicate existing infections and also move into new growth and resist weathering, thus avoiding repeated application, were sought after for many years but proved hard to find.

Eventually, in the late 1960s, no less than six different types of systemic fungicide with high levels of effectiveness appeared in rapid succession: benzimidzoles, carboxamides, morpholines, 2-amino-pyrimidines, and organophosphorus and antibiotic rice blast fungicides.

The reason for this sudden appearance of all these major systemics is not at all clear. Their chemical and biological properties, places of origin, and discovery processes were diverse. The adoption of *in vivo* screening tests and increased screening throughputs probably contributed, together with some luck. A second batch of highly active fungicides, with different degrees of systemicity, emerged in the mid-1970s: dicarboximides, phenylamides, triazoles, and fosetylaluminium.

After a lull of some 15 years, yet another wave has recently appeared, comprising strobilurins, anilinopyrimidines, and phenylpyrroles. These modern fungicides have attained new standards of disease control, often at relatively low doses, coupled with relatively low mammalian toxicity.

Some of them work against specific targets, for example dicarboximides against *Botrytis* and *Sclerotinia* spp. and 2-amino-pyrimidines against powdery mildews. Others have a broad spectrum of action, especially the strobilurins, which are unique in working on all the major classes of plant-pathogenic fungi (oomycetes, ascomycetes, and basidiomycetes).

Some are highly xylem-mobile, but others, notably the dicarboximides and strobilurins, have only limited mobility in the treated plant. Phloem mobility, permitting movement out of treated leaves and downward translocation to roots is generally absent but has been observed with fosetyl-aluminium. The biochemical mechanisms of action are diverse and cannot be described here individually.

As a general rule, the surface fungicides (sometimes called protectant fungicides—a misnomer because all systemic fungicides exert a protectant action) have a multiple action, affecting a number of different target enzymes in the target fungus, whereas the systemics affect either a single site or a relatively small number of sites.

Thus, the systemics are vulnerable to development of resistance in populations of the target pathogens through single-gene mutations that modify the site of action. In practice, many problems of resistance to almost all classes of systemic fungicides have occurred, whereas most of the surface fungicides have not encountered resistance development after many years of widespread use, often on schedules of many repeated applications.

Fungicides Today and Tomorrow

Over 200 fungicides have been introduced into agriculture to date, and, of these, some 140 are in current use. Many of these are described in the articles that follow. The number of formulated products, based on single fungicides or on mixtures of two or more, is several times larger. In 1998, the global sales value of fungicides was of the order of five and a half billion U.S. dollars.

While some fungicides have been withdrawn from commercial use for various reasons, the majority of those discovered are still in use. These include Bordeaux mixture and sulfur, whose use is currently increasing because their application is permitted under certain circumstances in organic crop production. When properly applied at the correct times, fungicides usually perform very well and have an acceptable margin of safety to humans, wildlife, and crop plants.

However, we still need new fungicides for several reasons. Despite the large total number of fungicides that are available, each particular crop disease typically is well controlled by only two or three marketed fungicides, each with its strengths and limitations.

A wider choice of treatment is desirable for many crop diseases. The development of resistant mutants of target pathogens has led to losses of effectiveness, in certain regions and uses, of most of the modern fungcide classes.

Further resistance problems seem likely to arise despite the considerable efforts of the agrochemical industry and farm advisory services to promote the use of countermeasures. New types of fungicide can act as effective replacements for these problem situations and also increase the diversity of treatment, which is a mainstay of fungicide resistance management.

Certain fungicides, for example, captafol, binapacryl, organomercurials, and ethylenebisdithiocarbamates, have been banned from or restricted in commercial application because of perceived toxicological risks or because manufacturers are not prepared to do additional toxicological or environmental safety evaluations required by regulatory authorities, and further withdrawals for these reasons are likely to occur. Fortunately, new fungicides with distinctive chemical and biological properties continue to be invented.

Fungicides introduced over the past two years or so, or known to be at an advanced development stage, include, among others, famoxadone (Du Pont), quinoxyfen (Dow-Elanco), fenhexamid (Bayer), iprovalicarb (Bayer), and MON6500 (Monsanto), together with several newer strobilurins. MON65500 is unusual in that it controls take-all disease of wheat, a root disease unaffected by almost all other fungicides.

Fluquinconazole also has reported recently to be effective against take-all. Acibenzolar-S-methyl (Novartis) has attracted special attention because it acts against a very wide range of plant diseases but does not directly affect the growth or metabolism of the causal organisms. It is known to act by stimulating systemic activated resistance (SAR) in the treated plant.

The well-established compound probenazole, which for some years has been the leading treatment for rice blast in Japan, is thought to act in a similar way. Pathogen resistance to probenazole has not developed, and such stability of performance may well be a general feature of SAR-inducing compounds. Thus fungicide invention continues very actively, despite the many mergers between agrochemical companies and consequent reduction in number of research centers and despite the ever-increasing standards of disease control and regulatory requirements with regard to margins of safety to the environment and human health.

Discovery still depends largely on high-throughput screening of candidate chemicals, now aided by target-enzyme tests, combinatorial chemistry, and computer modelling. However, the fully rational design of molecules that fit all the needs of a successful practical fungicide (biological activity, stability, environmental safety, crop safety, ease of manufacture, etc.) has yet to be achieved.

It is debatable whether the current pace of discovery can be maintained, especially because demand has plateaued or even declined in some areas. One critical factor, at present hard to assess, will be the extent to which plant varieties with improved, long-lasting disease resistance will emerge through genetic engineering and reduce the need for fungicide applications.

Use of biological control agents in crop disease management has achieved little significance up to now but may well increase in the future, particularly as seed treatments or in glass-house or produce-storage environments. Complementary to the inventive effort is the need for further research into the most cost-effective and safe ways of using fungicides.

At present, large quantities are wasted through inefficient application, inappropriate timing, or unnecessarily large doses. Computer-based decision-support systems are under development, and research continues on precision delivery systems designed to place pesticides where they are needed in the crop. Both of these approaches may help to optimize fungicide treatments.

The need for integration of fungicide applications with other crop management components, such as crop rotation, sowing date, choice of crop variety, fertilizer use, and soil cultivation, is increasingly recognized and researched. The integrated approach requires larger inputs of knowledge and observation by farmers and advisers and depends greatly upon their having a very good understanding of the nature and properties of the fungicides that are available to them.

FUNGICIDES, ANILOPYRIMIDINES

This group of antifungal compounds originated from studies conducted in East Germany (former GDR) by Krause et al.. The first fungicide to emerge from this class of 2-anilinopyrimidines was andoprim. It is highly toxic to many Peronosporales but does not inhibit *Pythium* spp. significantly. It acts mainly as a protectant, but it never reached the market.

On the other hand, the three related active ingredients, cyprodinil, mepanipyrim, and pyrimethanil, have been developed since the beginning of the 1990s. They are characterized by the absence of any substituent on the phenyl ring and by their efficacy toward several Ascomycetes or Adelomycetes but not toward Oomycetes. Cyprodinil, mepanipyrim, and pyrimethanil differ only by their substituents in the 4-position in the pyrimidine ring, which are, respectively, a cyclopropyl, a 1-propynyl, and a methyl group.

Physicochemical Properties and Synthesis

The main physicochemical properties of cyprodinil, mepanipyrim, and pyrimethanil are reported in Table elsewhere in this chapter. These various active ingredients are solids with melting points between 76°C and 133°C. They exhibit vapor pressure values below 1 mPa, except pyrimethanil, whose value is 2.2 mPa at 25°C.

The vapor activity of pyrimethanil could be partially explained by its volatility. Moreisover, pyrimethanil exhibits the highest water solubility It (121 mg/L) and the lowest n-octanol/water partition coefotficient (log K_{wo} = 2.8). The anilinopyrimidines are weak a bases with pKa values

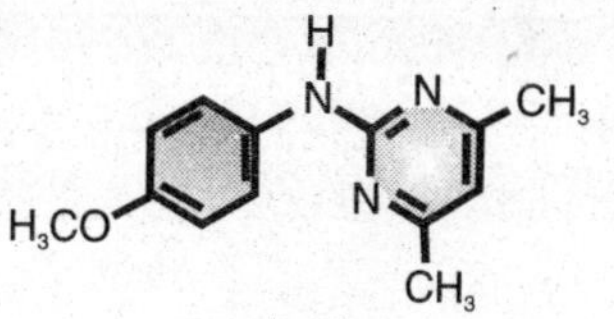

andoprim

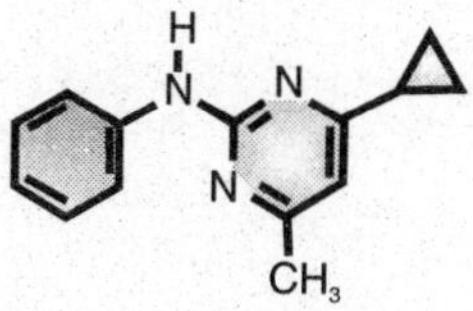

cyprodinil 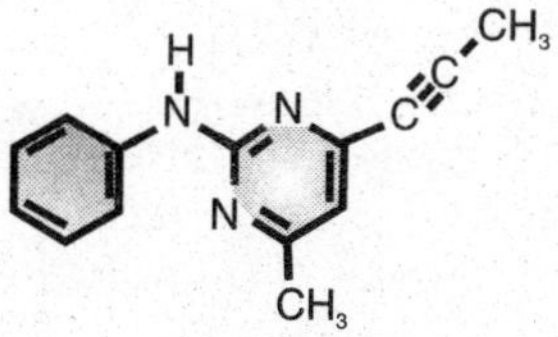 mepanipyrim 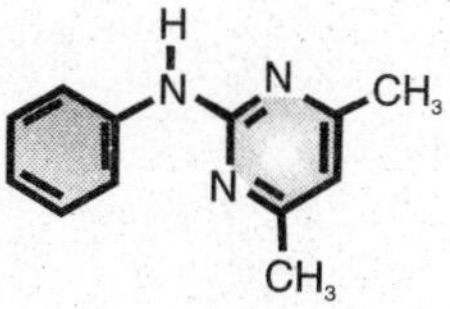 pyrimethanil

Figure 13.2: Structure of anilinopyrimidine fungicides.

between 3.5 and 4.5. Their solubility in organic solvents generally exceeds 100 g/L for acetone, dichloromethane, ethyl acetate, and alcohols; lower values are recorded for hexane.

The synthesis of cyprodinil can be accomplished by reacting phenylguanidine with the following diketone: 1-cyclopropyl-butane-1,3-dione. Another alternative can consist in cyclizing a guanidine salt with the previous diketone to give an aminopyrimidine and then reacting it with a monosubstituted halogen benzene.

Moreover, when formanilide and 2-methane sulfonyl-4-methyl-6 (propynil) pyrimidine are reacted in the presence of a base, in an organic solvent, they yield an intermediate formamide that, after hydrolysis, produces mepanipyrim (see patents).

Biological Properties and Uses

The fungal species sensitive to anilinopyrimidines belong only to Ascomycetes and Basidiomycetes. They include *Botrytis cinerea, Venturia spp., Alternaria spp., Monilinia* spp., and *Helminthosporium* spp.. The presence of the cyclopropyl side chain in cyprodinil enlarges its spectrum of activity to several major pathogens of cereals (e.g., *Tapesia spp., Erysiphe graminis, Rhyncosporium teres,* and *Septoria nodorum*). Pyrimethanil has additional activity against *Mycospaerella spp. in* bananas.

The anilinopyrimidines are not toxic to yeasts involved in the fermentation process of grape juice. Anilinopyrimidine fungicides are mainly used as foliar sprays, at 200 to 1000 g a.i./ha; cyprodinil and pyrimethanil are also used in mixture for seed dressing. They are formulated as wettable powders (e.g., mepanipyrim), suspension concentrates (e.g., mepanipyrim, pyrimethanil), or water dispersible granules (e.g., cyprodinil). Nowadays, the anilinopyrimidines can be used against scab on pome fruits, against anthracnose on peas, and against gray mold on various plants, including grapevine, fruits, vegetables and ornamentals.

To prevent the development of resistance in *B. cinerea* and *Ventura* spp., limitations of the anilinopyrimidine use are recommended. For instance, the FRAC (Fungicide Resistance Action Committee) recommends a maximum of two applications out of six treatments per season on grapevine against gray mold either with solo proucts (i.e., mepanipyrim, pyrimethanil) or mixtures (i.e., cyprodinil + fludioxonil).

Table 13.2: Nomenclature of Commercial Anilinopyrimidine Fungicides.

Common Name		*Cyprodinil*	*Mepanipyrim*	*Pyrimethanil*
Code number		CGA 219417	KUF-6201	SN 100309
KIF-3635		ZK 100309		
	IUPAC	N-(4-cyclopropyl-6-methylpyri midin-2-yl) aniline.	N-(4-prop-1-ynyl-6-methylpyri midin-2-yl) aniline.	N-(4,6 dimethyl-pyrimidin-2-yl) aniline.
Chemical names		4-cyclopropyl-6-methyl-N-	4-(1-propynyl)-6-methyl-N-	4,6-dimethyl-N-phenyl-2-pyridi
	CAS	phenyl-2-pyridinamine.	phenyl-2-pyridinamine	namine
CAS registry number		121552-61-2	110235-47-7	53112-28-0
Patent		EP 310550	EP 224339	DE 151404
			JP 63208581	
			US 4814338	
Manufacturer		Novartis now Syngenta	Kumiai/Ihara	AgrEvo, now Aventis
		(introduced by Ciba-Geigy)	(introduced by Schering)	
Trade-names		Unix, Chorus, Stereo, Koara, Radius, Switch	Frupica	Mythos, Scala, Sigarex, Walabi

In France, the advice is more restrictive, with only one spray per season, out of a maximum of three *Botrytis* treatments. Against *Venturia* spp., the proposal of FRAC is a maximum of four or five treatments per season, according to the use of solo products or mixtures.

Among the anilinopyrimidines, cyprodinil has a specific position as a foliar treatment on wheat and barley because it controls several major diseases such as powdery mildew, eyespot, leaf blotch, and net blotch. Pyrimethanil is used as a seed treatment in cereals in combination with sterol biosynthesis inhibitors (i.e., prochloraz, flutriafol), where it provides control of *Pyrenophora* spp. and *Fusarium* spp.

In order to broaden its spectrum of activity against cereal diseases, cyprodinil is also used in mixture with sterol biosynthesis inhibitors (e.g., fenpropidin, cyproconazole, propiconazole). This strategy is proposed on other crops (e.g., peas, pome fruits) and with the other anilinopyrimidines.

The companion fungicides can be multisite toxicants (e.g., chlorothalonil, thiram, ziram), sterol biosynthesis inhibitors (e.g., fluquinconazole, difenoconazole), and phenylpyrroles (e.g., fludioxonil). Under field conditions, the anilinopyrimidines have good protective and residual activities against fungal parasites. This preventive effect is due to the inhibition of penetration hyphae or appressoria.

In addition to this antipenetrant action, the anilinopyrimidines can reduce fungal infection stages that are formed inside the host tissues and consequently exhibit curative activities. Such a

Table 13.3: Physicochemical Properties of Anilinopyrimidine Fungicides.

Characteristics	Cyprodinil	Mepanipyrim	Pyrimethanil
Molecular formula	$C_{14}H_{15}N_3$	$C_{14}H_{13}N_3$	$C_{12}H_{13}N_3$
Molecular weight	225.3	223.3	199.3
Form	Beige powder with a weak odor	White crystals	Colorless crystals
Melting point in °C	76	133	96
Vapour pressure in Pa	5.1×10^{-4} (25°C)	2.3×10^{-5} (20°C)	2.2×10^{-3} (25°C)
Henry's constant in Pa m^3/mol	6.9×10^{-3} (calc.)	1.7×10^{-3} (calc.)	3.6×10^{-3} (calc.)
Log K_{ow}	4.0	3.28	2.84
Water solubility in mg/L	13-20(25°C)	3.1 (20°C)	121 (25 °C)
Solubility in g/L:			
—acetone	610	139	389
—ethanol or methanol*	160	15.4	176*
—ethyl acetate	—	—	617
—dichloromethane	—	—	1000
—hexane	30	2.1	24
pKa	4.44	4.5 (estim.)	3.52

phenomenon implies that the anilinopyrimidines can be transported within the treated organs. Furthermore, under experimental conditions, acropetal translocation of these fungicides has been observed.

Mode of Action and Resistance Phenomena

In vitro studies, conducted on minimal media, to examine the effects or anilinopyrimidines on *B. cinerea* have revealed a strong inhibition of germ-tube elongation and mycelial growth. A weaker and transient effect was recorded toward spore germination. On complex media, there may be a variability in response, caused by the ability of the fungus to obtain nutrients in a manner that circumvents the mode of action of anilinoprimidines.

Several amino acids, particularly methionine, have been shown to reverse the fungitoxicity of the anilinopyrimidines. Biochemical studies conducted with radiolabeled sulfate indicate that pyrimethanil inhibits the biosynthesis of methionine and suggest that the target could be cystathionine 6-lyase. Enzymatics studies, however, have shown only a weak inhibitory effect of the anilinopyrimidines on this enzyme.

Another common feature of the anilinopyrimidines their ability to prevent fungal secretion of hydrolytic enzymes such as proteases, cellulases, lipases, or cutnases, which play an important

Table 13.4: In Vitro Effects of Anilinopyrimidines and Other Fungicides Toward Botrytis cinerea.

Fungicides	Wild-Type Strains: EC50 Values in mg/L for: Spore Germination	Germ-Tube Elongation	Mycelial Growth	Resistant Strains: Resistance Levels for: Ani R1	Ani R2	Ani R3
Cyprodinil	0.01-0.50	0.008	0.010	13-250	10	2.5
Mepanipyrim	0.05-1.0	0.035	0.035	20-280	8.6	6.0
Pyrimethanil	0.1-2.0	0.055	0.075	15-90	5.5	5.5
Fludioxonil	0.055	0.017	0.003	<2	10	<2
Tebuconazole	>10	0.22	0.30	<2	<2	6.8
Fenhexamid	>10	0.050	0.014	<2	<2	9.0

role in the infection process. In *B. cinerea*, the same phenomenon also concerns laccase and could explain the reduced laccase activity in wines treated with pyrimethanil.

Additionally, in *in vitro* studies, this effect has been observed even in the absence of any inhibition of hyphal growth. The exact mechanism of action in the protein secretory pathway is not yet understood; it has been hypothesized that the target of anilinopyrimidines could be a step

Table 13.5: Toxicological Profiles of Anilinopyrimidine Fungicides.

Criteria	Cyprodinil	Mepanipyrim	Pyrimethanil
Acute oral LD_{50} for rats in mg/kg	>2000	>5000	4150-5971
Acute percutaneous LD_{50} for rats in mg/kg	>2000	>2000	>5000
Inhalation LC_{50} for rats in mg/L air (4h)	>1.2	>0.6	>2.0
NOEL in mg/kg body weight/day			
—rat	3 (2 years)	2.45 (2 years)	20 (2 years)
—dog	65 (1 year)	?	30 (1 year)
ADI in mg/kg body weight/day	0.03	0.024	0.17-0.20
Toxicity class: WHO (ai)/EPA (formulation)	III (Table 5)/?	III (Table 5)/?	III (Table 5)/IV
EC risk	Xi, R 43, R 50, R 53	?	R 51, R 53

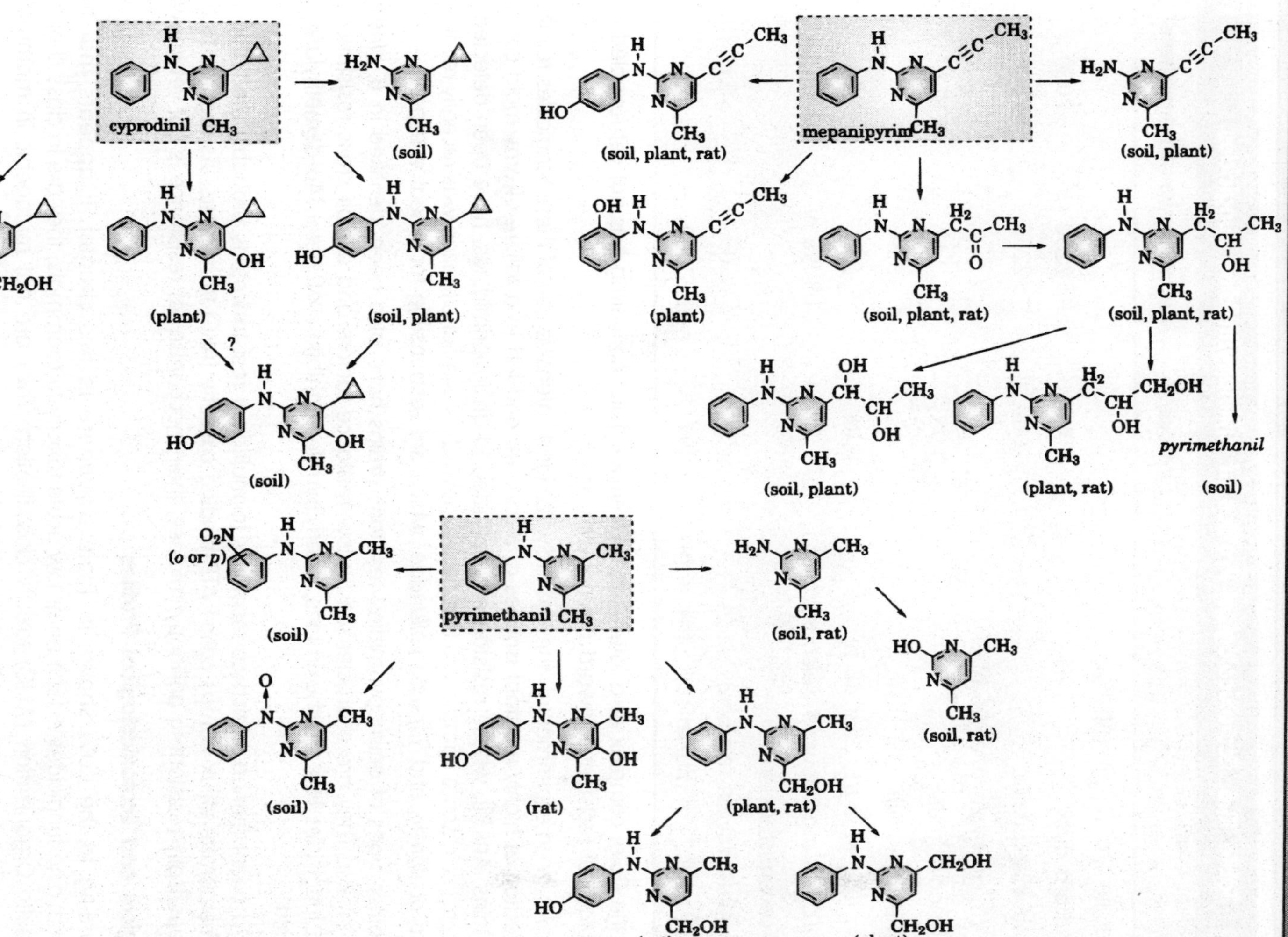

Figure 13.3: Metabolism of anilinopyrimidine fungicides.

Table 13.6: Ecotoxicological Profiles of Anilinopyrimidine Fungicides.

Criteria	*Cyprodinil*	*Mepanipyrim*	*Pyrimethanil*
Birds (acute oral LD_{50} in mg/kg)			
—Mallard duck	>2000	>2250	>2000
—Bobwhite quail	>2000	>2250	>2000
Fishes (LC_{50} in mg/L, after 96 h)			
—Rainbow trout	0.98	3.1	10.6
—Bluegill sunfish	1.07	3.8	
Daphnia (LC_{50} in mg/L, after 48 h)	0.10	5.0	2.9
Algae (EC_{50} in mg/L toward *Selenastrum sp.*, or *Scenedesmus sp.*	0.75 (72 h)	1.3 (96 h)	1.2 (96 h)
Bees (contact LD_{50} in mg/bee)	>0.1	>0.1	>0.1
Earthworms (LC_{50} in mg/kg soil, after 14 h)	192	>1000	625

involving the Golgi complex or a later stage. Extensive field monitoring has not yet revealed any case of practical resistance toward anilinopyrimidines.

However, in an experimental vineyard in Switzerland, intensive use of these fungicides resulted in the selection of highly resistant isolates of *B. cinerea* and led to a failure of gray mold control. Similar strains (Ani R1), whose resistance is restricted to anilinoprimidines, have been detected at low frequencies in some commercial European vineyards. A similar situation was recently recorded in *Tapesia acuformis* and *Tapesia yallundae*, which are responsible for wheat eyespot.

Two other types of anilinopyrimidine-resistant strains have also been identified in *B. cinerea* (Ani R2, Ani R3). They are characterized by low resitance factors and exhibit cross resistance to unrelated fungicides like dicarboximides, phenylpyrroles, inhibitors of sterol 14α-demethylase, or fenhexamid).

Such a phenomenon, which corresponds to multidrug resistances, is probably mediated by plasma membrane efflux pumps called ATP-binding cassette (ABC) tranporters. As regards to the strains specifically resistant to anilinopyrimidines, their mechanism of resistance is not yet known.

Toxicological and Ecotoxicological Profiles

According to the data shown in Table elsewhere in this chapter, it appears that the anilinopyrimidine fungicides exhibit no or low acute toxicity to mammals; they are in class III of the World Health Organization (WHO) toxicity classification. They are not mutagenic in mammals, bacteria, and fungi.

In long-term studies, the NOEL values range from 2.4 to 20-mg/kg body weight/day in the rat; mepanipyrim exhibits the lower value and can induce fatty liver in this rodent. This effect, which seems to be due to an inhibition of intracellular transport of hepatocytic very low density lipoproteins from the Golgi to the cell surface, shows similarities with that reported in fungi. Additionally, cyprodinil exhibits dermal sensitization (guinea pigs), whereas pyrimethanil can induce thyroid tumors in rodents with enhancement of the hepatic thyroid hormone metabolism and excretion. The anilinopyrimidines are not toxic to birds such as ducks or quails (acute oral LD_{50} > 2000 mg/kg).

On the other hand, they can have adverse effects toward aquatic organisms, including fishes, daphnia, and algae at concentrations below 1 mg/L. They seem to have limited noxious actions toward bees, beneficial arthropods, and earthworms.

Persistence and Metabolism

The anilinopyrimidine fungicides are stable to hydrolytic degradation in the pH range 5 to 9 for more than 1 year. They decompose rapidly in water when they are exposed to ultraviolet light; for instance, DT_{50} values of about 2 weeks have been recorded for cyprodinil and mepanipyrim. Under laboratory dark aerobic conditions, the soil DT_{50} values of anilinopyrimidines range from 2 weeks to 3 months.

Formation of unextractable bound residues represents the major route for dissemination of these fungicides in soils. Their degradation seems to be mainly mediated by soil organisms. Cleavage of the aniline–pyrimidine linkage represents the major degradation pathway of the anilinopyrimidines in soils, and it yields pyrimidine–amine derivatives.

The other reactions include hydroxylations, oxidations, and nitrations. The anilinopyrimidines show minimal movement into deeper soil layers, which is in agreement with their high Koc values (respectively, 3555944, 3510, 265-751 for cyprodinil, mepanipyrim, and pyrimethanil). When anilinopyrimidines are applied to the foliage of crops, the major residual components are the active ingredients.

Metabolism of these fungicides occurs mainly via hydroxylation at the phenyl ring, the pyrimidine, or the methyl moiety. Hydroxylated metabolites are recovered free or as carbohydrate conjugates. The studies conducted on grape give a wide distribution of the residue levels of anilinopyrimidines, depending on the number of treatments and the interval before harvest.

According to Cabras et al., pyrimethanil seems to be more persistent than cyprodinil (respective DT_{50} values: 57 and 12 days), and wine-making techniques cause residue reduction only with cyprodinil (c.a. 80%). In France, pyrimethanil is registered alone, with a delay between the last treatment and the harvest of 35 days, whereas a delay of 60 days has been given for cyprodinil (only used in mixture with fludioxonil).

The respective maximum residue level values have been established at 2 and 1 mg/kg of grape. The determination of anilinopyrimidine residues can be achieved by HPLC or by gas–liquid chromatography with a nitrogen–phosphorus detector. In the rat, after oral administration, the anilinopyrimidines are rapidly and almost completely eliminated in the urine and the feces.

There is no evidence for accumulation or retention of these fungicides or their metabolites in organs and tissues. The major metabolic pathways include hydroxylation of the methyl, the propynyl, the pyrimidinyl, or the phenyl moieties. Cleavage of the aniline–pyrimidine linkage is also of importance with pyrimethanil. Several metabolites can be recovered as glucuronide or sulfate conjugates.

FUNGICIDES, ANTIBIOTICS

Some conventional synthetic pesticides are reportedly hazardous to mammals through their biological concentration in the food chain, or their persistence in the natural environment, whereas naturally occurring pesticides are generally specific to target organisms and are inherently biodegradable because of their biogenesis.

Thus, use of antibiotics to protect crops against plant pathogens originated first by applying medicinal antibiotics such as streptomycin, oxytetracycline, and chloramphenicol. Then, by screening microbial products for the prime purpose of plant disease control, an epoch-making substance named blaticidin S was discovered as a good control agent against rice blast disease.

The success of blasticidin S in practical use inspired further research for new pesticides of micrbial origin, leading to the development of other excellent antifungal substances such as kasugamycin, polyoxins, validamycin A, mildiomycin, and so forth.

In addition, microbial products with insecticidal and herbicidal activity have also been found, e.g., polynactins, avermectins, and milbemycins as miticides, and bilanaphos as a herbicide.

Blasticidins S

Name Information

Common name: blasticidin S, CAS RN: [2079-00-7], Development codes: Bc S-BAB, Bc S-3 (S)-4-[[3-amino-5-[(aminoiminomethyl) methylamino]1-oxopentyl]amino]-1-[4-amino-2-oxo-1(2H)-pyrimidinyl]1,2,3,4-tetradeoxy-β-D-erythro-hex-2-enopyranuronic acid; 1-(4-amino-1,2-dihydro-2-oxopyrimidin-1-yl)-4-[(S)3-amino-5-(1-methylguanidino) valerylamino]-1,2,3,4-tetradeoxy-β-D-erythro-hex-2-enopyranuronic acid.

General Chemical/Physical Properties

Nucleoside antibiotic produced by *Streptomyces griseochromogenes. Biosynthesized* from *cytosine, glucose, arginine,* and *methionine.*

Colourless crystals, $C_{17}H_{26}N_8O_5$ (422.4), melting point: 235–236°C, $[\alpha]^{25}_D$ + 108.4 pKa1 2.4 (carboxyl), pKa_2 4.6, pKa_3 8.0, pKa_4 > 12.5 (*guanidino group*).

Solubility: In water, acetic acid > 30 g/L (20°C), Insoluble in *acetone, benzene, chloroform, diethyl* ether, *ethanol, ethyl acetate,* and *methanol.*

Unstable in alkaline solutions at >pH 8.

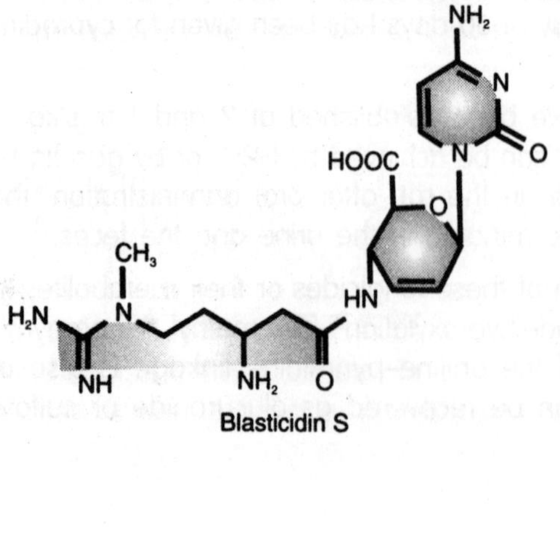

Blasticidin S

Uses

Trade name: Bla S. Control of rice blast caused by *Pyrycularia oryzae* by foliar application. Effective at 5–40 ppm (0.5–4 g of Bla S/10a). Benzylaminobenzene sulfonate of Bla S is least phytotoxic to rice plant.

Phytotoxic damage can be caused to alfalfa, beans, clover, egg plants, potatoes, tobacco, and tomatoes. Inhibits growth of many other fungi and bacteria. Formulation: DP, EC, WP. Mixes with fenitrothione and fenobucarb. Incompatible with alkaline materials.

Mode of Action/Toxicology

Inhibits protein synthesis both in eukaryotes and in prokaryotes. Interacts with ribosomal RNA in large subunit, interfering with the transpeptidation step. Inhibits cell-free protein synthesis in *P. oryzae* and *Escherichia coli*.

Acute oral LD_{50} for male rats: 56.8, female rats: 55.9, male mice: 51.9, and female mice: 60.1 mg/kg. Acute percutaneous LD_{50} for rats >500 mg/kg. Eye: severe irritation. Calcium acetate alleviates the eye problem.

Bacterial reversion tests: non-mutagenic. Toxicity class: World Health Organization (WHO) Ib (a.i.); U. S. Environmental Protection Agency (EPA) II (formulation), European Community (EC) risk T + (R28) Fish LC_{50} (48 h) for carp >40 mg/l. Daphnia LC_{50} (3 h) for Daphnia pulex >40 mg/L.

Metabolism/Transformation-Environment

^{3}H-blasticidin S administered to mice was excreted in the urine and feces within 24 h. Cytomycin and cytosin were identified as the main metabolites in and on rice plants, respectively. In *soil*, DT_{50} < 5 d. Metabolized to nontoxic deaminohydroxy blasticidin S *by Aspergillus sp.* and resistant *Bacillus cereus.* Novel deaminase and coding genes, *BSD* and *bsr*, were isolated as selectable marker genes for genetic engineering.

Cycloheximide

Name Information

Common name: Cycloheximide; Actidione, Naramycin 3-[2-(3,5-Dimethyl-2-oxocyclohexyl)2-hydroxyethyl-] glutarimide, CAS RN: [2163-69-9]

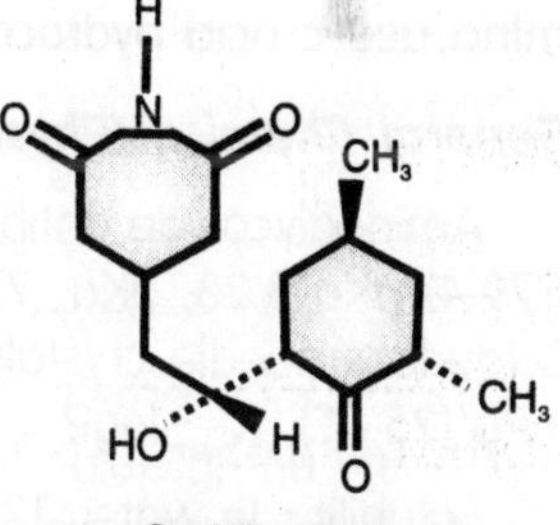

Cycloheximide

General Chemical/Physical Properties

Glutarimide-type antibiotic produced by *Streptomyces griseus.* Colourless crystals, $C_{15}H_{23}NO_4$ (281.4), melting point: 115.5–117 °C, weak acidic substance (pKa 11.2), Soluble in chloroform, isopropanol and methanol; water > 21 g/L (2°C). Stable in pH 3–5, but rapidly destroyed in alkaline solutions.

Uses

Effective for control of downy mildew on onion, shoot blight on larch, powdery mildew on roses and other ornamentals, rusts and leaf spots on lawn grasses, and azalea petal blight. Foliar application controls fungal diseases on turf and ornamentals (9,10). Inhibits growth of many plant pathogenic fungi. Potent rodent repellent. Also has plant growth regulatory properties.

Incompatible with alkaline materials.

Mode of Action/Toxicology

Strongly inhibits the growth of pathogenic fungi but no effects on bacterial growth, even at 100 mg/ml. Inhibits protein synthesis by interfering with the translocation step in eukaryotes, but not in prokaryotes. When ingested by animals, the agent causes excitement, tremors, salivation, diarrhea, and melena.

To remove toxicant from gut, activated charcoal and a catharitic dose of sodium sulfate are effective. Mechanisms of toxicity are not well defined, but hydrocortisone is antidotal, particularly in combination with the adrenergic agent *methoxyphenamine*.

Skin irritant. LD_{50} 2 mg/kg (rat, orl); 133 mg/kg (mice, lpr); 65 mg/kg (guinea pig); 60 mg/kg (monkey). Teratogenic effects.

Metabolism/Transformation-Environment

Rapidly inactivated at room temperature by diluted alkali with the formation of a volatile, fragrant ketone, 2,4-dimethylcyclohexanone. Hazardous to fish and wildlife.

Kasugamycin

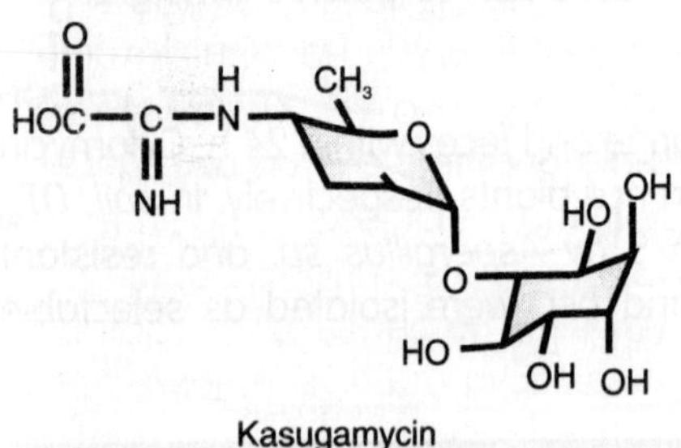

Kasugamycin

Name Information

Common name: kasugamycin, CAS RN: [6980-18-3]

3-O-[2-amino-4-[(carboxyiminomethyl)amino]-2,3,4,6tetradeoxy-α-D-arabino-hexopyranosyl]-D-chiro-inositol hydrochloride hydrate; 1L-1,3,4/2,5,6-1-deoxy-2,3,4,5,6 pentahydroxycyclohexyl 2-amino-2,3,4,6-tetradeoxy-4-(aiminoglycion)-α-D-arabino-hexopyranoside hydrochloride hydrate; [5-amino-2-methyl-6-(2,3,4,5,6-pentahydroxycyclohexyloxy) tetrahydropyran-3-yl] amino-α-imino acetic acid hydrochloride hydrate, CAS RN: [19408-46-9]

General Chemical/Physical Properties

Aminoglycoside antibiotic produced by fermentation of *Streptomyces kasugaensis.* $C_{14}H_{25}N_3O_9$ (379.4) pKa_1 3.23, pKa_2 7.73, pKa_3 11.0. Kasugamycin hydrochloride hydrate: $C_{14}H_{28}ClN_3O_{10}$ (433.8) Colourless needle crystals. M.p.: 202–204°C (decomp). $[c]^{25}_{D}$ + 120° (c 1.6 H_2O) V.p. $< 1.3 \times 10^{-5}$ mPa (25 °C). K_{ow}: logP < 1.

Solubility: In water, 125 g/L (25 °C); in methanol, 2.76; in acetone, xylene < 1 (all in milligrams/kilogram, 25°C). Stable in weak acids, but unstable in strong acids and alkalis. Very stable at room temperature. DT_{50} (50 °C) 47 d (pH 5), 14 d (pH 9).

Uses

Trade names: "Kasugamin," "Kasumin" Patent: JP 42006818; BE 657659: GB 1094566. Systemic fungicide and bactericide with protective and curative action. Controls rice blast caused *by Pyricularia oryzae* and other rice diseases, such as bacterial grain rot, bacterial seedling blight, and bacterial brown stripe caused by *Pseudomonas* spp.. Effective on other plant diseases, e.g., leaf mold and bacterial canker in tomatoes, bean halo blight, scab on apples, *Cercospora* spp. leaf spot on sugar beet, and bacterial soft rot on potatoes.

Nonphytotoxic to rice, tomatoes, sugar beet, potatoes, and other vegetables, but slightly toxic to peas, beans, grapes, citrus, and apples. Formulation: WP, DP, GR, UL, SL. Mixes with Bordeaux mixture; copper oxchloride; phthalide; phthalide + silafluofen; phthalide + silafluofen + tebufenozide; phthalide + validamycin + etofenprox; dichlofenthion + thiram. Incompatible with strongly alkaline pesticides.

Mode of Action/Toxicology

Inhibits hyphal growth of *P. oryzae* in acidic media (pH 5.0) but hardly inhibits it in neutral media. Interferes with binding of fMet–tRNA to the mRNA-30S ribosome complex, thereby preventing formation of initiation complex for protein synthesis.

It does not cause miscoding in protein biosynthesis. Kasugamycin-resistant strains of *P. oryzae* were reported, but the population of the resistant strains rapidly declined when application of kasugamycin was discontinued.

Acute oral LD_{50} for male rats >5 g/kg. Acute percutaneous LD_{50} for rabbits >2 g/kg. Nonirritating to eyes and skin (rabbits). Inhalation LD_{50} (4 h) for rats >2.4 mg/L. NOEL: (2 y) for rats 300, dogs 800 mg/kg diet. Nonmutagenic and nonteratogenin rats, and no effect on reproduction. Acute oral LD_{50} for male Japanese quail >4 g/kg. Fish LC_{50} (48 h) for carp and goldfish >40 mg/L. Daphnia LC_{50} (6 h) >40 mg/L. Bees LD_{50} (contact) >40 g/bee. Toxicity class: WHO (a.i.); EPA (formulation) IV.

Metabolism/Transformation-Environment

Kasugamycin orally administered to rabbits was excreted in the urine within 24 h. After oral administration to rats at 200 mg/kg, no residues were detected in organs or blood. When injected intravenously to dogs, it was mostly excreted within 8 h. In plants and soil, it was degraded to kasugamycinic acid and kasuganobiosamine, and finally to ammonia, oxalic acid, CO_2, and water.

Mildiomycin

Name Information

Common name: *Mildiomycin*; *Milanesin, Mildewmycin*; TF-138

4-amino-1-[4-[(2-amino-3-hydroxy-1-oxopropyl) amino]9-[(aminoiminomethyl)-amino]-6-C-carboxy-2,3,4,7,9-pentadeoxy-α-L-talo-non-2-enopyron-osyl]-5-(hydroxymethyl)2-(1H)-pyrimidinone.

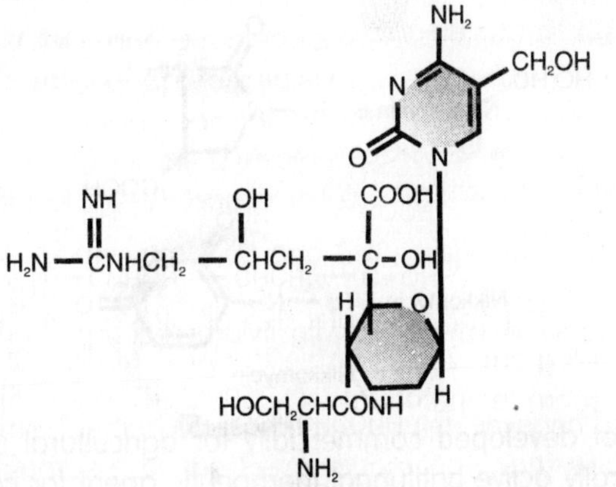

Mildimycin

General Chemical/Physical Properties

A nucleoside antibiotic isolated from the culture filtrate of *Streptoverticillium rimofaciens* B-98891. $C_{19}H_{30}N_8O_9$ (514.5). Basic and hygroscopic. Solubility: In water 580 g/L; in methanol 480; ethanol 160 (mg/L). Insoluble in ethyl acetate and ethyl ether. Susceptible to photodegradation by sunlight.

Uses

Specifically active against the powdery mildew pathogens; excellent curative activity on the disease of various plants *in vivo*. Used in practice for the control of powdery mildews on rose, spindle tree, and Indian lilac.

Mode of Action/Toxicology

Mildiomycin remarkably inhibits protein synthesis in *E. coli.* The synthesis of polypeptides in mammalian cell-free system from rabbit reticulocytes is less sensitive to *mildiomycin* than that from *E. coli.* The toxicity of mildiomycin is low; LD_{50} for acute toxicity in rats and mice is 500–1000 mg/kg by intravenous and subcutaneous injections, and 2.5–5.0 g/kg by oral administration. At a concentration of 1000 ppm, there is no irritation to the cornea and skin of rabbits for 10 days.

Metabolism/Transformation-Environment

Mildiomycin is highly susceptible to photodegradation on plants as well as to microbial degradation in soils. Its toxicity to Japanese killifish (*Oryzias latipes*) is not observed at a concentration of 20 ppm for 7 days.

Nikkomycin

Name Information

Common name: *Nikkomycin; Neopolyoxin*

4-(5-hydroxypyridin-2-yl)-4-hydroxy-3-methyl-2-aminobutyryl-(2-substituted-3,4-dihydroxy-tetrahydrofuran5-yl) glycine.

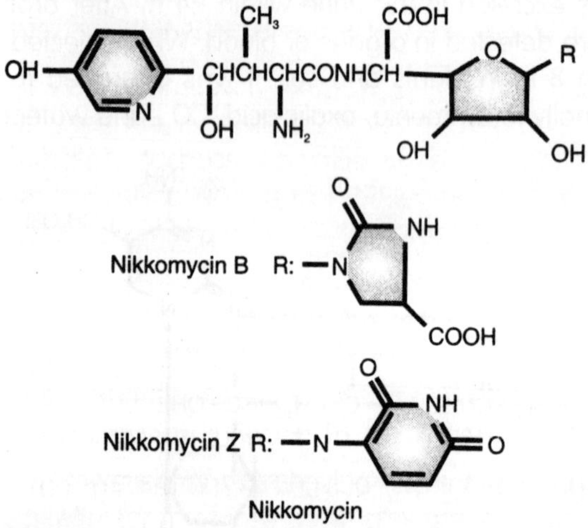

Nikkomycin B R:

Nikkomycin Z R:

Nikkomycin

General Chemical/Physical Properties

Nucleoside-type antibiotic isolated from culture filtrates of *Streptomyces tendae.* Some analogs were also prepared by mutasynthesis, utilizing a uracil *auxotroph* of *S. tendae.*

Nikkomycin Z (Neopolyoxin C): $C_{21}H_{20}N_5O_{10}$ (502.4), melting point: 194–197°C Nikkomycin B: $C_{20}H_{20}N_5O_{11}$ (506.2).

Uses

Nikkomycins inhibit the growth of various plant pathogenic fungi but are inactive against bacteria and yeasts.

Too susceptible to photodegradation by sunlight to be used in field application, and thus not developed commercially for agricultural use. Among the nikkomycins, nikkomycin Z is an orally active antifungal therapeutic agent for coccidioidomycosis. It is a competitive chitin synthase inhibitor that has been evaluated in mouse models for *coccidioidomycosis.*

Mode of Action/Toxicology

By inhibiting chitin synthetase in *fungi, nikkomycins* inhibit cell wall synthesis, ultimately causing fungal cells to swell and burst.

Metabolism/Transformation-Environment

There are no detailed reports on metabolism/transformtion of *nikkomycins* in the environment.

Oxytetracycline

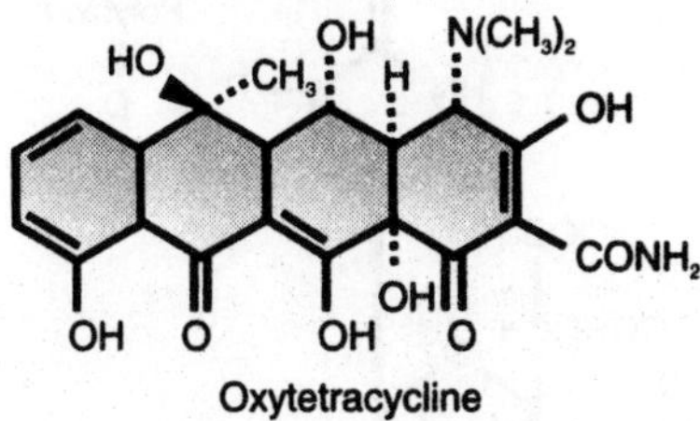

Oxytetracycline

Name Information

Common name: Oxytetracycline; terramycin, terramicin, oxymycin, terramistsin

CAS RN: [79-57-2]

4-(dimethylamino)-1,4,4a,5,5a,6,11,12a-octahydro-3,5,

6,10,12,12a-hexahydroxy-6-mehyl-1,11-dioxo-2-naphthacenecarboxamide; (4S, 4aR, 5S, 5aR, 6S, 12aS)-4-dimthylamino-1,4,4a,5,5a,6,11,12a-octahydro-3,5,6,10,12,12hexahydroxy-6-methyl-1,11-dioxonaphthacene-2-carboxamide.

General Chemical/Physical Properties

Isolated from culture filtrate of *Streptomyces rimoseus.* M.f. $C_{22}H_{24}N_2O_9$ (461.5), m.p. 184.5~185.5 °C, solubility: In water 250 mg/L; more soluble in acidic water. Stable in acidic solution but not in alkaline solution.

Uses

Antibacterial antibiotic used against some plant pathogens; also used as veterinary and human medicine. *Tetracyclines* are active against a wide range of *microorganisms* and effective in controlling some plant bacterial diseases caused by *Pseudomonas* spp. (Cucumber bacterial spot), *Xanthomonas* spp. (peach bacterial shot hole), and *Erwinia* spp.; protects against diseases caused by mycoplasma-like organisms.

In practice, it has been used to control bacterial diseases by mixing with streptomycin. The mixture occasionally causes chlorosis on the leaves of some crops when sprayed under high temperature and humidity.

Mode of Action/Toxicology

Tetracycline is a potent inhibitor of bacterial protein biosynthesis, with less activity on mammalian cells. It binds to the 30S and 50S bacterial ribosomal subunits, and it inhibits the binding of aminoacyl–tRNA and the termination factors RF1 and RF2 to the A site of bacterial ribosomes.

Acute oral LD_{50} for mice >7 g/kg; for rats >10 g/kg, acute intravenous 100 ~ 200 mg/kg. Tlm for black bass: 250 ppm (24 h).

Metabolism/Transformation-Environment

Oxytetracycline is easily taken up by plant leaves, especially through stomata, and rapidly translocated to plant tissues.

Polyoxins

Name Information

Common name: polyoxins CAS RN: [11113-80-7]. *Polyoxins* comprise 14 components (A–N); they are fungicidal except polyoxin C. B- and D-rich complexes are used in practice.

Polyoxin B: 5-[[2-amino-5-O-(aminocarbonyl)-2-deoxy-L-xylonoyl]amino]-1,5-dideoxy-1-[3,4-

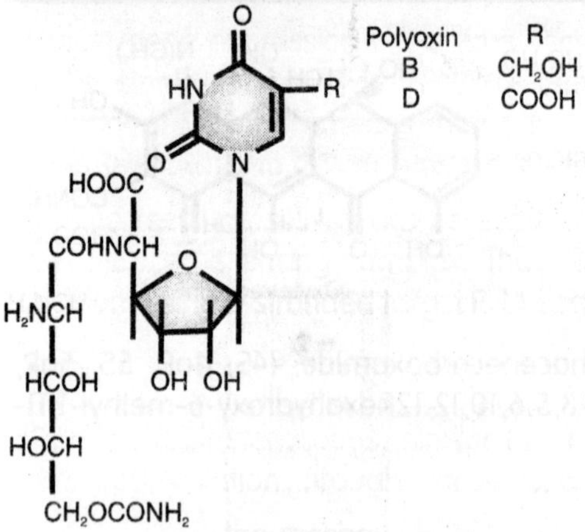

Polyoxin	R
B	CH_2OH
D	COOH

Polyoxins

dihydro-5-(hydroxy-methyl)-2,4-dioxo-1(2H)-pyrimidinyl]-β-D-allofuranuronic acid; 5-(2-amino-5-O-carbamoyl-2-deoxy-L-xylonamido)-5-dideoxy-1-(1,2,3,4-tetrahydro-5-hydroxymethyl-2,4-dioxopyrimidin-1-yl)-,6-D-allofuranuronic acid CAS RN: [19396-06-6] Polyoxin D; polyoxorim: 5-[[2-amino-5-O(aminocarbonyl)-2-deoxy-L-xylonoyl]amino]-1-(5-carboxy-β-4-dihydro-2,4-dioxo-1-(2H)-pyrimidinyl)-1,5-dideoxy-β-D-allofuranuronic acid; 5-(2-amino-5-O-carbamoyl-2-dioxy-L-xylonamido)-1-(5-carboxy-1,2,3,4-tetrahydro-2,4-dioxopyrimidin-1-yl)-1,5-dideoxy-β-D-allofuranuronic acid CAS RN: [22976-86-9]; zinc salt: [146659-78-1].

General Chemical/Physical Properties

Polyoxins are produced by fermentation of Streptomyces *cacaoi* var. *asoensis*. Polyoxin B and D isolated and their structure elucidated by K. Isono et al.. Polyoxin B complex: Consists of component B and several other polyoxins. Amorphous powder. M.f. $C_{17}H_{25}N_5O_{13}$ (507.4) M.p. > 160°C (decomp). Solubility: In water 1 kg/L (20°C); in acetone, methanol and common organic solvents <100 mg/L. Stability: Hygroscopic and thus should be stored under dry conditions. Stable between pH 1 and pH 8.

Specific rotation $[\alpha]^{20}_D$ + 34° (c = 1, water) pKa: pKal (carboxyl) 3.0, pKa_2 (amino) 6.9, pKa_3 (uracil) 9.4. Polyoxin D: M.f. $C_{17}H_{23}N_5O_{14}$ (521.4) Colourless crystals. M.p. > 190°C (decomp.) Solubility: In water <200 mg/L (20°C) (zinc salt); in acetone and methanol <200 mg/L (zinc salt). Stability: Hygroscopic; store under dry conditions. Specific rotation $[\alpha]^{20}_D$ + 30° (c = 1, water) pKa:pKa_1 (carboxyl) 2.6, pKa_2 (carboxyl) 3.7, pKa_3 (amino) 7.3, pKa_4 (uracil) 9.4.

Uses

The fungicidal activities of polyoxins reported on many plant pathogens, including *Alternaria* spp. and *Rhizoctonia zoctonia* spp..

Polyoxin B is used for control of Alternaria spp. and powdery mildews in apples and pears; *Botrytis cinerea* in vines and aubergines; powdery mildews in roses, chrysanthemums, and melons; blight of carnation; powdery mildew, brown spot, and gray mold in tobacco; powdery mildew and gray mold in strawberries; leaf mold, early blight, and gray mold in tomatoes; powdery mildew, gray mold, *Sclerotinia erotinia* spp. rot, and *Corynespora melonis* in cucumbers; *Alternaria* spp. blight in carrots; purple blotch in leeks. Patent: JP 493008 Formulation types: WP; EC; SG. Mixes with captan; oxine-copper; iminoctadine triacetate. Compatibility: Incompatible with alkaline materials. Tradename "Polyoxin AL."

Polyoxin D zinc salt is used for control of sheath blight (*Rhizoctonia solani*) in rice; canker in apples and pears; *Rhizoctonia solani, Drechslera spp., Bipolaris spp., Curvularia* spp., and *Helminthosporium spp.* in lawn turf. Formulation types: WP; PA. Mixes with thiram; propiconazole. Compatibility: Incompatible with alkaline materials. Tradenames "Kakengel" (zinc salt"Polyoxin Z" (zinc salt); "Stopit."

Application of polyoxins can be made at any growth stage of rice plants without causing phytotoxicity even at 800-ppm application. Foliar sprays of 200 ppm of polyoxins have produced no phytotoxicity on all the other crops tested.

Mode of Action/Toxicology

Polyoxin B causes a marked abnormal swelling on germ tubes of spores and hyphal tips in *Alternaria spp.*, and this makes the pathogen noninfectious. In a cell-free system of *Neurospora crassa,* polyoxin D inhibits the incorporation of N-acetyl-glucosamine (GlcNAc) into chitin in a competitive manner between UDP-GlcNAc and pollyoxin D; the pyrimidine nucleoside moiety of the antibiotics was shown to fit into the binding site of the enzyme protein, and the carbamoylpolyoxamic acid moiety of polyoxins stabilizes the polyoxin-enzyme complex.

Polyoxin-resistant strains of *A. alternata* were recognized in pear orchards after several years of intensive use of the antibiotics. The resistance was suggested to be caused by a lowered permeability of the antibiotic through the cell membrane into the site of chitin synthesis.

Acute oral LD_{50} of polyoxin B for male rats 21, female rats 21.2, male mice 27.3, and female mice 22.5 g/kg. Acute percutaneous LD_{50} for rats >2 g/kg. Nonirritant to mucous membranes and skin (rats). Inhalation LD_{50} (6 h) for rats 10 mg/L air. NOEL (2 y) for rats and mice >48-g/kg diet. Toxicity class EPA (formulation) IV.

Acute oral LD_{50} of polyoxin D for male rats and female rats >9.6 g/kg. Acute percutaneous LD_{50} for rats >750 mg/kg. Inhalation LD_{50} (4 h) for male rats 2.44, female rats 2.17 mg/L air. NOEL (2 y) for rats >50, mice >40 g/kg diet. Toxicity class EPA (formulation) III (WP).

Metabolism/Transformation-Environment

Product and residue analysis by bioassay using *Alternaria mali* ACI-1157 (for polyoxin B) and *Rhizoctonia solani* (= *Pelliculariasasakii*) ACI-1134 (for polyoxin D).

Polyoxin B: Fish LD_{50} (48 h) for carp >40 mg/L. Japanese killifish unaffected by 100 mg/L for 72 h. Daphnia LD_{50} (3 h) for *D. pulex* >40 mg/l. Other aquatic spp. LD_{50} (3 h) for *Moina macrocopa* >40 mg/l. *Soil/Environment*: In upland conditions at 25 °C, DT_{50} < 2 d (two soils, o.c. 6.2%, pH 6.3, moisture 23.3% and o.c. 1.1%, pH 6.8, moisture 63.6% respectively).

Polyoxin D: Fish LD_{50} (48 h) for carp >40 mg/L. Daphnia LD_{50} (3 h) for *D. pulex* >40 mg/L. Other aquatic spp. LD_{50} (3 h) for *Moina macrocopa* >40 mg/L. Soil/Environment: In flooded soil at 25°C, DT_{50} < 10 d (two soil types, o.c. 2.5%, pH 6.0, and o.c. 9.6%, pH 6.0 respectively).

In upland conditions at 25°C, DT_{50} < 7 d (two soil types, o.c. 0.6%, pH 6.4, moisture 10.7% and o.c. 6.2%, pH 6.3, moisture 61.9% respectively). In water DT_{50} 4 h (pH 5.5, 24°C), 8 h (pH 5.8, 26.5°C).

Streptomycin

Name Information

Common name: streptomycin; streptomycine O-2-deoxy-2-(methylamino)-α-L-glucopyranosyl-(1→2)-O-5-deoxy-3-C-formyl-α-L-lyxofuranosyl-(1→4)-N,Nbis(aminoiminomethyl)-D-streptamine; O-2-deoxy-2-metylamino-α-L-glucopyranosyl-(1→2)-O-5-deoxy-3-C-formyl-α-L-lyxofuranosyl-(1→4)-N3, N3-diamidino-D-strepamine; 1,1'-[1-L-(1,3,5/2,4,6)-4-[5-deoxy-2-O-(2-deoxy-methylamino-α-L-

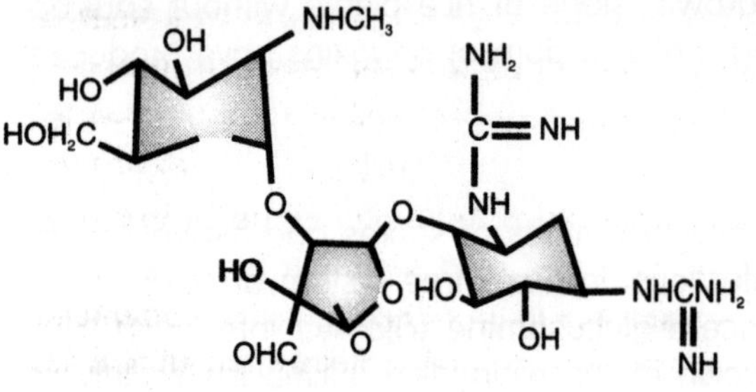

Streptomycin

glucopyranosyl)-3-C-formyl-α-L-lyxofuranosyloxy]-2,5,6-trihydroxycyclohex-1,3-ylene] diguandine. CAS RN: [57-92-1]

Streptomycin sesquisulfate CAS RN: [3810-74-0]

General Chemical/Physical Properties

Streptomycin is obtained by fermentatation of *Streptomyces griseus,* isolated as sesquisulfate.

Streptomycin: M.f. $C_{21}H_{39}N_7O_{12}$ (581.6) Stability: Unstable in strong acids and alkalis. Streptomycin sesquisulfate: M.f. $C_{42}H_{84}N_{14}O_{36}S_3$ (1457.3) Off-white powder. Solubility: In water >20 g/L (pH 7, 28 °C). In ethanol 0.9, methanol >20, petroleum ether 0.02 (all in g/L). Stability: hygroscopic. Specific rotation $[a]^{25}_{D}$ -84.

Uses

Control of bacterial shot-hole, bacterial rots, bacterial canker, bacterial wilts, fire blight, and other *diseas* caused by gram-positive species of bacteria in pome fruit, stone fruit, citrus fruit, olives, vegetables, potatoes, tobacco, cotton, and ornamentals. Chlorosis may occur on grapes, pears, peaches, and some ornamentals.

Formulation types WP; Liquid. Incompatible with pyrethrins and alkaline materials. A mixture of streptomycin and oxytetracycline is highly effective for the control of bactrial canker of peach, citrus canker, soft rot of vegetables, and various other bacterial diseases. Selected tradenam"Agrimycin 17" (sesquisulfate); "AS-50" (sesquisulfate).

Mode of Action/Toxicology

Streptomycin inhibits protein synthesis in bacterial cells by binding to the 30S ribosomal subunit and causes misreading of the genetic codes in protein synthesis.

Streptomycin-resistant strains are distributed in a wide range of plant pathogenic bacteria, such as *Xanthomonas oryzae, X. citri, Pseudomonas tabaci,* and *P.* lachrymans.

In agricultural use, the alternative or combined applictions of streptomycin and other chemicals with different action mechanisms is recommended in order to reduce the development of streptomycin-resistant strains in the field. Mutants of *E. coli* highly resistant to streptomycin are known to involve modification of the P10 protein of the bacterial ribosome 30S subunit.

Streptomycin: Acute oral LD_{50} for mice >10 g/kg. Acute percutaneous LD_{50} for male mice 400, female mice 325 mg/kg. May cause allergic skin reaction. NOEL: 125 mg/kg. Acute *i.p.* LD_{50} for male mice 340, female mice 305 mg/kg. Streptomycin sesquisulfate: Acute oral LD_{50} for rats 9, mice 9, hamsters 0.4 mg/kg.

Metabolism/Transformation-Environment

Streptomycin occasionally causes chemical injuries to vegetables and rice if it is applied at high concentrations. A mixture of streptomycin sulfate and iron chloride or citrate is effective to reduce the phytotoxicity of the antibiotic.

Validamycin A

Name Information

Common name: *validamycin; validamycin A*

[1S-(1α, 4α, 5β, 6α)]-1,5,6-trideoxy-4-O-β-D-glucopyranosyl-5-(hydroxymethyl)-1-[[4,5,6-trihydroxy-3-(hydroxymethyl)-2-cyclohexen-1-yl]amino]-D-chiro-inositol : 1L-(1, 3,4/2,6)-2,3-dihydroxy-6-hydroxymethyl-4-[(1S, 4R, 5S, 6S)-4,5,6-trihydroxy-3-hydroxymethylcyclohex-2-enyl-amino] cyclohexyl 6-D-glucopyranoside CAS RN: [37248-47-8].

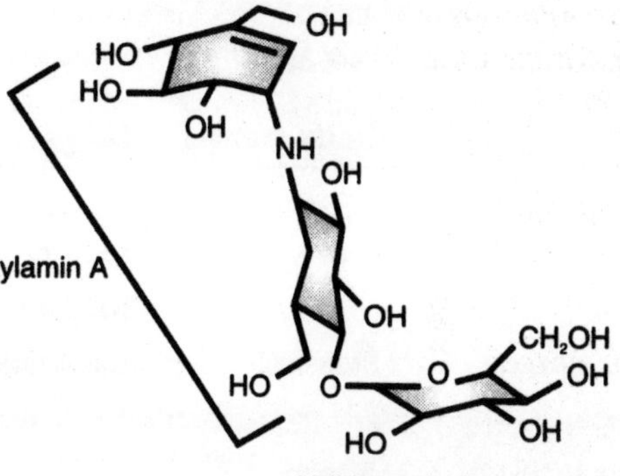

General Chemical/Physical Properties

Produced by the fermentation of *Streptomyces hygroscopicus* var. *limoneus* nov. var. Structure revised. M.f. $C_{20}H_{35}NO_{13}$ (497.5) Colourless, odorless, hygroscopic powder. M.p. 130–135°C (decomp). V.p. Negligible at room temperature. Solubility: Readily soluble in water, soluble in methanol, dimethylformamide and dimethyl sulfoxide.

Slightly soluble in ethanol and acetone. Sparingly soluble in diethyl ether and ethyl acetate. pKa 6.0 Stability $[a]^{24}_D$ + 110° (water).

Uses

Control of *Rhizoctonia solani* in rice, potatoes, vegetables, strawberries, tobacco, ginger, and other crops; dampingoff *diseases* of cotton, rice, and sugar beet. Applied as a foliar spray, soil drench, seed dressing, or by soil incorporation. No phytotoxicity was observed for over 150 species of plants sprayed with validamycin A even at a concentration of 1000 ppm. Validamycin A has been used to protect rice sheath blight, mainly in formulations of 3–5% solution or 0.3% dust. Formulation types DP, SL, DS, Liquid. Mixtures: fenobucarb, phthalide. Tradenames: "Validacin," "Mycin," "Solacol."

Mode of Action/Toxicology

Validamycin A specifically inhibits trehalase in *R. solani* AG-1 in a competitive manner between validoxylamine A (the possible active form of validamycin A) and the substrate, trehalose. Because trehalose is a storage carbohydrate in some fungi, trehalase is suggested to play an essential role for the digestion of trehalose to D-glucose and for its transportation to the hyphal tips.

Acute oral LD_{50} for rats and mice >20 g/kg. Acute percutaneous LD_{50} for rats >5 g/kg. Nonirritating to skin (rabbits). Not a skin sensitiser (guinea pigs). Inhalation LC_{50} (*4 h*) for rats >5 mg/L air. NOEL: In 90-d feeding trials, rats receiving 1 g/kg of diet and mice receiving 2 g/kg of diet showed no ill-effects. In 2-y feeding trials, NOEL for rats was 40.4 mg/kg daily. Toxicity Class WHO (a.i.) III; EPA (formulation) IV.

Metabolism/Transformation-Environment

Product and residue analysis by gas chromatography of derivatives. Residues in rice grains

Table 13.7: Aromatic Hydrocarbon Fungicide Compounds, Trade Names, Properties, and Origins.

Compounds	Trade Names	Usage	Acute Oral Doses mg/kg (Rats) LD_{50}	Vapor Pressure mm Hg	Patent	Introduction
Hexachlor--obenzene		Seed dressing	10,000	1.09×10^{-5}		1945
		Soil fungicide				
Quintozene	Brassicol	Soil fungicide	12,000	13.3×10^{-3}	IG Farben AG	1930
(Pentachloro-nitrobenzene)	Tritisan, Folosan	Seed dressing			(DRP 682048)	
Tecnazene	Fosolan	Soil fungicide	57	Volatile	Bayer AG	1946
(Tetrachloro-nitrobenzene TCNB)	Fusarex				USP 2615801	
Trichlorodi-nitrobenzene	Olpisan	Soil fungicide (*Plasmodiophora brassicae*, a. o. fungi)		Moderate Volatile		1950
Trichlorotri-nitrobenzene	Phomasan	Soil fungicide		Moderate		1953
Volatile Chloroneb	Demosan	Soil fungicide	11,000	3×10^{-5}	Du Pont de Nemour & Co. Inc. USP 3265564	1967
Dicloran	Allisan, Bortran	Fruit storage diseases, ornamental *Botrytis* and *Sclerotinia*, Rhizopus diseases	1,500–4,000	1.2×10^{-6}	Boots Co. Ltd (BP 845916)	1930
2-Phenylp--henol (OPP)	Dowicide	Fruit storage diseases		2,480	Moderate	1936
	Nectryl	Disinfectant		Volatile		
Biphenyl		Citrus storage	3,280	Volatile		1944

and straws were less than the detectable limit by glc.

Fish LC_{50} (*72* h) for carp >40 mg/L. Daphnia LC_{50} (24 h) for *D. pulex* >40 mg/L.

In animals, cleavage to glucose and an amine residue occurs. Soil/environment: Rapid microbial degradation in *soil;* $DT_{50} \leq 5$ h.

GENERAL COMMENTS

Antibiotics have an economic advantage over synthetic chemicals in that a variety of substances can be manufactured using one set of equipment and facilities. In addition, they are produced not from limited fossil resources, but from renewable agricultural products through fermentation by microorganisms. A negative is that, because of their highly specific mechanism of action, they are apt to suffer from the emergence of resistant strains.

Thus, it is sometimes necessary to combine the agents with other chemicals that have different mechanisms of action or to use them in rotation. Another limitation to the use of microbial products in agriculture would be a concern that their wide use might create resistant strains that may hinder medical treatment in humans. Fortunately, the use of agricultural antibiotics has not yet met any problem involving cross-resistance with medicinals.

Naturally, further precaution must be taken not to cause any resistance to human pathogens in the future. This approach, however, should not only be limited to pesticides of microbial origin; the important point is whether any pesticides may induce resistance to medicinals. Among the antibiotics used in practice, polyoxins and validamycins are safe fungicides; non-phytotoxic, and non-toxic to humans, livestock, and wildlife.

Such excellent characteristics seem to be due to their modes of action. Polyoxins selectively inhibit the synthesis of fungal cell-wall chitin, which does not exist in mammalian cells, and validamycin A is not fungicidal, but only disrupts the normal mycelial growth of pathogens in plants.

Probably because of such moderate activity, no occurrence of resistant strains has been reported in validamycin treatments, despite the large use of validamycin A in agriculture. Molecular biology is making rapid advances, and more efficient production of microbial products will become possible by modern gene engineering. Application of molecular biology to agricultural antibiotics often affords a unique feature for plant biotechnology. Biorational approaches will also become feasible in the design of new pesticides using microbial product templates as leads.

AROMATIC HYDROCARBONS

Aromatic hydrocarbon fungicides (AHF) represent an older and chemically heterogeneous group of fungicides consisting of an aromatic ring system, substituted by chloro-, nitro-, methoxy-, phenyl-, or amino-groups in various positions. They have been in use for a long time, but have lost their importance. Some of them are still valuable compounds in tropical countries and in use for controlling storage diseases.

Many of them could be replaced by newer compounds with a broader antifungal spectrum, better physical properties, and higher activity. Although chemically different, they are connected

Table 13.8: Antifungal Spectrum of Some Aromatic Hydrocarbon Fungicides (ED_{50} Values in mg/l for Inhibition of Radial Mycelial Growth on Malt Agar Medium).

Fungus	Chloroneb-	Dicloran	Quintozene	Diphenyl	Tolclofosmethyl	Etridiazole
Phytophthora cactorum	2	200	200	100	100	0.4
Pythium ultimum	1	200	150	100	100	0.3
Mucor mucedo	2	0.5	5	30	100	8
Botrytis cinerea	3	1	0.5	30	1	20
Sclerotinia sclerotiorum	3	2	5	30	1.4	10
Penicillium chrysogenum	—	0.5	20	100	—	90
Penicillium italicum	2	0.4	—	1	1.6	—
Aspergillus niger	—	100	200	100	—	100
Fusarium oxysporum	500	200	500	100	100	70
Ophiobolus graminearum	70	—	30	100	1.4	3
Pseudocercosp. herpotrich.	200	10	60	100	—	100
Rhizoctonia solani	15	200	200	30	0.1	70
Verticill albo-atrum	10	8	20	—	100	40
Schizophyllum commune	3	100	10	100	—	40

by a fungal cross-resistance based probably on a very similar mechanism of action. Etridiazole and especially tolclofosmethyl are somehow related to this group but will be dealt with in other chapters. Some properties of AHF are summarized in Table elsewhere in this chapter. Because of their high volatility, many compounds were used only as soil fungicides.

Table elsewhere in this chapter demonstrates that these fungicides exhibit unusual selective effects against some important fungi but are inactive for other ones. This specificity can be a disadvantage for their practical use; therefore, they were often mixed with other fungicides.

The value of this group lay in the low costs of synthes and the low mammalian toxicity (if no toxic impurities from the technical synthesis are present).

The detection of a fungal cross resistance with the structurally unrelated dicarboximides was very surprising. This points to a common mechanism of action. The biological conversion of some of these fungcides is summarized by Kaars-Sijpesteijn et al.. Some members of this group will be briefly characterized here. Further information can be found in the article by Lyr.

HEXACHLOROBENZENE (HCB)

HCB was an effective fungicide of the halogenated benzene class. It was discovered in France and exhibits a vapor-phase activity against some important soil-borne fungi and seed-borne diseases. HCB controls common bunt (*Tilletia foetida* and *T. caries*) and dwarf bunt of wheat (*T.*

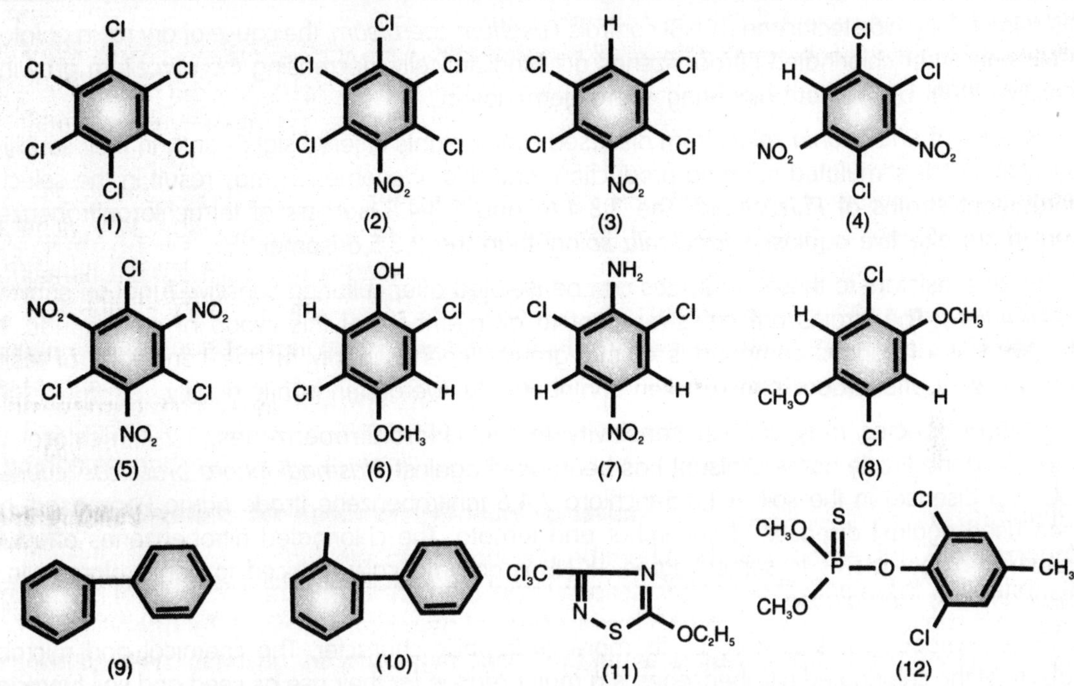

Figure 13.4: Structure formulae for compounds of the group of aromatic hydrocarbon fungicides. (1) Hexachlorobenzene; (2) Pentachloronitrobenzene (Quintozene); (3) Tetrachloronitrobenzene (Tectacene); (4) 1,2,4-trichloro-3,5-dinitrobenzene (Olpisan); (5) 1,3,5-trichloro-2,4,6-trinitrobenzene (Phomasan); (6) 2,4-Dichloro-3-methoxy-phenol (DCMP); (7) Dicloran; (8) Chloroneb; (9) Diphenyl; (10) o-Phenylphenol; (11) Etridiazole; and (12) Tolclophosmethyl.

controversa). It is also effective against the seed-borne fungus *Urocystis agropyri* (initiant of flag smut) but is inactive against other *Ustilago* spp. HCB had been used for more than 20 years as seed dressing and soil fungicide because of its low mammalian toxicity and low price.

In Greece it was used since 1958 as a seed dressing agent as an alternative to organic mercury compounds. However, in 1973 a marked decrease in effectiveness against *Tilletia* spp. was observed. It was reported that races of *T. foetida* were cross-resistant to HCB, PCNB, imazalil, and hydantoin.

Similar observations were made in Australia. The use of HCB as a seed dressing has ceased. It was replaced by other fungcides and fungicide combinations with broader spectra of activity against seed-borne pathogens.

CHLORINATED NITROBENZENES

The most widely used compound in this group is *pentachloronitrobenzene* (PCNB or quintozene). Quintozene exhibits an antifungal spectrum that is characteristic of this group, combined with low toxicity for mammals. It was extensively used as a soil fungicide to control diseases caused by *Rhizoctonia solani, Botrytis spp., Sclerotinia* spp., and *Sclerotium rolfsii.*

However, quintozene is completely ineffective against *Phytium, Phytophthora,* and Fusarium.

The related fungicide tectacene (TCNB) controls *Fusarium coeruleum*, the cause of dry rot in potatoes. PCNB and other chlorinated nitrobenzenes act fungistatically, decreasing the mycelium growth of sensitive fungi but without hindering spore germination.

Isolates of *Rhizoctonia solani* from diseased cotton plants differed significantly in their sensitivity to PCNB. PCNB stimulated sclerotia production, and this phenomenon may result in the selection of resistant strains of *Rhizoctonia*. The 2,3,4,6- and 2,3,4,5-isomers of tetrachloronitrobenzene were more effective against *Rhizoctonia solani* than the 2,3,5,6-isomer.

Strains resistant to these fungicides can be isolated after culturing sensitive fungi on sublethal concentrtions. The strains are cross-resistant to all members of this group of compounds. The effectiveness of various compounds of this group depends partly on the conditions of testing because very often vapor phase action contributes to overall fungistatic activity.

Fungal species may vary in sensitivity to the chloronitrobenzenes. 1,2,4-Trichloro-3,5-dinitrobenzene (trade name Olpisan) has been used against *Plasmodiophora brassicae* (cause of club root disease) in the soil. A 1,3,5-trichloro 2,4,6-trinitrobenzene (trade name Phomasan) had been used against diseases of cucumber and tomato. The chlorinated nitrobenzenes affect the growth of higher plants to some degree. PCNB is enzymatically reduced to pentachloroanilin by *Pseudomonas aeruginosa*.

This compound is less toxic by its more hydrophilic character. The chemical and microbial stability of the chlorinated nitrobenzenes is a major reason for their use as seed and soil fungicides over several decades. Also, the acute mammalian toxicity is quite low, but long-term effects have been noted in recent studies. Development of resistance of practical importance in soil fungi seems to require a longer time period compared with leaf pathogens, which have a higher sporulation capacity.

Therefore, the problems of resistance have never become a widespread problem. But the moderate activity of the chloronitrobenzenes against other important soil fungi has limited their practical use [further literature is cited by Corden]; very often combinations of PCNB with carboxin, carbendazim, mancozeb, captan, captafolacalixin, or thiram are used for seed dressing or as soil drench to control a broader spectrum of fungi, such as *Pythium, Rhizoctonia, Thielaviopsis, Diplodia*, and *Sclerotium*.

CHLORONEB

Chloroneb was developed by Du Pont de Nemours Co in 1967. It differs from the compounds described above in that it does not contain nitro-groups. In common with other chlorobenzene fungicides, chloroneb has a low mammalian toxicity and a significant vapor pressure so that it was used as a soil fungicide in the culture of beans, cucumber, and cotton.

In contrast with the nitrobenzene compounds, *Phytophthora* spp. are rather sensitive to chloroneb. *Pythium* spp. vary in sensitivity to chloroneb. Because of its low water solubility, chloroneb is only very weakly systemic.

The main targets of chloroneb were *Rhizoctonia solani, Pythium spp., Ustilago maydis*, and *Typhula spp.* and soil inhabiting *Phytophthora* spp. Chloroneb is not effective against *Fusarium* but has a relatively broad spectrum of activity compared with other compounds that are specifically active against oomycetes. By controlling *Rhizoctonia solani*, by seed-piece or in furrow applications, chloroneb increased potato yields in Texas. Like other fungicides of this group, resistant strains can be observed easily in the laboratory, and differences in sensitivity seem to exist also in the natural population.

DICLORAN

Dicloran (Botran) was discovered in 1959 by Boots Co. Ltd. Because of its high activity against *Botrytis, Mucor,* and *Rhizopus,* some *Penicillium spp., Monilia fructicola* and its low mammalian toxicity, dicloran is used against fruit and vegetable storage diseases, especially in peaches, sweet cherries, grapes, tomatoes, lettuce, and cabbage.

Table elsewhere in this chapter shows that dicloran has a selective spectrum of antifungal activity similar to that of PCNB. Both fungicides are highly active against *Mucor, Rhizopus,* and *Botrytis* but not active against *Phytophthora, Pythium* and *Fusarium.* Dicloran may be applied as spray or as fungicide-wax formulation.

It has been used as a seed treatment against *Sclerotium cepivorum* on onions, *Sclerotinia sclerotiorum* on beans, or on sunflowers. The chemical stability and low volatility of dicloran results in a prolonged disease control on leaves, fruits, and in the soil.

Similar to other compounds of this group, dicloran does not inhibit germination of spores but inhibits mycelium growth. Distortion and bursting of the germ tubes have been described.

Resistant isolates were obtained that are cross-resistant to other chloronitrobenzene fungicides and to dicarboximide fungicides but not to benzimidazololor triadimefon-fungicides. Some properties of dicloran- resistant isolates of *Monilinia fructicola* were similar to dicarboximide-resistant isolates. In cross-resistant isolates, mycelial growth on agar medium was inhibited more by dicloran than by the dicarboximides. For some practical applications, dicloran is replaced by the more effective dicarboximide fungicides, but cross-resistance remains a practical problem.

Application of dicloran reduces storage decay and sugar losses in sugar beets, when applied alone or in combination with carbendazim.

A combination of dicloran and imazalil effectively controlled *Penicillium sclerotigenum* and *Rhizopus spp. in* yellow yam.

A combination of iprodione and dicloran protected nectarines effectively against *Monilia fructicola* and *Rhizopus nigricans.*

O-PHENYLPHENOL (OPP)

This compound is unique among the AHF because it bears a free phenol group, but it has retained a limited use in plant protection. Its main application is the protection of stored fruits, especially citrus, and the disinfection of storage material. By vapor action it can protect packed fruits against decay by *Penicillium italicum* and *P. digitatum, Diplodia natalensis, Botrytis cinerea,* and other fungal species.

OPP is more selective than other free phenols but does produce phytotoxic effects. Sodium o-phenylphenate (SOPP) is used in practice because it is much less phytotoxic to fruits and has a greater water solubility. Tomkins tried to diminish the phytotoxic effect by esterification of the phenolic group. Acetate and butyrate esters gave efficient control of *Botrytis* on grapes and *Monilia* on peaches, but esters of o-phenylphenol are easily hydrolyzed into the free phenol.

The methyl ether was less active against *Diplodia* and virtually inactive against *Penicillia.* Eckert demonstrated that *Diplodia* can demethylate about 10% of the anisole, whereas *Penicillium digitatum* does not metabolize this compound. However, it is not likely that this small difference can account for the differences in the growth inhibition of these two fungi.

Apparently, biphenyl anisole, which is rapidly accumulated, is itself active against *Diplodia* but not against *Penicillium.*

This strongly resembles the situation with chloroneb, where the first demethylation product [DCMP = dichloromethoxyphenol] has the same spectrum of activity as chloroneb itself and does partly behave as a phenol.

Cross-resistance of SOPP and diphenyl against Penicillia became of practical importance in California in lemon packing houses after continuous use. 2-phenylphenol gave good protection of kiwifruits$_s$ against *Botrytis cinera* applied either by immersion or thermonebulization.

A reregistration for postharvest use was made for OPP and SOPP by EPA but only for pears and citrus fruits because of the high costs and extensive data requirements.

DIPHENYL (BIPHENYL)

Diphenyl is a vapor-phase fungistat that is used to control postharvest diseases in citrus caused by *Penicillium digitatum* and *P. italicum* inhibiting mycelial growth and sporulation. It is impregnated into paper sheets that are added to storage and transportation cartons.

Citrus fruits absorb this very stable compound in proportion to the vapor concentration in the surrounding atmosphere, which is related to storage temperature and storage time. The official tolerance limit for this compound on citrus fruits is 110 ppm (mg/kg) in the United States and 70 ppm in Europe and Japan..

Diphenyl, one of the weakest fungicides of this group because it is highly hydrophobic, is still a part of the protection strategy of citrus shippers.

MECHANISM OF ACTION OF AROMATIC HYDROCARBON FUNGICIDES

Several investigators have attempted to clarify the mechanism of action of this rather old group of selective fungicides. Although many effects at very different sites of the metabolism have been reported, no clear picture of the primary mechanism of action of these fungicides had emerged. Some of these effects are summarized by Lyr.

All authors agree that there should be a common mechanism of action in spite of the different chemical structures because fungi that developed resistance for one member of this group of fungicides are resistant to other members as well and also to dicarboximide fungicides, a more recently discovered group of fungicides, which share with AHF a 3,5-dichloro-benzene ring system.

RESULTS ON THE MECHANISM OF ACTION

With regard to the mechanism of action, the most thoroughly investigated compounds are chloroneb, quintozene (PCNB), and etridiazole, but there is at present no doubt that the other fungicides of this group act by a similar or identical mechanism. The reason for the different antifungal spectra is another problem that is not yet elucidated. The mechanism of action of chloroneb has been analyzed in detail in *Mucor mucedo*.

The ultrastructural effects of chloroneb are nearly identical with those of PCNB, which has a similar molecular space configuration compared with chloroneb. Therefore, they should be characterized together. Both compounds induce in *Mucor mucedo* a lysis of the inner mitochondrial membranes, beginning with a swelling of the cristae. The nuclear envelope vacuolizes, and the cell wall thickness increases dramatically within a few hours.

This stops the hyphal tip growth (radial growth) and, thus hinders the spreading of the fungus and a pathogenic attack. Simultaneously, the respiration decreases, but it is not specifically inhibited because the respiratory quotient of mitochondria remains nearly constant. Total dry matter accumulation also decreases. Chloroneb binds to mitochondrial proteins of sensitive, but not of resistant, strains in *Mucor.*

The latter do not show changes under the influence of chloroneb in the ultrastructure compared with the control. Comparing the amino acid composition of the proteins of isolated mitochondria of R- and S-strains of *Mucor mucedo,* Werner found nearly an equal percentage of all amino acids except for tyrosine, which was only 0.2% in resistant strains compared with 2.9% in S-strains.

This means that a tyrosine-rich protein has decreased in content or its composition has changed. It may be that a tyrosine molecule located in the active center of an enzyme has something to do with the binding capacity for chloroneb or PCNB. *Neurospora crassa,* a moderately sensitive species, binds significantly more chloroneb in its mitochondria than the insensitive *N. sitophila.* The tyrosine content of mitochondrial proteins of the former species is higher than in the latter.

Several biochemical investigations have established that tyrosine and tryptophane play an important role in the binding of flavin coenzymes to the apoenzyme. Tyrosine can form an intermolecular complex with charge transfer interaction, in contrast with hydroquinones.

The addition of tyrosine to the culture medium did not change the sensitivity of *Mucor* toward chloroneb [in contrast with the results of Kataria and Grover in Rhzoctonia solani]. This indicates that tyrosine synthesis not blocked but that a specific mutation reducing tyrosine rich sequences seems to have occurred in R-strains.

It should be mentioned that tyrosine is also an essential part of the active center of cytochrome P 450. Therefore, a point mutation at the active center could abolish the enzyme activity. AHF seems to interact with tyrosine sequences in some NADPH-dependent monooxygenases, which are capable of hydroxylation of benzene rings.

AHF seem to play the role of substrate analogs, which increase the rate of oxidation of NADPH

Table 13.9: Cross-Resistance of Mucor mucedo Comparing a Wild (S-strain) with a Resistant (R-strain) Selected on Chloroneb-Amended Medium.

Fungicide		*Concentrations (mg/l)*				
		1	**3**	**10**	**30**	**100**
Chloroneb	S	—	31	70	72	67
	R	0	0	0	0	0
PCNB	S	46	91	96	98	100
	R	0	18	32	47	49
Biphenyl	S	0	2	8	29	70
	R	0	0	0	3	8
Etridiazole	S	0	5	28	73	100
	R	0	-4 9	24	71	
Dicloran	S	69	85	100	—	—
	R	0	0	0	0	
Pentachlorophenyl-	S	0	43	83	91	100
Methylether	R	0	17	34	52	66
Pentachlorophenol	S	23	42	64	97	100
	R	22	45	65	96	100

Indicated is the inhibition of radial growth on malt agar (in percent of controls) by various concentrations of some fungicides.

but do not serve as normal substrates that can be hydroxylated (effector role). Resistance may be based on a less effective binding or on a lower enzymatic activity. According to Wang et al., chloroneb resistance in *Phytophthora cactorum* is controlled by a mitochondrial DNA in the cytoplasm, but the gene product is still unknown.

It should be mentioned here that Aeschbach et al. described the formation of diphenyl-bridge bonding between hydrophobic tyrosine sequences by a peroxidative action. This could be an interesting receptor structure for diphenyl or o-phenylphenol. DCMP, the demethylation product of chloroneb (Fig. 1.8), behaves nearly identical to chloroneb but has an additional weak uncoupling effect in isolated mitochondria, as can be expected because of its phenolic structure. Chlorinated phenols, to which chloroneb-resistant mutants do not show cross-resistance, have a much stronger uncoupling activity and never produce a cell wall thickening.

Further investigations revealed that chloroneb, as well as other members of AHF, induce a lipid peroxidation of mitochondrial and endoplasmic membranes in Mucor.A "cytochrome c-reductase" from *Mucor* was totally inhibited by chloroneb and other fungicides of this group, which were the most effective inhibitors of this enzyme presently known.

The primary toxic side effect of this inhibition seems to be a stimulated NADPH oxidation probably via a flavin peroxide, which could result in a lipid peroxidation of the phospholipids of such enzymes. This may start a cascade process in peroxidation of unsaturated fatty acids in the membranes of sensitive fungi. According to Kataria and Grover, cysteine, glutathione, hydroquinone, and tyrosine decreased or nullify the growth inhibition by chloroneb and PCNB in *Rhizoctonia solani*.

This mechanism of action can sufficiently explain all effects by AHF described until now. A lipid peroxidation by monooxygenases within mitochondria must destroy the structure of the inner membrane system and, by this, decrease overall respiration without a specific inhibition of the respiratory chain.

A production of free radicals and lipid peroxidation within the nuclear envelope [the occurrence of a "cytochrome c reductase" was demonstrated in nuclear membranes of mammals] not only impairs the membrane function and transport of RNA but can affect DNA, leading to strand scissions and chromosome aberrations. Desoxynucleosides are very sensitive to singlet oxygen or hydroxyl radicals. This could explain the genetic effects in fungi, observed by Georgopoulos et al..

"Cytochrome c reductases" are very often located in the endoplasmic reticulum. Therefore, after induction of lipid peroxidation by AHF, it is not surprising that protein synthesis at the ribosomal site is impaired. Orth et al. working with ribosomes of *Ustilago maydis* could not find a lipid peroxidation under the influence of AHF (chloroneb, tolclophos-methyl) or dicarboximides, nor an inhibition of a cytochrome-P450 reductase. This may point to the fact that although various species of "cytochrome c reductases" do exist, other properties modify the effect of AHF.

Membrane bound lipid synthesis can also be disturbed in many ways by lipid peroxidation. There still remains the phenomenon of pathological cell wall thickening under the influence of AHF as a typical effect, which contributes to the fungistatic effect of AHF.

According to Ulane and Cabib, chitin and glucan synthetase are present behind the hyphal tip in the cytoplasmic membrane in a dormant form under the influence of an inhibitor protein and can be activated by proteinases decomposing the inhibitor proteins.

This was demonstrated also in *in vitro* experiments. AHF also seem to activate chitin synthetase probably allostetically. Watanabe and Kondo found an activation of proteinases by interaction of lipid peroxidation with a proteinase inhibitor protein. An activation of proteinase by chloroneb was also described in *Mucor mucedo*.

Therefore, it is possible that, under the influence of a moderate lipid peroxidation in the cytoplasmic membrane, chitin and glucan synthetases are directly or indirectly activated, which can result in the observed pathological cell wall thickening and yeast-like growth. Other processes could support this effect.

The cytoplasmic membrane of *Mucor* seems not to be very sensitive to AHF because it is not lysed, and no extreme leakage of cell constituents occurs. But this does not mean that it is not affected at all.

The sensitivity of a membrane to lipid peroxidation is dependent on the presence of radical-forming enzymes, on the content of unsaturated phospholipids, and on the ratio of phospholipids to sterols within the membrane.

In *Mucor mucedo,* the rank of sensitivity appears to occur in the following order: inner mitochondrial membrane > outer membrane > nuclear membrane > EPR > cytoplasmic membrane.

The inner mitochondrial membrane seems to be most sensitive perhaps because of the abundance of radical generating flavin enzymes and/or because of a relatively high content of unsaturated fatty acids combined with a relative low sterol content, especially in some fungi.

In the case of etridiazole, similarities as well as differences in the mechanism of action do exist, when comparisons are made with AHF.

This is demonstrated in the cross-resistance values. Etridiazole selected R-strains of Mucor *mucedo* have a high factor of cross-resistance with most other AHF, but strains selected on chloroneb exhibit only a cross-resistance factor of 3,3 against etridiazole. The ultrastructural changes are somewhat different from those caused by chloroneb or PCNB in *Mucor,* especially in respect to the destruction of the inner mitochondrial membrane.

Here, a localized lysis of the membrane is typical, and this enlarges to a total destruction of the mitochondrial inner membranes. The overall effects on respiration, cell wall thickening, effects on the nuclear envelopment, and other parameters are rather similar to those of other AHF. The same is true for lipid peroxidation.

Resistance and cross-resistance in AHF has been described by several authors for various fungi. An open question remains, whether the AHF inhibited monooxygenases (target enzymes) are essential for normal growth and are still operating in the R-strains, though perhaps with small structural changes.

Resistant strains in most cases have similar growth rates as S-strains, and sporulation can be quite normal. As an exception, Van Tuyl observed that a chloroneb-resistant strain of *Penicillium expansum* was unable to grow and to sporulate normally and could not produce the typical green colour of mature spores.

The addition of chloroneb or PCNB restored growth and sporulation to normal. Werner confirmed these theses results and found that dichlorohydroquinone, p-dichlorobenzene, 2,4,5-trichlorophenol, 2,4,5-trichloroanisole, hexachlorobenzene, dicloran, or diphenyl did not reverse this defect, whereas chloroneb, DCMP, and PCNB at 100 and 1,000 ppm induced normal sporulation and a green colouring of the spores.

This means thatthe R-mutant needs chloroneb for induction of the target enzymes, perhaps for some peroxidative reactions necessary for normal development. On the other hand, these experiments could reflect differences in the receptor structure for the various AHF. The basis of this phenomenon is not yet elucidated.

CARBAMATES

Prothiocarb [N-(3-dimethylaminopropyl)thiolcarbamate hydrochloride] was first described in 1975 as a fungicide for control of soil-borne *Pythium* spp. and *Phytophthora* spp.. The marketing of this compound was stopped in 1978, and a close analogue, propamocarb hydrochloride [propyl 3(dimethylamino)propylcarbamate hydrochloride] was introduced in 1979.

Propamocarb hydrochloride is a versatile, effective, and safe carbamate fungicide with specific activity against numerous *Oomycete* species, causing seed, seedling, root, foot, and stem rots and foliar diseases in numerous edible crops and ornamental plants. Propamocarb hydrochloride was discovered and developed by Schering AG. The mode of action is different from that of other *Oomycete* fungicides, and propamocarb hydrochloride effectively controls strains that have developed resistance to other fungicides. A full data package is available for propamocarb, but that for prothiocarbhydrochloride is incomplete.

Toxicity of the Active Ingredients

The active materials are highly hygroscopic and are, consequently, very difficult substances to handle. For this reason, most toxicological investigations have been conducted with the products as an aqueous concentrate, containing 722 g/L propamocarb hydrochloride or 625 g/L prothiocarb hydrochloride.

Although propamocarb hydrochloride and prothiocarb hydrochloride are carbamates, they are, if at all, extremely weak cholinesterase inhibitors and should not be confused with carbamate insecticides.

Acute Oral and Dermal Toxicity

Acute Inhalation Toxicity

The determination of the 4-h LC_{50} of propamocarb hydrochloride liquid concentrate was not possible because at the highest administered concentration of 102 mg propamocarbhydrochloride/l air no effects were seen.

Primary Skin and Eye Irritation in the Rabbit

Propamocarb hydrochloride is not a primary dermal irritant. A very slight irritation was observed in a primary eye irritation study; such primary effects were completely reversible.

Sensitizing Properties

Propamocarb hydrochloride is not a skin sensitizer in guinea pigs.

Neurotoxicity

Propamocarb hydrochloride showed no signs of neurotoxicity at doses up to 1403 mg/kg/d in a 90-d study in the rat.

Ecological Properties of the Active Ingredient

Propamocarb hydrochloride does not persist in the soil. Following an adaptation phase, it is rapidly decomposed by microorganisms. The average half-lives are below 30 d. Ninety percent of the original material is decomposed within 70 d.

Propamocarb hydrochloride is very stable to hydrolys and photolysis in sterile aqueous media. However, aquatic microorganisms rapidly decompose propamocarb hydrochloride (up to 97% within 35 d).

Uses—Primary Uses

Primary Uses and Selectivity

Propamocarb hydrochloride is active on oomycetes fungi, including *Pythium spp., Peronospora spp., Phytophthora spp., Bremia* spp., and *Pseudoperonospora* spp., and is crop safe on a wide range of crops when applied as a seed, soil drench, or foliar treatment.

Propamocarb hydrochloride is registered in the United States as well as in a range of other countries to be used on potatoes and tomatoes to manage late blight (*Phytophthora infestans*), turf grass to control *Pythium* blight and damping-off (*Pythium* spp. and *Phytophthora* spp.), and herbaceous and woody ornamentals to control root rot and damping-off (*Pythium* spp. and *Phytophthora* spp.).

Mode of Action

The carbamate fungicides propamocarb and prothiocarb act by affecting the biosynthesis of a membrane constituent in *Pythiacious* species. Propamocarb hydrochloride exhibited a remarkable *low in vitro* effectiveness against mycelium growth of *Phytophthora infestans* strains (EC_{50} values 2481–6246 μg/mL active ingredient). On the other hand, production of sporangia (EC_{50} value <25 μg/mL) and oospores (916 μg/mL) was retarded by considerably lower concentrations of propamocarb hydrochloride.

As a consequence the compound inhibits all of the active growth stages in the development of the oomycete fungi with the consequence of a strong retardation of the spread of the pathogens under laboratory and field conditions.

COPPER COMPOUNDS AND SULFUR FUNGICIDES

Copper compounds and elemental sulfur are the two oldest of the agricultural chemicals used to control pests. The pest-averting properties of sulfur were known to the ancient Greeks as early as 1000 BC, but Forsyth was the first on record to suggest the application of sulfur for the control of plant diseases.

Elemental sulfur and a crude mixture of lime and sulfur became important for the control of powdery mildews of tree fruits, grapes and hops, and peach leaf curl.

The use of sulfur continued to increase until Millardet reported the fungicidal properties of Bordeaux mixture (a mixture of lime and copper sulfate) for controlling downy mildew on grapes.

Until copper was mixed with lime, it could not be used extensively because of its phytotoxicity. With the introduction of Bordeaux mixture, the use of sulfur declined until the early 1900s when sulfur was mixed with lime to form calcium polysulfide.

The use of sulfur then increased and peaked in the early 1940s when the organic fungicides were introduced. There was a reduction in use; however, sulfur and copper compounds continue to be important for controlling plant diseases. Factors that have prolonged the life of these products are low cost, lack of resistance, and safety to humans and the environment.

Copper also continues to be an important tool for controlling diseases, and its continued use is attributed to safety and low cost when compared with the organic fungicides. Except for copper and sulfur, inorganic chemistry has not had an important role in controlling plant diseases.

It seems remarkable that sulfur and copper were the first chemicals to be used for controlling diseases, and they continue to be important components of many current disease control programs.

Sulfur

Use

Sulfur is the oldest of the pesticides and is still sold in probably the largest quantities worldwide of all pesticides. On an active ingredient basis, it is used at rates ranging from 10 lbs to 30 lbs per acre, which is among the highest use rates per acre of any pesticide.

However, with the introduction of many new fungicides, its use declined from 172 million lbs in 1961 to 83 million lbs in 1995. Gammon et al. estimate the use of sulfur in California alone at 70 million lbs active ingredient, which is about one-third of the total weight of pesticides used in agriculture in California. Sulfur, while used mainly on grapes, tomatoes, and sugarbeets, is also used on numerous other crops.

In the common commercial form, elemental sulfur is a fine yellow powder. It is extremely insoluble in water; thus, particle size is important to its performance. The more finely ground formulations provide better performance and adherence to the plant surface, but residual activity is decreased. There are several allotropic forms of sulfur, some crystalline and others amorphous.

In addition to efficacy, sulfur has many advantages over other pesticides, including lack of chemical resistance and human and environmental safety factors. Because sulfur is a natural substance and toxicity is not an issue, it is exempt from USEPA registration requirements.

Environmental Fate

In soil, sulfur has been shown to be oxidized to the sulfite and further to the sulfate. It is sometimes added to fertilizers to reduce the pH of the soil when soils are too alkaline. When sulfur is used continuously in disease control programs, the soil pH can be lowered to levels below the ideal for plant growth. Overall, sulfur does not pose an environmental hazard.

Mammalian Toxicity

One of the advantages of sulfur is its low mammalian toxicity. An excellent summary of the acute mammalian toxicity for several sulfur formulations is by Gammon. Following is a brief summary of the acute toxicity data for sulfur. In nearly all categories (EPA Guidelines 81-1 through 81-6) acute toxicity for most of the sulfur formulations was either category III or IV. Only one formulation was category I (guideline 81-4), and it contained copper hydroxide, a compound that is an eye irritant. In summary, acute mammalian toxicity of sulfur formulations is very low, showing little or no acute toxicity.

Mechanism of Action

The mechanism of sulfur toxicity to fungi has been the subject of several investigations. Several theories have been presented, and these can be categorized as:

1. the *oxidation theory,* which suggests that elemental sulfur is oxidized to sulfur dioxide, sulfur trioxide, pentathionic acid, or other compounds and that one or more of the oxidized products kills the fungus
2. the *hydrogen sulfide theory,* which suggests that elemental sulfur is reduced to hydrogen sulfide, which acts as the toxicant
3. the *direct-action theory,* based on the toxic action of the sulfur molecule itself

Wilcoxon and McCallan disproved the oxidation theory. The results of their study showed the oxidized products were nontoxic when neutralized by forming salts. The hydrogen sulfide theory was widely accepted for several years, and McCallan and Wilcoxon showed that leaf tissue of 26 species of higher plants and the conidia of 16 species of fungi were able to evolve hydrogen sulfide from elemental sulfur, thus supporting the hydrogen sulfide theory.

Miller et al. critically reexamined the hydrogen sulfide theory and presented evidence that toxicity could not be attributed to hydrogen sulfide. They found that a preparation of colloidal sulfur was from 5 to 50 times more toxic on an equivalence bas than hydrogen sulfide, depending on the species of fungi. The fungal species susceptible to the toxic action of sulfur generally produce hydrogen sulfide.

The mechanism for this reaction has been the subject of several investigations. de Rey-Pailhade in 1888 showed that an ethanol extract from fungal spores would reduce sulfur to hydrogen sulfide and considered it to be an 'enzyme,' which he designated as philothion, later shown to be glutathione. Several other mechanisms have been reported to form hydrogen sulfide but do not explain the fungitoxic action of sulfur.

Tweedy and Turner reported that sulfur interferes with many metabolic pathways while being reduced. Data obtained by them and Byrde et al. showed that sulfur stimulates the activity of NADH2 and inhibits the reduction of cytochrome c. Tweedy and Turner further showed that reduction of sulfur to form hydrogen sulfide was not inhibited by cyanide but is inhibited by methylene blue and electron transport inhibitors acting on the substrate side of cytochrome c.

It was hypothesized that sulfur accepts electrons from the electron transport system, possibly from cytochrome b. The transfer is most likely directly from cytochrome b, but the supply of electrons is enzyme-dependent. Oxidative phosphorylation would be decreased, and adenosine diphosphate and inorganic phosphate begin to accumulate. The accumulations of these metabolic regulators enhance the oxidation of endogenous substrates to form more adenosine triphosphate (ATP). Toxicity of sulfur is due to the depletion of energy-yielding substrates and the ATP required for metabolism.

Copper

Most of the heavy metals display some activity in controlling the growth of fungi. According to Horsfall, the best estimation of toxicity from studies reported in the literature is as follows:

Ag>Hg>Cu>Cd>Cr>Ni>Pb>Co>Zn>Fe>Ca. Horsfall also reported that this order is similar to the order of stability of the chelate forms by metal ions. The only heavy metal that plays a major role in the commercial control of plant diseases is copper.

Uses

Copper's first reported use for disease control was in 1761 when it was used as a seed treatment for controlling bunt of wheat. In 1839, copper sulfate was used in French vineyards as a wood treatment for preserving the posts used for trellises.

Copper compounds continue to be extremely important as wood preservatives. Because of its phytotoxicity, copper sulfate was not used extensively until Millardet reported the efficacy of Bordeaux mixture, a mixture of lime and copper sulfate, for controlling downy mildew on grapes.

By adding lime to copper sulfate, phytotoxicity was decreased, and efficacy was not affected. The copper fungicides quickly gained worldwide recognition for controlling plant diseases.

Although Bordeaux mixture was being used in most disease control programs, it was somewhat phytotoxic to many plants. This detrimental effect led to efforts to develop other copper compounds, which were less phytotoxic, yet maintained disease control. The technologically advanced copper fungicide formulations currently provide effective control on a vast number of crops, and many of these formulations contain copper hydroxide.

Copper hydroxide has a low solubility in water, and the new formulations allow for the slow release of Cu^{+2} ions providing a residual control without phytotoxicity. The use of copper products for disease control in 1995 is estimated at more than 8 million lbs. In recent years, copper salts of rosin acids and fatty acids have been introduced as fungicides.

Mechanism of Toxicity

The biological performance of fixed copper formulations is very dependent on the availability of the Cu^{+2} ion. Therefore, the successful use of copper as a fungicide on higher plants must be based primarily on a relatively greater exposure of the fungal protoplast to the cupric ion than those of the host plant. Toxic effects include:

1) the blocking of functional groups of biologically important molecules (enzymes and transport systems for essential metal ions);
2) the displacement and/or substitution of essential metal ions from biomolecules and functional cellular units;
3) conformational modification, denaturation, and inactivation of enzymes; and
4) disruption of cellular and organocellular membrane integrity. Cupric ions have been shown to cause leakage of cellular metabolites, which is probably due to damage to the cell membrane.

Toxicology

There is no established acceptable daily intake (ADI) value for copper hydroxide compounds. The USEPA and other regulatory bodies have determined that there is no known evidence of chronic adverse human health effects from dietary exposure, and none would be expected except in a case of massive intake. The copper ion is a trace element essential for the growth and well being of man.

Humans have a natural, efficient homeostatic mechanism for regulating the body levels of copper ions over a wide range of dietary intakes. This mechanism integrates the absorption, retention, and excretion of copper to stabilize the body burden. Most of the absorbed copper is excreted. The adult human body burden of copper is typically 80 mg to 150 mg.

Because of this characteristic and the acute toxicity data, the USEPA and other regulatory groups have determined that chronic, genotoxicity, mutagenesis, carcinogenesis, and reproductive studies are not required for copper hydroxide. The acute LD50 values are between 1,00 to 2,000 mg/kg. The acute skin absorption for most copper products ranges from 2,000 to 5,000 mg/kg. Generally, copper products are not highly toxic but can cause irritation to the eyes.

CYMOXANIL FUNGICIDES

Nomenclature

Common name: cymoxanil (BSI, ANSI, ISO)

IUPAC name: 1-(2-cyano-2-methoxyiminoacetyl)-3-ethyl-urea

Chemical Abstract name: 2-cyano-N-[(ethylamino)carbonyl]-2-(methoxyimino)acetamide

CAS Registry number: 57966-95-7

Development code: DPX-T3217

Trade name: Curzate

Patent

U.S. 3957847

Manufacturer

Discovered and commercialized by DuPont Crop Protection Chemicals.

Physical/Chemical Properties

Form: Off-white to light pink, odorless crystals

Molecular formula: $C_7H_{10}N_4O_3$

Molecular weight: 198.18

Melting point: 159–160°C

Vapor pressure: 1.1×10^{-6} mm Hg or 1.5×10^{-4} Pascals at 20°C

Density: 1.32 g/cm^3 at 25°C

Solubility (in mg/L at 20°C): Water: 890 at pH 5.0, 780 at pH 7.0; acetone: 62,400; hexane: 37; methanol: 22,900; 1-octanol: 1,430; toluene: 5,290; ethyl acetate: 28,000; acetonitrile: 57,000; methylene chloride: 133,000

Stability: Stable in acidic environment. Decomposes in alkaline environment. Light promotes decomposition.

Analysis: Product analysis by HPLC. Residue analysis by GC. Details available from DuPont.

Mode of Action

Preventive, curative, translaminar (penetrant), and local systemic actions. Inhibits nucleic acid synthesis, amino acid synthesis, and cellular processes in the target plant pathogen.

Used in combination with other fungicides with different modes of action to control downy mildew diseases of grapes, hops, sunflower, tobacco, sugar beet, plantation crops, cucurbits, lettuce, onions, and several other vegetable crops. Also used to control the late blight disease of potatoes and tomatoes.

Mixtures

Sold as a wettable powder (WP) or wettable granule (WG) or suspension concentrate (SC) in mixtures with one or more of the following fungicides: famoxadone, mancozeb, metiram, zineb, dithianon, fosetyl, folpet, copper hydroxide, copper oxychloride, copper sulfate, chlorothalonil, zineb, propineb, and oxadixyl. These mixtures may be synergistic in their action against the target pathogen. They provide a broader spectrum of activity, longer residual action (thus extending spray intervals), and help prevent the development of fungal resistance to cymoxanil and the other fungicide in the mixture.

Compatibility

Incompatible with alkaline materials.

Crop Tolerance

Nonphytotoxic when used according to label directions.

Mammalian Toxicity

Rat oral LD50: 960 mg/kg, mouse oral LD50: 860 mg/kg. Rabbit dermal LD50: >2,000 mg/kg. Mild eye irritant to rabbit (clears at 48 h). Mild, transient dermal irritation to rabbit (clears at 48 h). Inhalation LC50 (4 h) for male and female rats >5.06 mg/L. Nononcogenic and nonteratogenic.

Ecotoxicology

Oral LD50 for bobwhite quail and mallard ducks >2,250 mg/kg. Eight-day dietary LC50 for bobwhite quail and mallard ducks >5,620 mg/kg diet. Fish LC50 (in mg/L at 96 h): rainbow trout 61, bluegill sunfish 29, common carp 91, Zebra fish >47.5 mg/L. Earthworm LC50 (14 d) >2,208 mg/kg soil. *Daphnia magna* LC50 (48 h) 27 mg/L. Algal growth inhibition LC50 (72 h) 5.2 mg/L. Honeybee contact LD50 >25 μg/bee.

Degradation and Metabolism

Animals

Radiolabeled cymoxanil is metabolized in the goat to natural products, including fatty acids, glycerol, glycerin, and other amino acids, lactose, and acid-hydrolysable formyl and acetyl groups.

Plants

Rapid degradation to naturally occurring amino acids, particularly to glycine, with subsequent incorporation into constituent sugars, starch, fatty acids, and lignin.

Soil

In laboratory soils, DT50 0.75–1.5 d (5 soils, pH range 5.7–7.8, o.m. 0.8–3.5%). In the field, DT50 (bare soil) 0.9–9 d.

Precautions

Use as per label directions. Avoid contact with eye and skin. Keep away from food, and feed and out of reach of children.

DICARBOXIMIDES FUNGICIDES

This group of antifungal compounds originated from studies on related chemicals with herbicidal activities, by Sumitomo. The first fungicide to emerge from this class of cyclic imides, now designated dicarboximides, was dichlozoline in 1967. On account of toxicological problems, this active ingredient was not pursued further.

The search for other cyclic imides, substituted on the nitrogen atom by a 3,5-dichlorophenyl group, exhibiting high fungicidal activity and devoid of major secondary effects was successful. Nowadays, four dicarboximides are used worldwide, and according to their chemical structure, they can be classified into three classes: oxazolidinedione (e.g., vinclozolin, chlozolinate), hydantoin or imidazolinedione (e.g., iprodione), succinimide or pyrrolidinedione (e.g., procymidone). Other experimental antifungal dicarboximides are shown in Figure elsewhere in this chapter.

Physicochemical Properties and Synthesis

The main physicochemical properties of chlozolinate, iprodione, procymidone, and vinclozolin are reported in Table elsewhere in this chapter. These various active ingredients are solids with melting points between 108°C and 166°C. They exhibit vapor pressure values below 1 mPa, except procymidone whose value is 18 mPa at 25°C.

Their water solubility ranges from 2 mg/L for chlozolinate to 13 mg/L for iprodione, whereas they exhibit similar n-octanol/water partient coeficients (log Kow between 3.0 and 3.15). Their solubility in organic solvent exceeds 100 g/L for acetonitrile, acetone, dichloromethane, and ethyl acetate; lower values are recorded for heptane or hexane and alcohols.

These dicarboximide fungicides can be obtained either by cyclization of N-(3,5-dichlorophenyl) carbamates under basic conditions or dehydration of N-(3,5-dichlorophenyl) dicarboxylic acid amides. These intermediate compounds can be prepared by reacting 3,5-dichloroaniline with acid anhydrides (amides) or chloroformates (carbamates). An alternative way to synthesize oxazolidinedione fungicides (i.e., chlozolinate, vinclozolin) can consist in mixing 3,5-dichlorophenyl isocyanate with hydroacids or hydroxyesters.

In other respects, the reaction of 3,5-dichlorophenyl isocyanate with glycine leads to an hydantoic acid that, after cyclization, gives 3-(3,5-dichlorophenyl) hydantoin. Then iprodione is obtained by reacting the previous hydantoin with isopropyl-isocyanate.

Biological Properties and Uses

Although the dicarboximides are mainly used against Helotiaceae (*Botrytis*, *Monilinia*, and *Sclerotinia*), they are also effective toward many other fungi. For example, they can inhibit fungi in the Basidiomycetes (*Corticium*, *Rhizoctonia*, *Tilletia*, *Typhula*, *Ustilago*), the Zygomycetes (*Mucor*, *Rhizopus*), and the Ascomycetes (Cochiobolus, *Didymella*, *Glomerella*, *Leptosphaeria*, *Penicillium*).

Among the Adelomycetes, species of *Alternaria*, *Cercospora*, *Colletotrichum*, *Fusarium*, *Helminthosporium*, *Phoma*, and *Thielaviopsis* are sensitive to dicarboximides. However, these fungicides are not toxic to phytopathogenic Oomycetes and to yeasts involved in the fermentation process of grape juice.

They are mainly used as foliar sprays at 500-1000 g a.i./ha. For example, toward *B. cinerea* on grapevine, the application rates are 750 g a.i./ha for iprodione, procymidone, and vinclozolin or 1000 g a.i./ha for chlozolinate. They are formulated as wettable powders (WP), suspension concentrates (SG), or water-dispersible granules (WG).

The other sprayed crops are fruits (including top fruits, strawberries, raspberries), vegetables

Table 13.10: Nomenclature of Commercial Dicarboximides.

Common Name		*Chlozolinate*	*Iprodione*	*Procymidone*	*Vinclozolin*
Code number and other name		M 8164 dichlozolinate	26019 **RP** glycophene	S 7131 dicyclidine	BAS 352F Vinclozoline (Fr.)
Chemical names	IUPAC	Ethyl (+/-)-3-(3,5-dichloro-phenyl)-5-methyl-2,4-dioxo-oxazolidine-5-carboxylate.	3-(3,5-dichlorophenyl)-N isopropyl-2,4-dioxo-imidazolidine-1-carboxamide. 2,4-dione.	N-(3,5-dichlorophenyl)-1,2-dimethylcyclopropane-1,2-dicarboximide. 1,3-oxazolidine	(RS)-3-(3,5 dichlorophenyl) 5-methyl-5-vinyl
	CAS	(+/-)-ethyl 3-(3,5-dichloro-phenyl)-5-methyl-2,4-dioxo-5-oxazolidinecarboxylate.	3 (3,5-dichlorophenyl)-N-(1-methylethyl)-2,4-dioxo-1-imidazolidine-carboxamide.	3-(3,5-dichlorophenyl)-1,5-dimethyl-3-azabicyclo hexane-2,4-dione.	(+/-)-3-(3,5 dichloro-phenyl) 5-ethenyl-5 methyl-2,4 oxazolidinedione.
CAS registry number		282-714-4	36734-19-7	32809-16-8	50471-44-8
Patent		BE 874406 DE 2906574 IT20579178	GB 1312536 US 3755350 FR 2120222	GB 1298261 US 3903090	DE 2207576
Manufacturer		Isagro (introduced by Montedison)	Rh̄one Poulenc, now Aventis	Sumitomo	BASF
Trade names		Serinal	Rovral, Kidan, Verisan	Sumisclex, Sumilex, Kimono	Ronilan, Flotilla

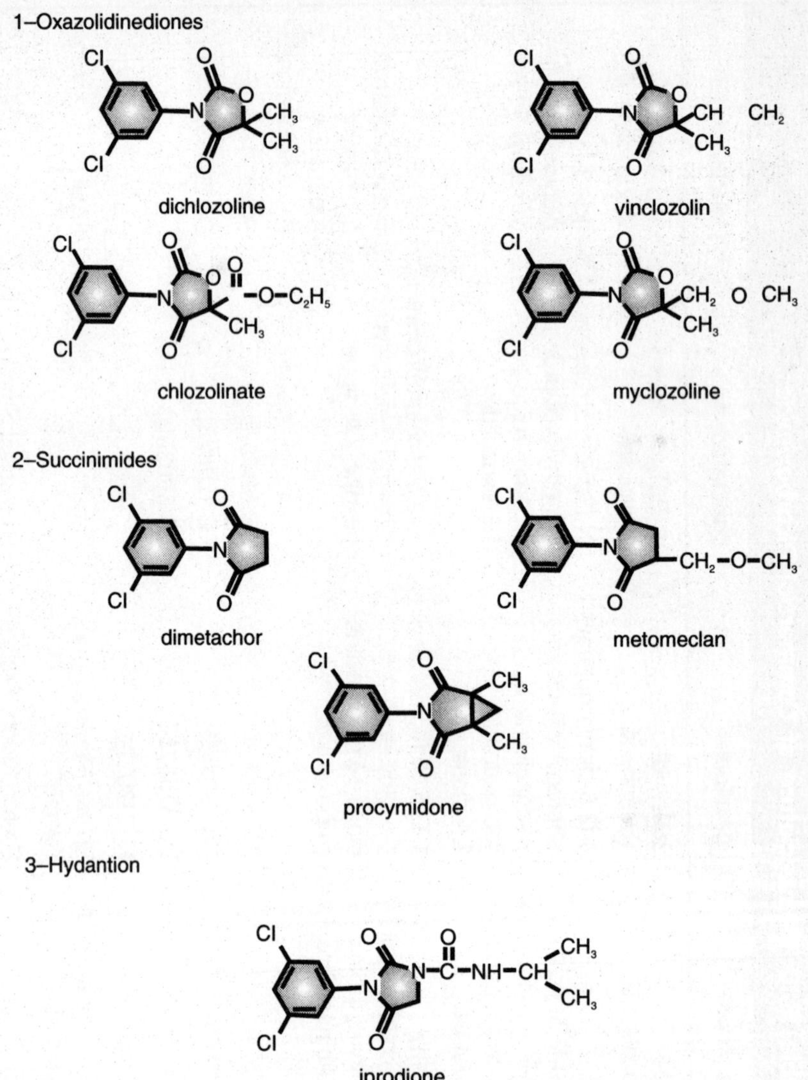

Figure 13.5: Structure of dicarboximide fungicides.

(including aubergines, cabbages, beans, peas, tomatoes, lettuces, chicories), ornamentals, cereals, sunflowers, oilseed rapes, soya beans, and peanuts. On turfs and lawns, at 3000-12000 g a.i./h, they can control *Sclerotinia homoeocarpa, Corticium fuciforme,* and some *Fusarium spp. Soil* drenches, seed treatments, (e.g., 150-g/q seeds on allium against *Sclerotium cepivorum* or on beet against *Phomaetae*), and dips or sprays (on ornamental bulbs or potato tubers) are other ways of dicarboximide uses.

In order to broaden the spectrum of antifungal activity of dicarboximides, or in connection with anti-resistance strategies, mixtures with other fungicides have been developed. The companion fungicides can be multisite inhibitors (e.g., chlorothalonil, thiram, maneb), sterol biosynthesis

Table 13.11: Physicochemical Properties of Dicarboximide Fungicides.

Characteristics	*Chlozolinate*	*Iprodione*	*Procymidone*	*Vinclozolin*
Molecular formula	$C_{13}H_{11}C_{12}NO_5$	$C_{13}H_{13}C_{12}N_3O_3$	$C_{13}H_{11}C_{12}NO_2$	$C_{12}H_{19}C_{12}NO_3$
Molecular weight	332.1	330.2	284.1	286.1
Form	Colorless and	White, odorless	Colorless crystals	Colorless crystals,
odorles Solid	crystals		slight aromatic odor	
Melting point in °C	112	134	166	108
Vapour pressure in Pa	1.3×10^{-5} (25°)	5×10^{-4} (25°C)	1.8×10^{-2} (25°C)	1.3×10^{-4} (20°C)
Henry's constant in				
Pa m³/mol	2.3×10^{-3} (calc.)	5.5×10^{-2} (calc.)	1.1 (calc.)	1.4×10^{-2} (caL)
log K_{ow}	3.15	3.0	3.14	3.0
Water solubility				
in mg/L	2.0 (25°C)	13.0 (20°C)	4.5 (25 °C)	2.6 (20 °C)
Solubility in g/L:				
—acetone	>250	342	180	334
—acetonitrile	—	150	135	—
—ethanol or	methanol	13	30	20* 15*
—ethyl acetate	>250	225	90	233
—dichloromethane	>250	450	=200	475
—hexane or heptane	2	0.6	—	4.5

inhibitors (e.g., bromuconazole, diniconazole, imazalil), or tubulin-binding compounds (e.g., carbendazim, thiabendazole, thiophanate-methyl).

Under field conditions, the dicarboximides exhibit mainly preventive activities against fungal parasites. However, curative and eradicative properties are sometimes reported. Furthermore, under experimental conditions, translocation of these fungicides has been observed, but this systemicity is probably not involved for their practical efficacy.

Mode of Action and Resistance Phenomena

The dicarboximides inhibit conidial germination and mycelial growth, but the latter process is the most sensitive one. Stunting, swelling, and bursting of germ tubes and changes in hyphal branching have been reported. In *B. cinerea,* similar morphological alterations are induced by phenylpyrroles and aromatic hydrocarbons.

Furthermore, in laboratory mutants of various fungal species, positive cross resistance occurs between these three families of fungicides, and it is associated with an increased susceptibility to high osmotic pressures. In laboratory mutants of *Ustilago maydis* and *Neurospora crassa,* the genes conferring dicarboximide resistance encode protein kinases that could be implied in the osmoregulation. It is suggested that inhibition of such enzymes may activate a MAPK (mitogen-activated protein kinase) cascade of reactions resulting in increased glycerol synthesis. In other respects, it has been proposed that the toxicity of dicarboximides resulted from the overproduction of reactive oxygen forms that cause lipid peroxidation, membrane destruction, and other nonspecific toxic effects.

Table 13.12: In Vitro Effects of Dicarboximides and Other Fungicides Toward Botrytis Cinerea.

	Wild-type Strains			*Resistant Strains*	
	EC_{50} Values in mg/L for:			*Resistance Levels for:*	
Fungicides	*Spore Germination*	*Germ-tube Elongation*	*Mycelial Growth*	*Field Strains*	*Laboratory Mutants*
Chlozolinate	3.5	1.5	0.9	11.1	>25
Iprodione	2.2	0.9	0.15	8.0	>200
Procymidone	2.5	0.8	0.12	16.7	>200
Vinclozolin	0.8	0.5	0.15	8.0	>200
Dicloran	8.0	4.0	0.7	5.0	>35
Fludioxonil	0.05	0.015	0.003	1.5	>200

Table 13.13: Toxicological Profiles of Dicarboximide Fungicides.

Criteria	*Chlozolinate*	*Iprodione*	*Procymidone*	*Vinclozolin*
Acute oral LD_{50} for rats in mg/kg	>4500	3500	6800	>15000
Acute percutaneous LD_{50} for rats in mg/kg	>5000	>2500	>2500	>5000
Inhalation LC_{50} for rats in mg/L air (4 h)	10	>5	>1.5	>29
NOEL in mg/kg body weight/day				
— rat	10 (90 d)	6 (2 yr)	15 (2 yr)	1.4 (2 yr)
— dog	5 (1 yr)	18 (1 yr)	75 (90 d)	2.4 (1 yr)
ADI in mg/kg body weight/day	?	0.06	0.1	0.01
Toxicity class: WHO (a.i.) / EPA (formulation)	III/III	III/IV	III/?	III/IV
EC risk	Not classified	Xn, R40, R50, R53	Not classified	Xn, R40, R43, R50, R53, R62, R63

The fact that the scavenger α-tocopherol protects fungi against dicarboximide toxicity is supporting for a toxic mechanism of this type. The target could be a plasma-membrane–bound NADPH-dependent flavin enzyme, inhibition of which would initiate pathological oxidative processes. Enhanced levels of catalase in dicarboximide-resistant strains of *B. cinerea* could therefore be a resistance mechanism not related to the target site of these fungicides.

Table 13.14: Ecotoxicological Profiles of Dicarboximide Fungicides.

Criteria	*Chlozolinate*	*Iprodione*	*Procymidone*	*Vinclozolin*
Birds (acute oral LD_{50} in mg/kg):				
— Mallard duck	>4500	>10000	>4000	>2500
— Bobwhite quail	>9000	>2000	—	>2500
— Japanese quail	>4500	—	>6600	—
Fishes (LC_{50} in mg/L, after 96 h):				
— Rainbow trout	27.5	4.1	7.2	22
— Bluegill sunfish	>80	3.7	10.3	50
Daphnia (LC_{50} in mg/L, after 48 h)	1.18	0.25	>1.8	4.0
Algae (EC_{50} in mg/L toward *Selenastrum* sp or *Scenedesmus sp**).	30 (96 h)	1.9 (120 h)	2.6* (96 h)	>1 (120 h)
Bees (contact LD_{50} in mg/bee)	>0.1	>0.4	>0.1	>0.2
Earth worms (LC_{50} in mg/kg soil, after 14 d)	—	>1000	>1000	1000

The intensive use of dicarboximides led to the development of practical resistance in *B. cinerea*, especially on grapevine. In field strains, the resistance levels toward dicarboximides range from 8 to 20 and positive cross resistance occurs with aromatic hydrocarbons but not with phenylpyrroles. They differ from laboratory mutants, which are highly resistant to dicarboximides, aromatic hydrocarbons, and phenylpyrroles.

However, in both cases, dicarboximide resistance seems to be determined by the same polymorphic major gene that encodes a two-component histidine kinase probably involved in fungal osmoregulation. Under field conditions, in the absence of *dicarboximide* treatments, the proportion of resistant strains decreased gradually.

Consequently, the restricted use of dicarboximides is justified, and for example, in French vineyards, the actual recommendation is one application per season; temporary interruptions are needed in the case of more intensive use. Mixtures of dicarboximides with multisite toxicants (e.g., *thiram, chlorothalonil*) are also proposed as an anti-resistance strategy. Field-resistant strains have also been detected within the following species: *Fusarium nivale, Sclerotinia homeocarpa, Monilinia fructicola, Sclerotium cepivorum*, and *Alternaria* spp. In several cases, positive cross resistance is recorded among *dicarboimides, aromatic hydrocarbons*, and *phenylpyrroles*. But the practical problems remain limited.

Toxicological and Ecotoxicological Profiles

According to the data from Table elsewhere in this chapter, it appears that the dicarboximides exhibit no or low acute toxicity to mammalians; they are in class III of World Health Organization

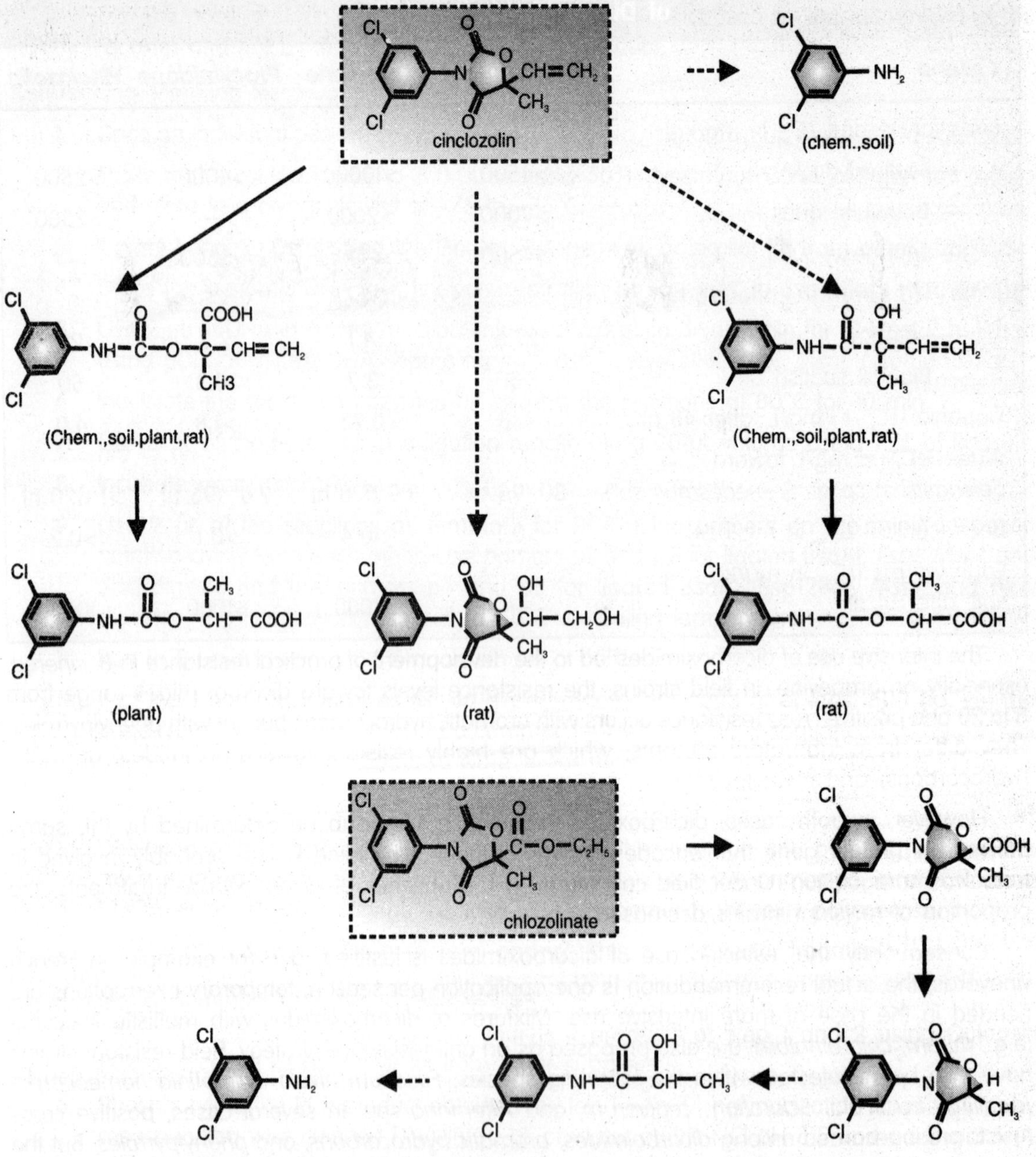

cinclozolin

(chem.,soil)

(Chem.,soil,plant,rat)

(Chem.,soil,plant,rat)

(plant)

(rat)

(rat)

chlozolinate

chem.: chemical degradation (especially under basic conditions)

Figure 13.6: Metabolism of chlozolinate and vinclozolin.

(WHO) toxicity classification. They are not mutagenic in mammals or in bacteria, whereas such a phenomenon has been recorded in the filamentous fungi *Aspergillus nidulans;* it could be related to effects of activated oxygen forms on chromosomes. In long-term studies, the NOEL values range from 1.4 to 15 mg/kg body weight/day in rats and from 2.4 to 75 mg/kg body weight/day in dogs.

Vinclozolin exhibits the lower values and shows anti-androgenic properties in laboratory animals; it also shows dermal sensitization (guinea pigs). The ADI values are respectively 0.005, 0.06, and 0.1 mg/kg body weight/day for vinclozolin, iprodione, and procymidone. The dicarboximides are not toxic to birds such as ducks or quails (acute oral LD_{50} > 2000 mg/kg).

The effect of dicarboximides toward aquatic organisms including fishes, daphnia, and algae are reported in Table elsewhere in this chapter; only iprodione and vinclozoline are considered to be dangerous (EC risk: R 50 and R 53). The dicarboximides seem to be devoid of any noxious action toward bees, beneficial arthropods, and earth worms.

Persistence and Metabolism

One of the common features of dicarboximides is their instability in alkaline solution, with DT_{50} values below 1 h at pH above 8. An attack of the hydroxide ion at either one of the carbonyl moities of the dicarboximide heterocyclic ring leads to its cleavage, and it generally yields anilides. Such a reaction can also occur when aqueous solutions of dicarboximides are exposed to ultraviolet and 3,5-dichloroaniline is often produced. Dechlorination is reported as a photodegradation reaction when the dicarboximides are in organic solvents. Among the dicarboximides, procymidone is the most persistent in soils with DT_{50} values of several months.

For iprodione, the mean soil DT_{50} under laboratory dark aerobic conditions is about 1 month. However, their mobility is slight, which is in agreement with their high K_o values (respectively, 1300 and 975–1200 for procymidone and iprodione). The oxazolidinediones (e.g., chlozolinate, vinclozolin) are less persistent in soils than are the previous dicarboximides, and they are also less adsorbed to soil organic matter (Koc, respectively, 300 and 100–375 for chlozolinate and vinclozolin).

The main degradation mechanism seems to be the cleavage of the cyclic imide, and subsequently, 3,5-dichloroaniline is produced. In the case of procymidone, dechlorination and hydroxylation at the methyl group and at the 4-phenyl position are also reported. Furthermore, an enhanced degradation of dicarboximides can take place after successive applications and leads to reduced performance.

This phenomenon is well documented for iprodione or vinclozolin, with DT_{50} values from 22 to 30 days after the first application to only 1–2 days after the third application. It is probably related to the selection of adapted *Pseudomonas spp.*

When the dicarboximides are applied to the foliage of crops, the major components identified are the active ingredient and metabolites resulting from the cleavage of the cyclic imide. 3,5-dichloroaniline is not detected, except after treatments with procymidone, whose methyl groups can also be oxydated. In the case of iprodione, N-dealkylation and cleavage of N-carboxamide linkage are also observed.

Numerous studies have been conducted on grape, and they show a wide distribution of the residue levels (generally between 0.5 and 5 ppm), depending on the number of treatments and

chem.: chemical degradation (especially under basic conditions)

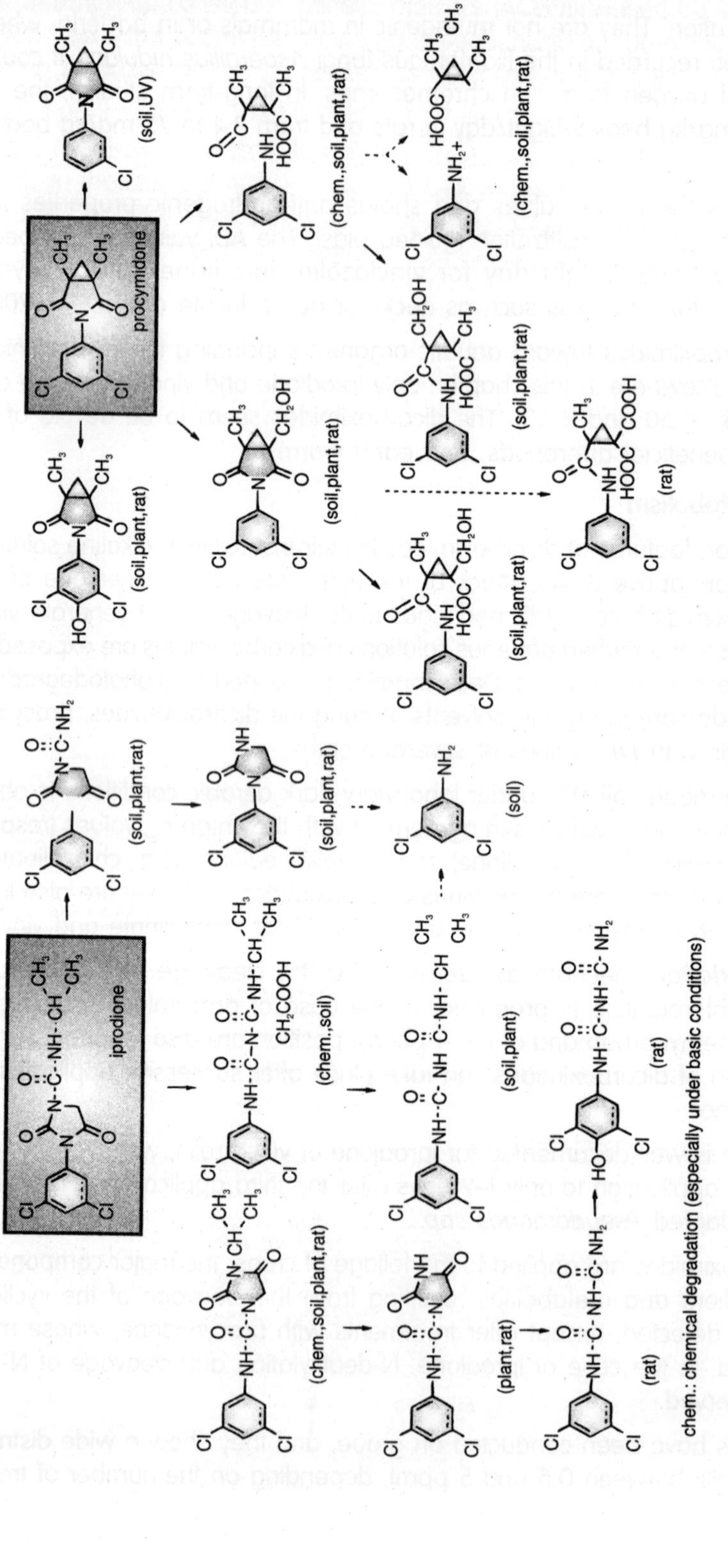

Figure 13.7: Metabolism of iprodione and procymidone.

the interval before vintage. The content of dicarboximides greatly decreases during wine making; further reduction is obtained by clarification with charcoal.

The final concentrations of dicarboximides in wine are normally below 0.5 ppm, and the parent compounds can be chemically degradated (opening of the heterocycle ring). Kinetic data indicate a degradation rate order: chlozolinate > vinclozolin > procymidone > iprodione. The determination of dicarboximide residues is generally achieved by gas–liquid chromatography; it consists in estimating (with an electron-capture detector) the amount of 3,5-dichloraniline produced after the hydrolysis of dicarboximides.

In the rat, after oral administration, the dicarboximides are rapidly eliminated in the urine and the feces. The tissue residue distribution profile confirms the previous observation and shows a slight retention of dicarboximide residues in the liver. The major metabolic pathways include opening of the cyclic imide, alkyl, or aryl hydroxylations. Several metabolites can be recovered as glucuronide or sulfate conjugates.

DIMETHOMORPH FUNGICIDES

Dimethomorph, marketed, in chronological order, by Shell, American Cyanamid, and BASF, is the common name for the fungicide developed under code names CME 151, WL 127 294, AC 336379, and CL 336379. The chemical names for dimethomorph are (E,Z)-4-[3-(4-chlorophenyl)-3 (3,dimethoxyphenyl) acryloyl]morpholine or (E,Z)-4-[3-(chlorophenyl)-3-(3,4-dimethoxyphenyl)-1-oxo-2-propenylmorpholine. Dimethomorph has been described as a cinnamic acid derivative.

Physicochemical Properties

Dimethomorph has a molecular formula of $C_{21}H_{22}ClNO_4$ and a molecular weight of 387.9. The active ingredient consists of (E)- and (Z)-isomers, in a ratio of approximately 1 : 1. Dimethomorph forms colourless crystals at room temperature and has a melting point of 127-148°C, with a melting point of 135.7-137.5 °C for the (E)isomer, and 169.2-170.2 °C for the (Z)-isomer. At 25 °C, the vapor pressure of the (E)-isomer is 9.7 × 10-4 mPA and that of the (Z)-isomer is 1.0 × 10-3 mPa.

The KowlogP-values of the (E)-isomer and the (Z)-isomer are 2.63 and 2.73, respectively (20 °C). The bulk density of dimethomorph is 1318 kg/m3 (20°C). At 20-23°C, the solubility of dimethomorph in water is less than 50 mg/L. Dimethomorph is hydrolytically and thermally stable under normal conditions.

It is stable for more than 5 years in the dark. The (E)- and (Z)-isomers are interconverted in sunlight. USESDimethomorph is effective against members of the Peronosporaceae spp. and Albuginaceae spp. (downy mildews and white rusts). It is also active against Phytophthora spp., but it is inactive against Pythium spp.. It has good protectant, curative, and antisporulant activities.

Dimethomorph has shown translaminar activity after foliar application and systemic uptake via the roots following soil drench. Dimethomorph is used worldwide for control of downy mildew and Phytophthora spp. diseases in vines, potatoes, vegetables, and other crops.

It is used in combination with contact fungicides such as dithianon, mancozeb, copper compounds, and fentin hydroxide. The fungicide is available as an emulsifiable concentrate, dispersible concentrate, wettable powder, or a water-dispersible granule.

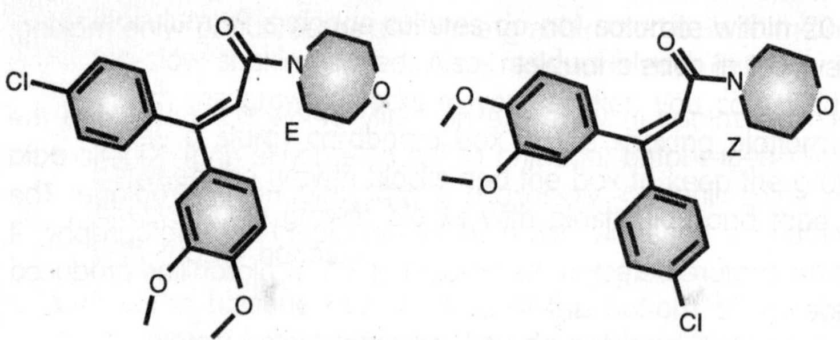

Figure 13.8: The (E)- and (Z)-isomers of dimethomorph.

Mode of Action

Only the (Z)-isomer is intrinsically active, but because of rapid interconversion of isomers in light, it has no advantage over the (E)-isomer in practice. The primary binding site of the molecule is unknown. In the presence of dimethomorph, there is loss of control of the biochemical processes involved in normal cell wall biogenesis of sensitive Oomycete fungi, which results in cell lysis. Mycelial growth and formation of sporangia are development stages most sensitive to the action of dimethomorph. Oomycete target fungi resistant to phenylamides are not cross-resistant to dimethomorph.

Toxicology

Dimethomorph has a World Health Organization (WHO) classification III, indicating that the product is slightly hazardous. The acute oral LD_{50} values for male and female rats were 4300 and 3500 mg per kilogram of bodyweight, respectively. Acute oral LD_{50} values for male and female mice were greater than 5000 and 3700 mg per kilogram of bodyweight, respectively.

The acute percutaneous LD_{50} for rats exceeded 5000 mg per kilogram of bodyweight. Dimethomorph is not an irritant to the skin or the eyes of rabbits and not a skin sensitizer in guinea pigs. The No Observed Effect Level (NOEL) over 2 years was 200 mg per kilogram bodyweight in male and female rats, and 450 mg per kilogram of diet for dogs. The LC_{50} for rats by inhalation was higher than 4.2 mg/L.

Dimethomorph was not oncogenic in studies in rats and mice over a 2-year period. The acceptable daily intake was established at 0.09 mg per kilogram of bodyweight. The acute intraperitoneal LD_{50} for male and female rats was 327 and 297 mg per kilogram bodyweight, respectively. Dimethomorph was tested for bird toxicity on mallard ducks.

The acute oral LD_{50} was greater than 2000 mg per kilogram of bodyweight. With oral and contact LD_{50} values greater than 100 μg per bee, dimethomorph is not hazardous to bees. Dimethomorph was tested for fish toxicity on carp and bluegill sunfish.

To carp, the LC_{50} after 96 h of exposure was 14 mg/L. To bluegill sunfish, the LC_{50} after 96 h of exposure was greater than 25 mg/L. The EC_{50} of dimethomorph to earthworms was greater than 1000 mg/kg soil. The EC_{50} of dimethomorph to Daphnia was 49 mg/L after 48 h of exposure, and to algae 29 mg/L after 96 h of exposure.

Metabolism/Transformation

In rats, the major route of metabolism is demethylation of one of the dimethoxy groups or by oxidation of one of the CH_2 groups (*ortho*- or meta-position) of the morpholine ring. In plants, the only significant component of the residue, when present, is dimethomorph.

FERIMZONE FUNIGICIDES

Fermizone

Ferimzone was introduced by Takeda Chemical Industries Ltd. Ferimzone is the commonly accepted name for this compound. The chemical name is (Z)2'-methylacetophenone 4,6-dimethyl-2-pyrimidinylhydrazone.

General Chemical and Physical Properties

Ferimzone has a molecular weight of 254.3. Ferimzone has a specific density of 1.185 g/ml and forms colourless crystals at room temperature. It has a melting point of 175–176°C. The vapor pressure of ferimzone at 20°C is 4.11×10^{-3} mPa. At 25°C, the K_{ow} logP-value is 2.89. It dissolves in water in amounts up to 162 mg/L. The compound is stable in sunlight and in neutral and alkaline aqueous solutions.

Uses

Ferimzone has systemic properties. It is mainly used in rice-producing areas in Asia for the control of brown spot (*Helminthosporium oryzae*), narrow brown leaf spot or Cercospora leaf spot (*Cercospora oryzae*), and rice blast (*Pyricularia oryzae*).

Ferimzone is marketed as a driftless formulation, a suspension concentrate, or a wettable powder formulation. It is also marketed as a mixture with phthalide.

Mode of Action

Research to elucidate the mode of action of ferimzone was conducted on the rice blast pathogen, *Pyricularia oryzae*. The results of these studies did not indicate that ferimzone affects the synthesis of RNA, DNA, protein, cell walls, or lipid components. Ferimzone inhibited mycelial growth of *P. oryzae*, but it did not inhibit spore germination until a germ tube had formed, consisting of several hyphal cells with a normal nucleus.

However, cytoplasm of both spores and hyphae was granulated and localized after exposure to ferimzone. No effect of ferimzone was observed on the cell wall architecture of mycelia. Ferimzone exhibited a fungistatic effect, as spores or mycelia resumed normal development processes once ferimzone was removed from their environment.

Ferimzone did not affect respiratory activity of mycelia. Ferimzone caused leakage of some electrolytes from mycelia, which suggests that the compound affected or disrupted membrane function.

Toxicology and Metabolism

Ferimzone has a World Health Organization (WHO) classification III, which indicates that the product is slightly hazardous. The acute oral LD_{50} values for male and female rats were 725 and 642 mg per kilogram of body weight, respectively. Acute oral LD_{50} values for male and female mice were 590 and 542 mg per kg of body weight, respectively.

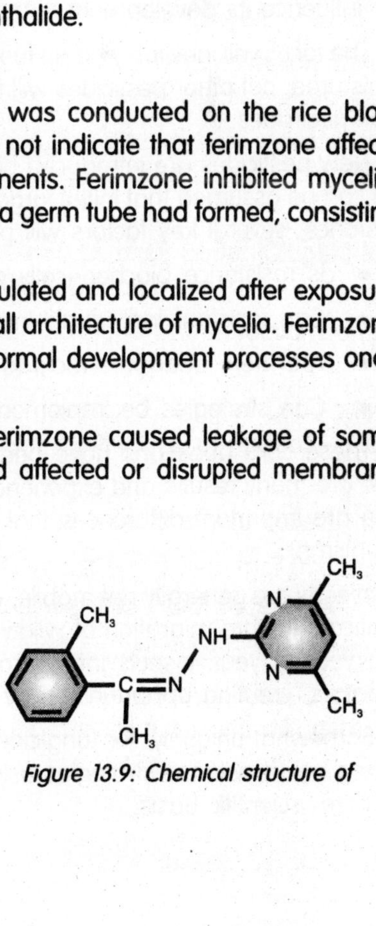

Figure 13.9: Chemical structure of

The acute percutaneous LD_{50} for rats exceeded 2000 mg per kilogram of body weight. The LC_{50} for rats by inhalation was higher than 3.8 mg/L after 4 h of exposure. Ferimzone was tested for bird toxicity on bobwhite quail and mallard ducks. Acute oral LD_{50} values for these two species exceeded 2250 and 292 mg per kilogram of body weight, respectively.

Ferimzone is not hazardous to bees with oral LD_{50} values greater than 140 µg per bee. Ferimzone was tested for fish toxicity on carp and *Moina macrocopa.* To carp, the LC_{50} after 72 h of exposure was 10 mg/L. For *M. macrocopa,* the LC_{50} after 24 h of exposure was greater than 40 mg/L. In soil, the half-life of ferimzone (DT_{50}) ranges from 3 to 14 days, depending on soil type.

FUNGICIDES, FUNGAL RESISTANCE TO CHEMICAL CONTROLS

Pesticides will remain an essential component of crop protection for the foreseeable future, ensuring high quality yields free from pests and diseases. The long-term viability of pesticides is always threatened by resistance, which not only affects performance, causing unnecessary impact on the environment, but it also means loss of biochemical modes of action that are valuable resources generated with considerable private and public investment.

Experience shows that the resistance problem can be managed, but it first requires assessment of the risks involved. This review examines the phenomenon of resistance, its causes, and factors that influence its development.

The focus will inevitably be on fungicidal resistance because of the authors' research experience in this area, but other pesticides will be included where this may identify generic factors applicable to all pesticides.

New pesticides are introduced only after extensive field trials establish their effectiveness, and so it can be assumed that initial target populations are sensitive. If intensive selection is to generate resistance, several key factors will play a part in determining the possibility of this happening.

- Is resistance biochemically possible?
- Can resistant individuals increase in frequency?
- Are there adequate tools to monitor resistance?
- Can strategies be implemented to maintain effectivenes ness?

These core questions have been the subject of extensive research over many years, and there are many results and experiences to draw on for all pesticides, as well as for antibiotics. But there are important differences that influence approaches to dealing with resistance in different organisms.

Weeds are generally not mobile, whereas insects and fungi can travel great distances, allowing significant inward migration of wildtype sensitive individuals, so that resistance levels can change from year to year. Insects incorporate behavioural factors into their life cycles, and often have predators, creating opportunities for Integrated Pest Management strategies.

Somewhat uniquely, for fungicides, crossresistance patterns invariably follow modes of action placing anti-resistance strategies using fungicide mixture partners with different modes of action on a firm scientific basis.

Table 13.15: Resistance Caused by Different Alleles of the CYP51A1 Gene (Sterol 14α Demethylase).

Mutant	Resistance Factor		
	Allele Fluconazole	Ketoconazole	Itraconazole
G129A	1	1	1
S405F	4	4	2
Y132H	4	16	2
G464S	4	4	2
R467K	4	4	2
Y132H; S405F	>64	32	8
Y132H;G465S	32	32	4
G464S; R467K	8	4	2
G129A G464S	16	4	1

Resistance Factor (RF) = MIC value resistant strain/ MIC sensitive strain. Mutant alleles were identified in clinically resistant strains of *Candida albicans* and expressed in yeast (*Saccharomyces cerevisiae*). Table derived from data presented in Sanglard et al.. Amino acid codon 132 is equivalent to codon 136 in CYP51A1 in plant pathogenic fungi.

Is Resistance Biochemically Feasible?

The answer to this question is generally yes. Chemical pesticides are small molecules able to bind to specific sites within their target proteins, and in some way disrupt normal function. Small changes in protein structure can easily affect pesticide binding, and indeed, a single amino acid change is usually all that is required.

In β-tubulin, the target of benzimidazole fungicides, substitution of glutamate at amino acid codon 198 with neutral amino acids, such as glycine or alanine, is sufficient to reduce binding of carbendazim to this target, yet increase binding of the phenylcarbamate fungicide diethofencarb

Substitution of this same glutamate with either lysine, valine, or histidine has more drastic effects on protein conformation and fungicide binding. Alteration of phenylalanine at the neighboring codon 200 for tyrosine also generates benzimidazole resistance, but not negative cross resistance.

Just how changes at codon 198 affect the dimensions of the benzimidazole binding site have yet to be resolved, but recent developments in electron crystallography have produced a three-dimensional structure of tubulin of sufficient resolution (3.7 A) to identify binding sites.

Much early work on benzimidazole resistance equated this major gene resistance with high levels of resistance. It is now clear that these high levels of resistance, coupled with negative cross-resistance, were only associated with alleles E198G, E198A, or E198V. Other point mutations at either amino acid codons 200 or 198 gave lower resistance levels. So linking major gene resistance with high levels of resistance is not always correct.

Indeed, phenotypes with different levels of resistance are now recognized in many pathogens. Although several different mechanisms clearly contribute to resistance to demethylation inhibitors

Table 13.16: Current Status of Strobilurin (QOI) Resistance in Practice.

Pathogen	Disease	Distribution
Erysiphe (=Blumeria) graminis f.sp. tritici	Wheat Powdery mildew	N. Europe
Erysiphe (=Blumeria) graminis f.sp. hordei	Barley Powdery mildew	Scotland
Sphaerotheca fuligenea	Cucurbit Powdery mildew	E. Asia, Spain
Pseudoperonospora cubensis	Cucurbit Downy Mildew	Japan
Plasmopara viticola	Vine downy mildew	France, Italy
Mycosphaerella fijiiensis	Banana black sigatoka	Costa Rica
Venturia inaequalis	Apple scab	Germany
Didymella bryonae	Cucurbit gummy stem blight	United States
Corynespora cassiicola	Cucurbit leaf spot	Japan

(DMIs), point mutations in the target sterol 14a demethylase (14DM; CYP51 gene) are linked with resistance in some pathogens. Amino acid substitutions can affect fungicide binding, and modeling studies suggest that substitutions alter the polar environment in the substrate pocket, so that DMI fungicides no longer enter easily.

The sterol substrate must not be obstructed, however, because these resistant strains are seemingly as fit as wild-type ones. This change differs significantly from the first point mutation reported to be associated with DMI resistance in *Saccharomyces cerevisiae* (G243A), which as a result, thrust a histidine into a position that could link it with the sixth coordination position of the heme iron, and this blocked DMI access to this site.

This point emphasizes a common feature of resistance in that mechanisms selected in the laboratory are seldom those found in natural populations. High levels of resistance to strobilurin and related fungicides (QOIs, formerly the STAR group) have been reported in several fungi, and sequence analysis of at least partial fragments of the mitochondrial encoded target cytochrome bc-1 gene suggests that resistance is also linked to a point mutation [G143A].

In this case, glycine is replaced by the slightly larger alanine in the region of the protein known to be involved in the binding of QOI fungicides and close to the Qo center. How this affects the conformation of the target cytochrome, or indeed the neighboring iron–sulphur protein is not clear.

Nor is there any biochemical evidence as yet to confirm that this G143A mutation alters fungicide binding. At least some of these resistant fungi seem as fit as wild-type ones, and so presumably, electron transport through the Qo center is not affected. In insects, too, single point mutations in target receptors or proteins cause resistance, and always, it seems, it is the same few amino acids that are replaced in different resistant insect species.

γ-aminobutyric acid (GABA) gated chloride ion channel is the target of cyclodiene insecticides, and replacement of just a single amino acid (alanine 302) with glycine or serine confers resistance in a wide range of insects, including mosquitoes, whiteflies, and cockroaches. The voltage-gated sodium channel is the target for DDT and pyrethroid insecticides, and here just two amino acid substitutions in a conserved region of the protein confer resistance.

Three point mutations (I129V, G227A, F288Y) in acetylcholine esterase confer resistance to organophosphorus and carbamate insecticides, whereas three point mutations in a cytochrome P450 monooxygenase (CYP6A2) reduce enzyme activity sufficiently to cause resistance in several insect species.

Unfortunately, there are no similar molecular studies published at present on target site resistance to herbicides, but it does seem that generally the same point mutation confers resistance to a pesticide group in different organisms.

Although many different mutants may be generated in the laboratory, those occurring in resistant field or clinical populations are limited tc just a few sites in the target protein. It is surprising that these sites are in highly conserved regions, because these are presumably conserved across different organisms because the function attributed to that part of the protein is essential. More than one mutation may occur, increasing the level of resistance.

In *Candida albicans*, the F132Y mutation does not occur alone, but only in combination with at least one other mutation, and this generates much higher levels of resistance to DMI fungicides. But point mutations at amino acid codons 198 and 200 in β-tubuiin, which both cause resistance to benzimidazole fungicides in several fungi, have never been found together in field strains, although they were recently combined by sitdirected mutagenesis, and appeared to confer benomyl resistance in *Erysiphe* (*=Blumeria*) *graminis*.

Other mechanisms can cause resistance. Metabolic detoxification of fungicides is seldom encountered, perhaps because fungi lack excretory mechanisms and enzymes needed to generate polar metabolites that can be expelled through aqueous routes. But some examples have been reported recently, including resistance of *Venturia inaequalis* to kresoxim methyl and *Botrytis cinerea* to fenhexamid.

This is in contrast with the situation in plants or insects, in which metabolic detoxification by upregulated cytochrome P450s and glutathione-Stransferases can cause resistance. One consequence of these nonspecific resistance mechanisms is that cross resistance extends across unrelated chemistries, and consequently, anti-resistance strategies are difficult to devise.

Cells possess a battery of membrane spanning proteins involved both in nutrient uptake and efflux of unwanted molecules. Efflux requires energy, especially where lipid soluble pesticides are expelled across a lipid membrane and against a concentration gradient. Energy may be derived from adenosine triphosphate (ATP) and mediated through ATP-binding cassettes (ABC transporters), or through an electrochemical gradient in the case of the major facilitator superfamily (MSF) of proteins. The requirement for an ATP-binding site ensures that they are highly conserved and, consequently, that their function can be explored by a variety of molecular and biochemical techniques.

Furthermore, their activity is often induced by the presence of foreign molecules. The first report of fungicide resistance due to increased efflux was before the role of ATP-binding casettes was fully understood. Increased efflux of fenarimol from resistant *Aspergillus nidulans* mutants was reversed by the uncoupler of oxidative phosphorylation CCCP, and the mutants became sensitive.

Further characterization of one of these ABC transporters in *Aspergillus nidulans* (AtrBp) revealed an affinity to all major classes of agricultural fungicides and some natural products. Deletion mutants all showed normal growth in the absence of toxicants, but displayed increased sensitivity to a range of fungicides.

Overexpression mutants were less sensitive to a wide range of chemicals. In the human pathogen *Candida albicans*, benomyl resistance is not caused by an altered 6-tubulin; instead resistance was traced to overexpression of an MFS-type multidrug resistance (MDR) transporter. But not only was benomyl efflux increased, so too were fluconazole and other molecules, so that the resistant strains were cross-resistant to many unrelated compounds.

MDR is recognized as an important factor in resistance to anti-cancer drugs, but its expression is not always constitutive and resistance can be lost when foreign molecules are withdrawn. How far MDR contributes to pesticide resistance in the field is not clear. It is the cause of resistance to DMIs in some *C. albicans* strains, where target-site changes are not so important.

There are many ABC transporters that expel different groups of compounds, and Northern analysis has revealed increased expression following fungicide treatment. However, upregulation of the expression of ABC transporters has not been conclusively linked with pesticide resistance in practice, although these mechanisms may contribute to underlying low levels of resistance.

Unlike with MDR, in no cases in which fungicides have failed against plant pathogens because of resistance has cross-resistance extended beyond compounds with the same mode action. The plasma membrane proton-pumping ATPase is an essential cell-surface transporter that generates the electrochemical driving force for multidrug efflux. Selective inhibition of this target may counter resistance, at least to fungicides, through inactivating efflux pumps.

It will be interesting to see if this concept, which implies action outside the cell, avoiding detoxification and limiting side effects in the host, emerges as a practical solution to MDR. Overexpression of a target is a further resistance mechanism. For instance, selection with ketoconazole led to overexpression of 14DM in S. *cerevisae* and to the cloning of the CYP51 gene.

Prolonged laboratory selection of *C. albicans* for more than 300 generations with fluconazole produced one resistant population with distinct overexpression of the target CYP 51 gene [Erg 11]. Although this mechanism does not commonly cause resistance in practice, in DMI-resistant field strains of *Penicillium digitatum*, constitutive expression of CYP 51 was about 100-fold higher than in wild-type sensitive strains. This was linked to a tandem repeat of a transcriptional enhancer in the promoter region of the gene. Unfortunately, these molecular studies were not backed up with spectroscopy to show that enhanced transcription did indeed increase the amount of the target

14DM protein. But in the aphid *Myzus persicae*, resistance to organophosphorus and carbamate insecticides occurs through gene amplification. Amplified genes are dispersed around the genome and cause increased production of two almost identical carboxylesterases (E4 and FE4), although it is rare to find both genes amplified in an individual aphid.

Resistance is largely due to the greater amount of enzyme protein sequestering more insecticide and lowering its effective concentration. Some aphid clones with amplified E4 lose resistance spontaneously and cease overproduction of the esterase. In *Myzus persicae* revertant aphids retain their full complement of amplified genes, but produce wild-type levels of the corresponding mRNA, indicating that regulation of esterase production is at the transcription level.

Unusually, transcription of amplified genes enhanced by methylation, both in the coding and upstream promoter regions, and silenced when methylation is lost. Similar amplification of two carboxyl esterase genes causes organophosphorus and carbamate resistance in the mosquito (*Culex pipiens*), although in this case, each gene has a number of allozymes leading to a complex mixture of different alleles and resistance levels in different mosquito populations.

A feature of gene amplification is that it is often reversed when selection is withdrawn, but because the genome changes associated with resistance remain in place, resistance quickly reappears if these insecticides are reintroduced. Gene amplification does not seem to be a factor in fungicide or herbicide resistance. P450 mono-oxygenases are a superfamily of ubiquitous enzymes involved in the metabolism of many xenobiotics.

They catalyze a variety of reactions, including hydroxylations, dealkylations, and demethylations, which often lead to detoxification of pesticides. Both overproduction, and enlarged substrate specificity contribute to insecticide resistance in many insects, and treatment with pipeonyl butoxide, a P450 monoxygenase inhibitor, synergizes insecticide action and lessens the impact of resistance.

Increased enzyme activity is generally due to overexpression of the genes encoding these mono-oxygenases, but enhanced stability of mRNAs and proteins may play a part. Although P450 mono-oxygenases may play a role in some herbicide resistance mechanisms, there is no evidence that this mechanism is involved in fungicide resistance.

Metabolic changes to avoid a target site are seldom a cause of resistance. DMI fungicides block sterol 14 demethylation, but the resulting 14 methyl fecosterol functions normally as a bulk membrane component. A later enzyme in the ergosterol biosynthesis pathway, sterol 5-6 desaturase, normally inserts the 5-6 double bond, but when 14α methyl fecosterol is its substrate, it generates a toxic 3-6 diol instead Resistance is achieved through a suppressor mutation, which inactivates the 5-6 desaturase.

However, there is no evidence that this mechanism can account for DMI resistance in plant pathogens. Electron transport is a basic component of respiration and energy production in all higher organisms, and so it is somewhat surprising that it has proved a useful selective target for insecticides (rotenone) and fungicides (carboxanilides, QOIs).

Although resistance to QOI inhibitors may involve a point mutation (G143A) in the target site gene, in one highly resistant field isolate of *V. inaequalis*, the target cytochrome b-c_1 of complex III was unchanged. Evidence of synergy with SHAM (salicylhydroxamic acid), a fairly specific inhibitor of terminal alternative oxidase, suggests that a shift in metabolism to alternative respiration, avoiding complex III, may cause resistance, especially during spore germination. Yet during later mycelial growth, SHAM did not synergize strobilurins, and NADH consumption by submitochondrial particles from both wild-type and resistant isolates was inhibited by trifloxystrobin to almost the same extent.

It seems that resistance may also involve some, as yet unknown, mechanism that compensates upstream of NADH dehydrogenase in the respiratory chain, for the energy deficiency generated by strobilurin treatment. Most fungi utilize branched respiratory chains, but until the complex regulation for each branch point is fully understood, an explanation of how organisms develop resistance by redirecting respiration to avoid targets such as complex III will remain unclear.

Can Resistant Strains Increase in Frequency?

When a new pesticide is being developed, resistant individuals must be rare; otherwise performance would not be effective enough to gain registration. Because pest and disease populations are extremely large, it is reasonable to expect that all possible resistance mutations will occur Many mutations will be lethal, but some will survive and increase in frequency through Darwinian selection when treatments are applied.

Unless the relative fitness of these mutations is high, their frequency will decline when selection is relaxed, and competition with wild-type -individuals re-established. Rarely does this happen once resistance has become a practical problem, and it may take many years for the original wild-type population to -re-establish. Resistance generated in the laboratory is not constrained by limitations in fitness, and it is possible to a obtain resistant mutants that are never isolated from field populations.

Point mutations at 10 different sites in the β-tubulin gene confer resistance to benzimidazole fungicides, but only two of these are commonly encountered in practice, and then only involving a few of the possible amino acid changes. Most mutations are associated with some functional penalty, although it is not easy to predict in advance which mutations will be successfull and cause practical resistance.

Knowledge of ED50 values, or some other measure of fungicide sensitivity, is not sufficient to predict the evolution of resistance under different pesticide treatments. Biochemical fitness must -be combined with pathogenic fitness before resistance can become a problem, but experience suggests that this can happen surprisingly quickly depending on the biology of the organisms involved.

Changes in the frequency of mutations that confer resistance to strobilurin fungicides is particularly difficult to understand. A mutation might confer resistance in one mitochondrion, but other mitochondria in the same cell remain sensitive and generate reactive oxygen species (ROS), which is presumably the damaging factor in strobilurin action. Somehow, cells must protect themselves against the effects of ROS sufficiently to stay alive to allow the resistant mitochondria to divide.

This may be achieved through alternative oxidase or some other antioxidant system. This is likely to be a key factor in the development of strobilurin resistance, and perhaps not all pathogens operate effective defense mechanisms against ROS; in which case, strobilurin resistance may not become a problem in these diseases.

It is possible that some biochemical targets have a low resistance risk. Certainly there are pesticides in which resistant mutants can be generated in the laboratory, or even can be found in field populations, but their frequency does not increase. The new fungicide quinoxyfen interferes in some way with this G-protein signaling process, which controls the early steps in powdery mildew infection.

Components of this signaling pathway are often multifunctional, and mutations may have many pleiotropic effects. Quinoxyfen-resistant strains can be generated in barley powdery mildew (*E. graminis f.sp. hordei*) in the laboratory, and they can be isolated from field populations. This resistance is associated with a change affecting a GTP-ase activating protein, which locks the G-protein signaling pathway in an "off" position.

These mutants have defects in sporulation and sexual reproduction, and they are of low pathogenic fitness. Targeting later steps in signaling pathways broadens the spectrum of activity, and it may bring a greater risk of resistance. Phenylpyrrole and dicarboximide fungicides target the osmotic signal transduction probably at a kinase step.

Although resistance can be a problem, especially with dicarboximides, careful use of antiresistance strategies has managed resistance and has maintained their effectiveness. It seems that signaling pathways may be associated with a low resistance risk. Regardless of the biochemical and pathological fitness, resistant strains must survive in populations where other selective forces operate.

Where resistance is controlled by many genes, stabilizing selection serves to focus the sensitivity of the population around a mean, which does not necessarily reflect the highest level of resistance. Initially, selection generates a broad range of sensitivity, but subsequently, the extremes of the population distribution are lost. Stabilizing selection has been observed in many systems, including in barley powdery mildew in response to selection with the hydroxypyrimidine and DMI fungicides.

Are there Adequate Tools to Monitor Resistance

Bioassay, coupled with probit and related methods for data analysis, has long been the cornerstone of resistance monitoring. Its main advantage is that it measures resistance regardless of the mechanisms contributing to it. Bioassays are extremely adaptable and can be applied to most systems, with the proviso that the measured response is relevant to the action of the pesticide.

Effects on spore germination are a quick way to monitor fungicide sensitivity, but they can be misleading if the fungicide has no effect on germination, but works at some later stage of development. Media components used in any bioassay must not affect the outcome. Anilinopyrimidines fungicides interfere in some way with the biosynthesis of certain amino acids, and including these in any bioassay medium results in a decrease in sensitivity of up to 50-fold in the eyespot fungus *Pseudocercosporella herpotrichoides* (=*Tapesia yallundae*,).

This could lead to incorrect diagnosis of resistance. The main disadvantages of bioassays are twofold. They can be very resource demanding and do not identify the genotype (allele) causing resistance. Where several mechanisms contribute to resistance, a series of doses are required to obtain sufficient information to generate a meaningful dose/response relationship and ED50, ED95 values, or similar measures of sensitivity.

If the level of resistance is large, a single discriminating dose may be sufficient, although this

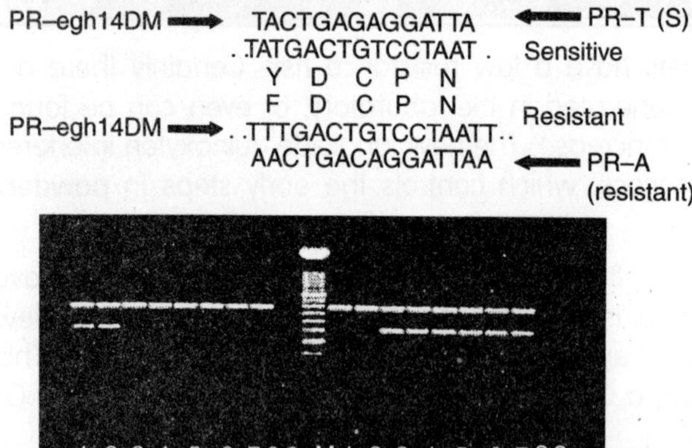

1–23D5 (S), 2–212 (R), 3–DE67 (R), 4–DE68 (R), 50MB27 (R), 6–MB28 (R), 7–19–18 (R), 8–AP2–19 (R), 9–HALCYON

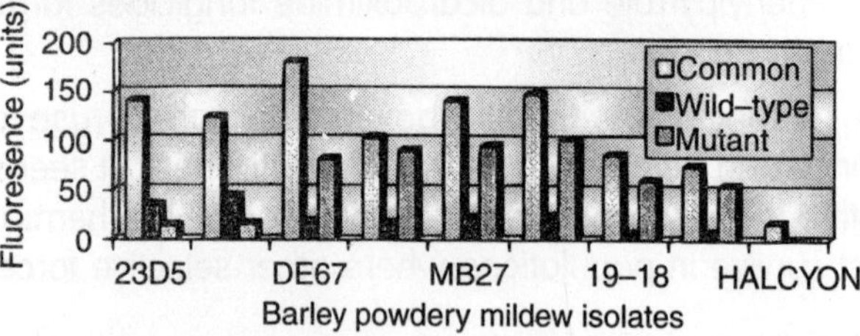

Figure 13.10: Allele specific PCR detection of the Y136F point mutation linked to DMI resistance in barley powdery mildew (Erysiphe graminis f.sp. hordei) DNA extracted from infected barley leaves infected and 100-ng DNA used in PCR reaction. (A) Allele-specific primers (PR-T specific for sensitive allele; PR-specific for resistanct allele) and PCR products separated on an agarose gel. Upper, common band results from nonspecific primers and confirms absence of inhibitors. Eight isolates plus healthy barley (cv. Halcyon). S = Sensitive. R = Resistant. (B) Fluorescence measurements on some of the same isolates. 150-µl thiazole orange (1 : 200 dilution) added to 5-µl PCR reaction.

must be established from a dose-response relationship initially, and it requires the appearance of field resistance to validate the dose level. Bioassay will not necessarily identify the mechanism involved. Indeed, bioassay is the only practical solution where mechanisms are unknown, and it currently is the most widely used monitoring method.

Biochemical and DNA-based methods offer ways to overcome the disadvantages of bioassay, providing mechanisms are known. Increased esterase activity in organophosphorus and carbamate-resistant aphids can be measured easily and cheaply, and the kinetics of the reaction identify which of the two resistant esterase forms is present. Assays are easily automated, allowing hundreds of aphids to be tested daily.

Many polymerase chain reaction (PCR) techniques are available to detect point mutations, and these methods are now being applied to monitor pesticide resistance. Where a resistant mutation generates a new restriction site, digestion of a PCR product with this enzyme identifies resistance. But perhaps the simplest PCR method uses specific primers (allele-specific oligonucleotides, ASO), which only generate a PCR product if there is a perfect match with the target sequence.

Figure elsewhere in this chapter provides an example detecting the Y136F mutation in the CYP51 gene in barley powdery mildew, which is linked to certain levels of DMI resistance. Designing two primers with the mismatch base pair at the 3' end, and using a common upstream primer, resistant and sensitive alleles can be identified using just a 2-cm infected leaf piece as the source template. This leaf piece can be stored frozen prior to use. Because only a single PCR product is formed, cyanine dyes such as PicoGreen and thermostable SYBR1 Green, which only fluoresce when bound to double-stranded DNA, allow fluorescence measurements to quickly identify a positive outcome, avoiding further manipulation involving gel electrophoresis.

These PCR methods are not easily quantified, but recent developments in real-time PCR, and specific hybridization techniques involving Taqman Applied Biosystems probes or molecular beacons, allow accurate measurement of the frequency of resistance mutations, and changes as

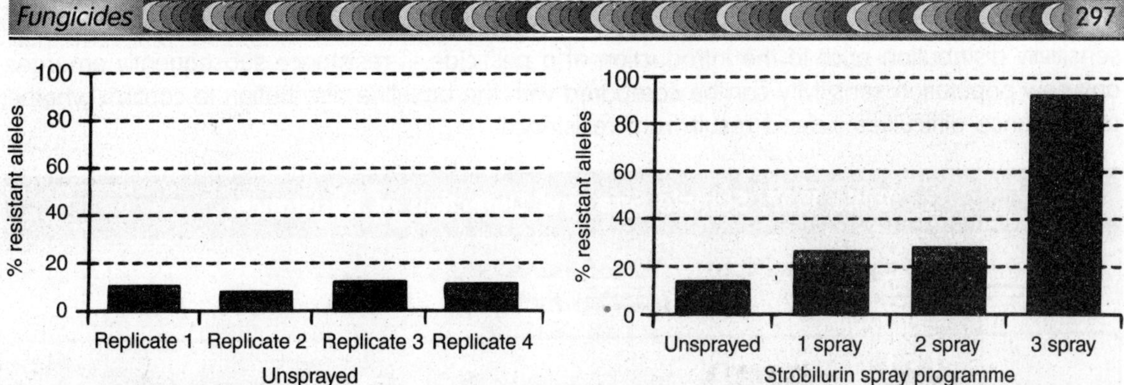

Figure 13.11: Selection for strobilurin resistance in wheat powdery mildew (Erysiphe graminis f.sp. tritici). A wheat crop was sprayed with Amistar (a.i. Azoxystrobin) at 0.66 L/Ha (two-thirds recommended rate) on three occasions. The first spray (T1) was applied at GS 31; the second (T2) at GS 45; and the third (T3) at GS 59. Samples were taken from nine sites within the crop just before the first spray was applied, and 2 weeks after subsequent treatments were applied.

a result of selection. Different fluorophores incorporated into each probe, or beacon, offer the potential to monitor more than one mutation in a single PCR reaction.

These techniques are extremely specific and provide information on allele frequencies that can be directly incorporated into population genetic models that predict the outcome of different treatment strategies.

However, Figure elsewhere in this chapter emphasizes the caution needed in interpreting these data, because strains that are bioassay resistant, but either have a different mutation to that included in the primer/probe, or have a different mechanism entirely, will be identified as sensitive. One difficulty surrounding all monitoring methods aimed at following the development of resistance is the sampling protocol needed to provide accurate pictures of the resistance level in any population.

Sample size is also important, but that depends very much on the aim of the monitoring exercise. The initial frequency of resistance is one of several key parameters in the prediction of how resistance might spread within a population. Table elsewhere in this chapter shows that attempts to identify, by bioassay, an initial frequency of $1:10^{-3}$ requires a significant resource input, and looking for frequencies of $1:10^{-6}$ or $1:10^{-8}$ is impossible.

PCR technologies offer the prospect of detecting resistant mutations much earlier than by bioassay. For example, we can detect the strobilurin-resistant mutation in wheat powdery mildew at frequencies down to $1:10^{-4}$, and conduct assays in a single day, compared with up to a month needed to grow and bioassay mildew isolates.

Despite this ability to detect the G143A mutation, the relationship between its frequency and its effect on field performance of QOI fungicides remains unclear, whereas apparently unexposed populations can contain the G143A mutation, albeit at low frequencies. Data from PCR assays relate to allele frequencies, and not to phenotypic frequencies of individual isolates. The sensitive nature of PCR assays may also reveal a background level of the wild-type allele in otherwise resistant populations.

An essential prerequisite for monitoring resistance development is to determine the baseline sensitivity distribution prior to the introduction of a pesticide. If resistance subsequently emerges, any new population sensitivity can be compared with the baseline distribution to confirm whether performance difficulties indeed result from resistance.

Table 13.17: Sample Size and Detection of Resistance.

Frequency of Resistant Mutation	*No. of Individuals Needed for Assay*	*Sample Area (Ha)*
1 in 10^{-3}	3×10^{3}	0.2
1 in 10^{-5}	3×10^{5}	23.1
1 in 10^{-6}	3×10^{-6}	230.8
1 in 10^{-8}	3×10^{8}	23,077.0

The sample size needed to establish the baseline and to make these comparisons depends on several factors, including the magnitude of the expected resistance, and the experimental error associated with the bioassay method. For example, a tenfold shift in sensitivity in the apple scab pathogen (*V. inaequalis*) to the fungicide flusilazole can be confirmed with a sample size of just 50 individuals.

Can Use Strategies be Implemented to Maintain Effectiveness?

Resistance to agrochemicals has been a practical problem for over 50 years, but no pesticides have been totally lost from the market solely because of resistance. In part this is due to successful innovative chemistry, which has solved potential problems through development of products with new modes of action. This approach to combating resistance wil! no doubt continue, along with ways to prolong effectiveness of existing products through anti-resistance strategies.

These manipulate the population dynamics of the target organisms by placing several different obstacles in the way of their normal development. The framework for this approach lies in predictive modeling of the response of target organisms to selection by different control strategies. Although there are shortcomings to these models, they identify many key factors, including fitness parameters, which have so far been impossible to measure.

But molecular technologies that monitor changes in alle frequencies provide an opportunity to seriously estimate for the first time the fitness of different resistance alleles in natural populations exposed to selection. Given the limitations of predictive modeling, there is inevitably an empirical component to devising anti-resistance strategies. At the cornerstone of all anti-resistance strategies lies the use of different obstacles to disrupt development of the target organism, integrating where possible nonchemical control measures with pestcides.

Products with different modes of action can either be mixed together or alternated, although the success of alternations depends largely on a loss of fitness of resistant types in the absence of the selective pesticide. Without this, there would be no reversion to sensitivity when the alterntive product was being used. The success of mixtures also rests in the reduced frequency of any combined resistance and in its impact on fitness.

For example, dual resistance to DMI, and the hydroxypyrimidine fungicide ethirimol in U.K. barley powdery mildew populations, was much lower than expected from the individual resistance frequencies in the same populations. This probably accounted for the effectiveness of a mixed seed treatment formulation involving ethirimol and the DMI fungicide flutriafol. Usually, antiresistance strategies restrict the number of treatments of an "at risk" pesticide, minimizing the extent that a target organism is exposed to selection.

Even so, treatments are generally applied during the early phase of an epidemic when populations are small and most vulnerable to selection. The effectiveness of anti-resistance strategies may not, therefore, increase in proportion to the reduction in treatments. Dose rate is another factor that may influence the effectiveness of anti-resistance strategies, but the relationship between dose and selection is a complex one.

In practice, strategies based on recommended rates are difficult to implement, especially in crops such as cereals since growers reduce rates to save costs where they can achieve adequate disease control when infection levels are not too high. Lowering dose rates may favour the buildup of partially resistant individuals, especially where resistance is polygenic, whereas high dose rates would kill all individuals and, in effect, exert no selection.

Alternatively, low dose rates, or leaving areas untreated (refugia), should slow the evolution of resistance, because more sensitive and fitter individuals survive, although the commercial reality of poorer control might be unacceptable. Critical data derived from field experimentation that might distinguish between these two possibilities are hard to come by, and what is available is conflicting.

Indeed, over the range of dose rates that are agriculturally acceptable, and where dose rates seldom go below one quarter of the recommended rate, dose rate does not seem to be a major factor affecting selection.

Where several pesticides with the same mode of action are available, the most effective will exert the strongest selection Physicochemical properties, such as vapor movement, may contribute to uneven distribution within a crop, which may encourage diversity and slow any spread of resistance. Repeated treatments with an at-risk pesticide over a wide geographic area must be avoided. Importantly, all mixture partners should be effective against the target, and when the dose rate of a mixture is reduced, this should not result in at least one of the partners becoming ineffective.

This may have been the case recently when a mixture of kresoxim-methyl and epoxiconazole fungicides failed to prevent rapid development of resistance to the strobilurin fungicide in wheat powdery mildew. Growers applied the mixture repeatedly, but at dose rates reduced to levels such that epoxiconazole no longer controlled mildew. Despite the recommendations by the industry-sponsored Resistance Action Committees and other groups over the years, examples of effective anti-resistance strategies are few.

This may, in part, be due to the fact that there is generally no true comparison with a no-strategy "control." There is no cross-resistance between DMI and morpholine fungicides, and alternations between the two groups has largely been successful in extending the life of at least the DMI for control of banana sigatoka disease [*Mycosphaerella fijiensis*].

A mixture of benzimidazoles and phenylcarbamates, which both target 6-tubulin yet select different resistance alleles, has been applied as an anti-resistance strategy against gray mold (*B. cinerea*) in grapes, taking advantage of the negative cross resistance between these two groups.

Unfortunately, dual resistance is known, and so this otherwise attractive strategy can fail. More recently, an anilinopyrimidine (cyprodinil) and phenylpyrrole (fludioxinil) mixture has also been deployed as an anti-resistance strategy against gray mold disease, although limitations on the number of applications restrict its usefulness as an anti-resistance strategy.

CONCLUSIONS

~~Resistance threatens to weaken our ability to control many important pests, diseases, and weeds. With the possible exception of a few insect pests, cooperative research between private and public organizations has managed the problem without serious losses in control. New chemistries have certainly helped, but so too has improved understanding of resistance mechanisms, and the inherent factors associated with particular pests, diseases, and weeds, and each pesticide group, which contribute to resistance.

Adding to this knowledge base is an ongoing process. Indeed, discovery of resistance to a new pesticide does not necessarily mean its demise. Rather, it provides the knowledge impetus needed to develop sustainable anti-resistance strategies.

Hymexazol Fungicides

Hymexazol was originally introduced by Sankyo. Common names used for this fungicide are hymexazol and hydroxyisoxazole. The chemical name for hymexazol is 5-methylisoxazol-3-ol, 5-methyl-1,2-oxazol-3-ol, 5-methy3(2H)-isoxazolone. The fungicide is marketed under the tradename Tachigaren.

General Chemical/Physical Properties

The active ingredient can be isolated to 99% purity. Hymexazol has a molecular weight of 99.1, and it forms colourless crystals at room temperature. It has a melting point of 86 °C to 87 °C. The vapor pressure at 25°C is fairly high at 182 mPa. With a K_{ow} log P value of 0.48, it is relatively water soluble. At 20°C, up to 65.1 g of pure grade hymexazol dissolve in water of neutral pH, 58.2 g dissolve in water of pH 3, and 67.8 g dissolve in water of pH 9.

In aqueous solutions, the compound is a weak acid with a PKa value of 5.92. Hymexazol is stable under alkaline conditions and relatively stable under acidic conditions. The compound is stable to light and heat.

Uses

Hymexazol controls Fusarium wilt disease, caused by *F. oxysporum f. sp. cucumerinum*, in cucumber plants. It is also used worldwide as a systemic soil and seed fungicide for the control of diseases caused by *Fusarium, Aphanomyces, Pythium*, and *Corticium spp. in* rice, sugarbeet, fodderbeet, vegetables, cucurbits, and ornamentals. Depending on the region of use, hymexazol is available as a dispersible powder for soil incorporation, a wettable powder for seed treatment, and an aqueous liquid for soil drenching applications.

Hymexazol is applied as a soil drench at 30g to 60g active ingredient per hectoliter or by soil incorpo-ration. It is also used as a seed dressing or as a commercial seed treatment on pelleted seed for sugar and fodder beet at 5g to 90g per kg of seed.

Mode of Action

Among the fungicides with activity against Oomycete fungi other than propamocarb, hymexazol is unique in that it provides activity against certain *Aphanomyces spp.*. In addition, hymexazol is active against *Pythium*, *Fusarium*, and *Corticium* spp. Hymexazol has been used as an experimental tool to isolate *Phytophthora* spp.; however, certain *Phytophthora* species are inhibited by hymexazol.

Hymexazol does not provide activity against Oomycetes of the family of Peronosporaceae, such as downy mildews. Hymexazol is rapidly translocated and has locally systemic distribution properties. It also exhibits moderate apoplastic (xylem-mediated) transport properties, but no symplastic (phloem-mediated) transport properties. Investigations into its mode of action were conducted on *Fusarium oxysporum f. sp. cucumerinum* and *Pythium* spp. In *Fusarium*, hymexazol was suggested to interfere with RNA and DNA syntheses.

In Pythium, hymexazol was suggested to interfere only with RNA synthesis. Hymexazol was shown to affect mycelial growth and sporulation of *Pythium* but not zoospore mobility and germination. Upon entry into the plant, hymexazol is rapidly transformed into glucosides.

The O-glucoside has fungitoxic activity, whereas the N-glucoside is not fungitoxic. The N-glucoside has been associated with certain plant growth promoting effects, such as stimulation of lateral root hair development in seedlings.

Toxicology

Hymexazol has a World Health Organization (WHO) classification III, indicating that the product is slightly hazardous. It is neither mutagenic, nor carcinogenic, nor teratogenic. The acute oral LD_{50} values for male and female rats were 4,678 and 3,909 mg per kg of bodyweight, respectively. Acute oral LD_{50} values for male and female mice were 2,148 and 1,968 mg per kg of bodyweight, respectively. The acute percutaneous LD50 exceeded 10,000 mg/kg in rats and exceeded 2,000 mg/kg in rabbits.

Hymexazol is not a skin irritant, but it may irritate eyes and mucous membranes. The No Observed Effect Level (NOEL) over 2 years was 19 mg and 20 mg per kg bodyweight in male and female rats, respectively, and 15 mg per kg of bodyweight in dogs. The LC_{50} for rats by inhalation was higher than 2.47 mg/L.

Hymexazol was tested for bird toxicity in Japanese quail and mallard duck. The acute oral LD50 value in mg per kg of bodyweight in Japanese quail was 1,085. The acute oral LD_{50} values in mg per kg of bodyweight in malard duck was greater than 2,000. Hymexazol is not haardous to bees with oral and contact LD50 values greater than 100 µg per bee. Hymexazol was tested for fish toxicity in carp and rainbow trout. After 96 h of exposure, the LC50 was 165 mg/L for carp and 460 mg/L for rainbow trout.

Metabolism/Transformation

In animals, following oral administration, *hymexazol* is metabolized to glucuronides. In plants, *hymexazol* undergoes transformation to O- and N-glucosides. In soil, hymexazol is degraded to 5-methyl-2-(3H)-oxazolone with a half-life (DT_{50}) ranging from 2 days to 25 days.

Inhibitors of Mitochondrial Energy Production Fungicides

Adenosine triphosphate (ATP) is the common cellular currency that provides chemical energy, directly or indirectly driving the bulk of cellular biosynthetic and transport processes in fungi. Modulators of mitochondrial function that block ATP production are commercially effective fungicides, insecticides, miticides, and herbicides.

Although the mitochondrial respiratory pathway is highly conserved, both broad spectrum and taxon-specific inhibitors have been identified. Currently, the strobilurins and other molecules acting at respiratory complex III are a particularly important new class of commercially successful compounds controlling fungal diseases.

In order to understand the diversity of activities of those compounds affecting mitochondrial ATP production, a general description of the respiratory machinery involved is beneficial. Synthesis of ATP in mitochondria is a complex process involving transfer of energy through a series of proteins embedded in the inner mitochondrial membrane.

Douce estimated that a single respiratory chain in plant mitochondria, similar to many fungal mitochondria, may have an aggregate MW of approximately 1.5×10^6, being comprised of as many as 40 oxidation-reduction centers and 50 polypeptides. The components of the electron transport chain are grouped into four respiratory complexes (I, II, III, and IV) based on function and physical association in the membrane. Reductant, in the form of NADH produced in the enzymatic reactions of the Kreb's cycle in the mitochondrial matrix, enters the respiratory chain through the NADH dehydrogenase function of complex I.

Respiratory complex II contains a succinate dehydrog-enase activity that can capture energy by oxidizing succinic acid, also generated by the Kreb's cycle. Either of these respiratory complexes can then transfer reductant to a pool of ubiquinone (Q) in the membrane. The reducing power subsequently moves from the reduced ubiquinone through complex III to cytochrome c.

Cytochrome c is oxidized by complex IV, which transfers the reducing equivalents to O_2, the ultimate reductant acceptor. Transport of reducing power down this electron transport chain is linked to transfer of protons from one side of the inner membrane (the mitochondrial matrix) to the other (the intermembrane space) at complexes I, III, and IV. Because protons do not readily cross the lipid bilayer of biological membranes, this unidirectional flux generates a proton gradient across the inner membrane.

The electrochemical gradient resulting from this unequal distribution of protons provides the energy to drive synthesis of ATP by ATP synthase, also located in the inner mitochondrial membrane. A generalized schematic of these processes is shown in Figure elsewhere in this chapter. A number of fungicides interfere with the generation of the electrochemical gradient and its use to generate cellular ATP.

Compounds that interfere with the transfer of electrons from one component of the electron transport chain to another are referred to as mitochondrial electron transport inhibitors (METIs). They block production of ATP by preventing generation of the electrochemical gradient needed to drive ADP phosphorylation.

This usually occurs by the inhibitor binding to a protein component of the electron transport chain, preventing further transfer of reductant, thus blocking establishment of the proton gradient. Normally, the electrochemical gradient established by flow through the respiratory chain is dissipated as protons flow back to the mitochondrial matrix through a channel in the ATP synthase.

Energy stored in the gradient is captured by the ATP synthase and used to phosphorylate ADP to ATP. However, it is also possible to dissipate the proton gradient through artificial mechanisms, which do not allow capture of stored energy in ATP. Compounds that act by this mechanism are said to "uncouple" electron transport from ATP synthesis.

The most common mechanism by which uncouplers dissipate the proton gradient is by acting as proton ionophores. That is, they facilitate transfer of a proton from one side of the membrane to the other, effectively reducing the proton gradient. This is essentially a physical process of movement through the lipid-protein membrane and does not involve specific interactions with enzymes or membrane proteins.

Because uncouplers do not interact with specific target proteins, they are thought to be less susceptible to the development of certain types of resistance compared with other inhibitors (se Fungicides: 2-Aminopyrimidines). Although the end result of preventing ATP synthesis is similar, mechanistically the METIs and uncouplers are distinct classes.

Complex I Inhibitors

Fenaminosulf

Introduced in 1955, fenaminosulf (1; Lesan; Dexon) was one of the first selective, systemic fungicides. It is active against Phycomycetes such as Pythium, *Phytophthora*, and *Aphanomyces* but has little or no effect on higher fungi belonging to the Ascomycetes or Basidiomycetes. Fenaminosulf has a relatively high water solubility (20,000 mg L^{-1}), a feature often associated with fungicides that specifically inhibit Oomycetes. Another important physicochemical property is photolytic instability, one of the reasons that fenaminosulf was developed as a seed dressing and soil fungicide.

As a consequence of acute toxicity (see below), limited spectrum, and photo-instability, the use of fenaminosulf decreased, and it is now listed in *The Pesticide Manual* among products that have been superseded. Although no longer used, it is included here for completeness, being the only example of a commercial fungicide that acts on complex I.

During its period of use, fenaminosulf, as wettable powder (WP), dust, or granular formulations, was employed as a seed dressing and soil-applied fungicide to control damping off and foot/root rots. Fenaminosulf-based products were used on a range of plants including vegetables, field crops, ornamentals, and turf.

Structurally, fenaminosulf is closely related to azo dyes. The latter are well known carcinogens, and fenaminosulf itself tests positive in mutagenicity assays. In addition, it has a relatively low acute oral LD50 of 60 mg kg^{-1} in the rat.

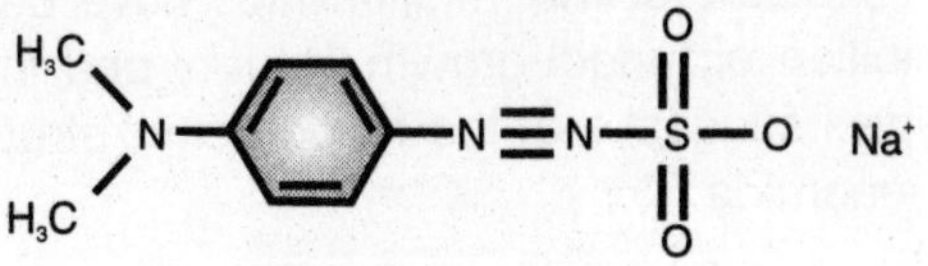

(1) Fenaminosulf

At the biochemical level, the site of action of

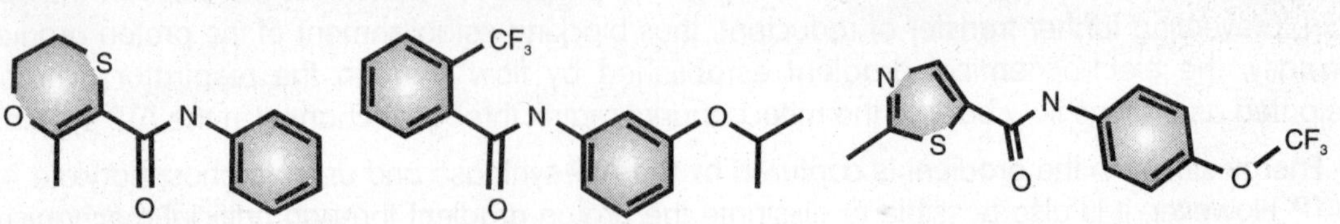

(2) Carboxin (3) Flutolanil (4) Triflumazid

fenaminosulf is complex I (NADH : ubiquinone : oxidoreductase). This was first indicated by Tolmsoff using mitochondria from *Pythium* and supported by later experiments using electron transport particles from beef heart (7,8). In the latter studies, the I50 for NADH oxidase was 1.4 μM, while comparable inhibition of succinate oxidation required a fungicide concentration of 2 mM.

The natural product and insecticide, rotenone, is also a complex I inhibitor, though it differs from fenaminosulf in having no effect on the externally facing NADH dehydrogenase found on the inner membrane of plant mitochondria.

Based on the observations that fenaminosulf inhibits not only NADH-ubiquinone reductase but also oxidation of NADH by menadione and ferricyanide, it has been proposed that the site of action within complex I involves interference with the flavin component of the dehydrogenase. In contrast with fenaminosulf, rotenone is considered to act on the O2-side of both the flavin and iron-sulfur centers.

In general, specificity to fungicides may be determined by differences between fungi with respect to uptake, metabolic detoxification (or activation), the active site, or to the existence in insensitive species of a pathway whose operation circumvents the biochemical "lesion" that would otherwise result from action of the fungicide. For fenaminosulf, the first two explanations have been invoked.

Tolmsoff proposed that sensitivity was attributed to the absence of barriers to penetration that exists in insensitive fungi such as *Fusarium* and Rhizoctonia. Alternatively, specificity may reflect differential detoxification. This mechanism may also explain plant safety to fenaminosulf because mitochondria from sugar beet have been shown to metabolize the fungicide.

Complex II Inhibitors

Carboxamides

The enforced withdrawal of mercury-based seed dressings left a significant gap in the portfolio of products for plant disease control that has been filled, in part, by the carboxamide fungicides. Following the introduction of carboxin in the 1960s, a range of related compounds has been described, more recent examples including flutolanil and triflumazid.

Structure-activity relationships have been explored in detail with respect to disease control, inhibition of fungal growth *in vitro,* and intrinsic activity on isolated biochemical systems. These studies suggest that the fundamental structural unit required for biological activity is N-phenyl-2-butenamide.

The carboxamides are primarily active against Basidiomycete fungi and provide control of smuts, bunts, and rusts, and diseases caused by *Rhizoctonia solani.* Important crop outlets are small grain cereals and cotton. Carboxamides, as WP or dispersible liquid formulations, are typically applied as seed dressings in combination with other fungicides, such as thiram, guazatine, and imazalil. The partner fungicides enhance activity and broaden spectrum to non-Basidiomycete fungal diseases, including damping-off (*Fusarium nivale*) and leaf spot (*Drechslera graminea*).

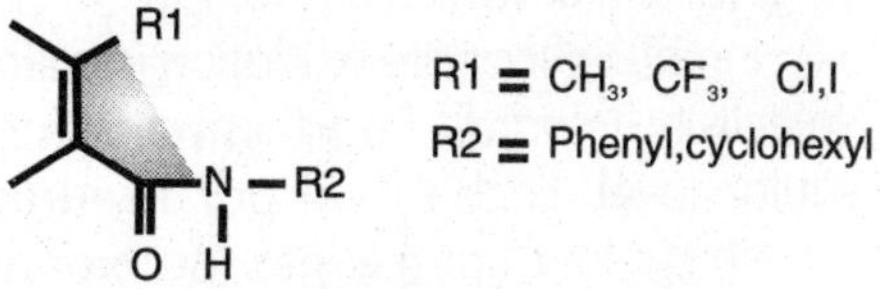

R1 = CH_3, CF_3, Cl,I
R2 = Phenyl,cyclohexyl

(5) *N*–Phenyl–2–butenamide

As well as providing control of important diseases, carboxamides may also exhibit benificial plant growth regulating effects that result in growth stimulation. The most important physicochemical properties for carboxamides are those that confer systemicity, such that a product applied as a seed dressing is taken up and redistributed in the plant to provide disease control.

The relevant properties are water solubility and octanol water partition coefficient. For carboxin, a representative carboxamide, the values are 199 mg L^{-1} (25°C) and 158 (logP 2.2; 25°C), respectively. These values presumably give an appropriate balance between lipophilicity and hydrophilicity to confer apoplastic movement.

In general, the carboxamides have favourable toxicological and environmental profiles. For a detailed account, which focuses on carboxin and oxycarboxin, the reader is referred to a recent review by Kulka and von Schmeling.

Complex II (succinate : ubiquinone reductase) is a membrane-bound enzyme that catalyzes the oxidation of succinic acid to fumaric acid. The complex consists of two subunits, a flavoprotein and an iron-sulfur protein, together with two smaller polypeptides for quinone binding and anchoring of the enzyme to the inner mitochondrial membrane.

There is a high degree of homology in the enzyme from fungal, plant, mammalian, and bacterial systems, and much of the work on carboxamide mode of action has employed mitochondria and submitochondrial fractions isolated from beef heart. Effects of carboxamides on complex II have been investigated by kinetic studies, photoaffinity labeling, and electron paramagnetic resonance (epr) spectroscopy.

Collectively, the results indicate that carboxamides act by interfering with electron transfer from the iron-sulfur protein to ubiquinone. Although details of the mechanism remain to be clarified, two recent studies, each comparing genes from wild-type and carboxamide-resistant strains, have shed light on the nature of the fungicide binding site.

The first study indicated that the *oxr-1B* mutation in the smut fungus *Ustilago maydis* results in a single amino acid substitution in the iron-sulfur protein. In the second study, using genes and gene products from the bacterium *Paracoccus denitrificans*, resistance was again conferred by a single amino acid, change but in a membrane-anchor polypeptide.

Based on these results, Matsson et al. concluded that the mutations in *U. maydis* and *P. denitrificans* defined a region in complex II that represented the catalytic center for quinone reduction and to which carboxamides bind to interfere with electron transfer. The high degree of conservation in complex II between fungi and other organisms raises the question of the basis for selectivity.

With respect to mammalian systems (e.g., beef heart mitochondria), where succinate-ubiquinone reductase is as sensitive to carboxamides as the enzyme from Basidiomycete fungi, the explanation for low acute toxicity in the whole organism is probably excretion. The finding that carboxin inhibits succinate oxidation in mitochondria from *Botrytis cinerea*, an Ascomycete, indicates that selectivity towards Basidiomycete fungi is not wholly based on differential sensitivity at the subcellular level. Lack of efficacy towards *B. cinerea in vivo* might instead have an agrikinetic basis, such as low uptake into the organism.

In view of the intrinsic activity of carboxamides to non-Basidiomycetes, Leroux speculated that it may be possible to discover inhibitors of complex II with a broader spectrum than that exhibited by the carboxamides.

Complex III Inhibitors

Inhibition of mitochondrial electron transport at complex III has emerged as an important mode of action in the control of fungal plant diseases. Complex III, or the bc1 complex, is composed of multiple proteins that collectively transfer electrons from ubiquinol, contained in the ubiquinone (Q) pool, to cytochrome c oxidase. Three of these (cytochrome b, cytochrome c1, and the Rieske Fe-S protein) have electron transfer capabilities.

Cytochrome b transfers electrons between its two redox centers that form a transmembrane electrical circuit. These redox centers are Qi (b562; Q_n; center N) on the negative side of the inner mitochondrial membrane and Q_o (b_{566}; Q_p; center P) on the positive side. On this protein, electrons are transferred from the low potential cytochrome b_{566} (b_L) at the Q_o reaction center to the high potential cytochrome b_{562} (bH) at the Qi reaction center.

The Q_o center is formed by cytochrome b, the iron-sulfur protein and, possibly, another small protein subunit. It carries out the oxidation of ubiquinol, whereas the Qi center catalyzes the ubiquinone reductase reactions. Inhibitors at complex III likely act as quinone/quinol analogs and are grouped into three classes.

Class I includes natural products such as myxothiazol and fungicides such as the strobilurins, the oxazolidinediones (famoxidone), and the imidazolinones (fenamidone). These compounds block two reactions at the Q_o site. Class II inhibitors, including the hydroxyquinone analogs, block electron transport between the iron-sulfur protein and cytochrome c_1, plus reduction of the b_{566} heme group.

Class III inhibitors bind to the Q_i site, preventing electron transport from quinone to the high-potential b562 heme, thus preventing reoxidation of the enzyme. This class includes antimycin, funiculosin, HQNO (2-n-heptyhydroxyquinoline-N-oxide), and the recently introduced fungicide IKF-916. The most significant class of fungicidal inhibitors of mitochondrial energy production in current use is the strobilurins.

These compounds were derived from the natural products strobilurin A and B, first isolated in the 1960s from wood-decaying Basidiomycetes, including *Strobilurus tenacellus*, found on pine cones. Sixteen different natural strobilurins have been isolated and characterized.

Excellent reviews of the process by which the commercial strobilurins from Zeneca and BASF were discovered have been published by Clough & Godfrey and Sauter et al..

It should be noted that, for many of the strobilurins, an increase in plant vitality, known as the "greening effect," has been found to increase yield, delay senescence, enhance water utilization, and alleviate environmental stress.

Grossman et al. suggest that kresoximmethyl treatment shifts the plant hormonal balance in favour of cytokinins and abscisic acid rather than ethylene. In these studies, kresoxim-methyl blocked production of ethylene in wheat by inhibiting the induction of enzymes involved in ethylene synthesis. The biological effects have been confirmed in the field and are considered important aspects of overall strobilurin performance.

Kresoxim-methyl

Kresoxim-methyl (6; Discus; Juwel (mixtures); Stroby; BAS 490 F; methyl (E)-2-methoxyimino-[2-(o-tolyloxmethyl)phenyl]acetate, CAS 143390-89-0) was introduced in 1996 by BASF as the first broad-spectrum strobilurin fungicide.

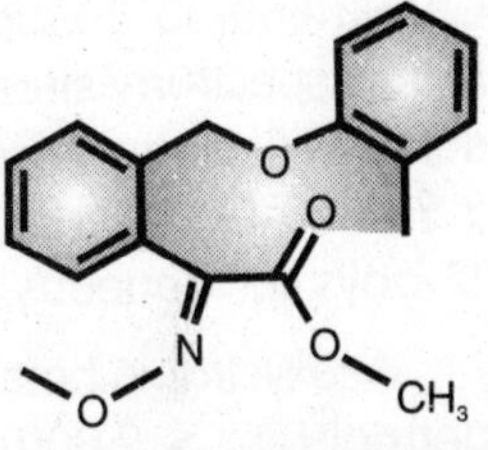

(6) Kresoxim–methyl

Kresoxim-methyl forms white, mildly aromatic crystals and has the following physicochemical properties: mp = 97.2 to 101.7°C; vp = 2.3×10^{-3} mPa at 20°C; logP = 3.4 at pH 7, 25°C; water solubility = 2 mg/L at 20°C; hydrolytically stable at pH 7 (20°C) for 24 h.

Kresoxim-methyl is used as a protectant fungicide to control a broad range of diseases caused by Ascomycete, Basidiomycete, and Oomycete pathogens. Major us include: scab (*Venturia inaequalis*) on top fruit; powdery mildews on apples (*Podospheara leucotricha*), vines (*Uncinula necator*), cucurbits (*Sphaerotheca fulginea*), and sugar beets (*Erisyphe betae*); cereal diseases, including mildews (*Blumeria graminis*), rusts (*Puccinia recondita* and *P. striiformis*), scald (*Rhynchosporium secalis*), and glume blotch/leaf spot (*Septoria nodorum* and *S. tritici*); blast (*Magnaporthe grisea*) and sheath blight (*R. solani*) on rice; and downy mildew of vines (*Plasmopa viticola*) and vegetables.

It is formulated as a suspension concentrate (SC), suspension emulsion (SE), or wettable granule (WG). Mixtures with fenpropimorph (Juwel) are preferred for cereal mildew control. It is also marketed in a three-way mix with fenpropimorph and epoxiconazole (Juwel Top) to broaden the disease control spectrum and aid in prevention of resistance development. It has eradicant activity on powdery mildews and scab.

Kresoxim-methyl has an acute oral LD50 > 5,000 and an acute dermal LD50 > 2,000 mg/kg in rats. The NOEL for rats is 800 ppm, and the ADI = 0.4 mg/kg bw (body weight). Kresoxim-methyl is not a skin or eye irritant, is nonmutagenic and nonteratogenic. It shows toxicity to aquatic organisms (fish 96-h LC_{50} = 0.681 to 1 mg/L) but does not cause permanent damage.

Other nontarget organisms show the following levels of sensitivity: bird: 14-day LD_{50} = 2,150 mg/kg; bee: 4h LD_{50} > 20 μg/bee; worm: LC_{50} > 937 mg/kg; *Daphnia:* 48-h EC_{50} = 0.186 mg/L; and algae 0- to 2-h EC50 = 63 μg/L.

Kresoxim-methyl is rapidly metabolized in mammalian systems to the virtually inactive carboxylic acid, accounting for its low toxicity and high level of selectivity. At-harvest residues in cereals and top fruit are <0.05 mg/kg and <1 mg/kg in grapes and vegetables. The soil DT_{50} = <3 days, and the Koc is 219 to 372. For the main metabolite, the Koc is 17 to 24.

Hydrolytic stability tests indicate a DT_{50} of 34 days at pH 7 but only 7 h at pH 9. Kresoxim-methyl has a low, but detectable, vapor pressure and a relatively high lipophilicity. Due to these properties, redistribution on the leaf surface is thought to occur due to the compound partitioning between the plant's waxy cuticle and the gas phase above the leaf surface.

Thus, although it is not readily transported within the plant, it is effectively redistributed on the plant surface by a mechanism termed "quasi-systemic".

Azoxystrobin

Azoxystrobin (7; ICI A5504; Methyl (E)-2-{2-[6-(2-cyanophenox-y)pyrimidin-4-yloxy]phenyl)-3-methoxyacrylate; CAS 131860-33-8; Quadris, Heritage, MW = 403.4) is a second broad-spectrum strobilurin fungicide (28,29). It forms white solids with a mp = 116°C; vp = 1.1×10^{-7} mPa at 25°C; logP = 2.5 at 20°C; water solubility = 6 mg/L; and a half-life of 11 to 17 days for aqueous photolysis.

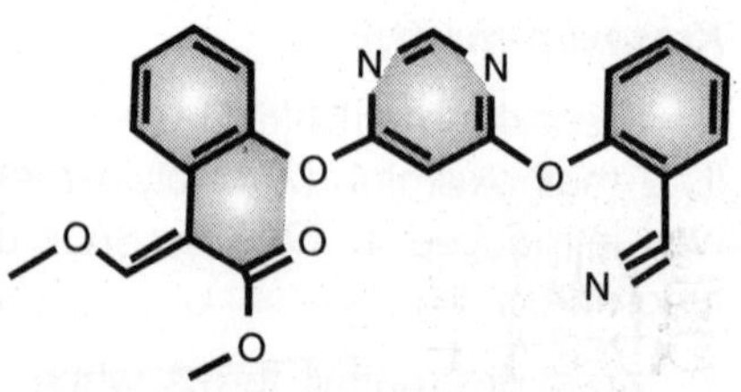

(7) Azoxystrobin

Azoxystrobin has an acute oral LD_{50} > 5,000 and an acute dermal LD_{50} > 2,000 mg/kg in rats. Azoxystrobin gives only slight skin and eye irritation and is non-mutagenic and nonteratogenic. Its NOEL is 18 mg/kg bw/day. It has no toxicity to birds in acute studies (LD_{50} > 2,000 mg/kg). It is harmless to other nontarget organisms (honey bees, earthworms, beneficial arthropods) due to low toxicity and rapid degradation in the environment, which minimizes exposure.

Environmental fate studies show rapid degradation in soil, with a DT_{50} of 1 w to 4 w. Metabolites are also rapidly degraded in soil. Photolysis studies show a DT_{50} of 11 days. Due to the rapid degradation and low soil mobility, no leaching is found in field studies. As a result, ground water contamination is unlikely.

Azoxystrobin controls key diseases on major crops with a high level of crop safety. It is active against all the major groups of fungi, including Ascomycetes, Basidiomycetes, Deuteromycetes, and Oomycetes. In addition to the major diseases listed above for kresoxim-methyl on cereals, rice, grape vines, and cucurbits, azoxystrobin is also used for turf diseases such as brown patch (*R. solani*), Pythium blight (*Pythium aphanidermatum*), and Fusarium patch (*Microdochium nivale*).

It is also used against black and yellow Sigatoka on banana (*Mycosphaerella fijiensis* and *M. musicola*). Azoxystrobin's lower vapor pressure gives it less cereal powdery mildew activity than its competitors, but its lower metabolic breakdown rate by the plant and greater water solubility provides it with relatively greater acropetal translocation and redistribution within the plant compared with kresoxim-methyl.

Azoxystrobin shows good translaminar movement, as well as systemic movement providing foliar protection when applied as a root drench, nursery box granule, paddy granule, or seed treatment. Although it shows excellent curative and eradicant activity, it is most effective when used as a protectant fungicide.

Trifloxystrobin

Trifloxystrobin (8; CGA 279202; Flint; (E,E)-methoxyimino-2-[1-(3-trifluoromethylphenyl)-ethylideneaminooxymethyl]-phenyl-acetic acid methyl ester; CAS 141517-21-7, MW = 408.8) is an oximinoacetate strobilurin introduced in 1998. Like kresoxim-methyl, the pharmacophore is an oxime ether ester.

Trifloxystrobin exists as an odorless white powder, with the following physicochemical characteristics: mp = 72.9°C; bp = 312°C; vp = 2.3 × 10^{-3} Pa at 25°C; logP = 4.5 at 25°C; water solubility = 610 μg/L at 25°C.

Trifloxystrobin is a broad-spectrum protectant fungicide that is particularly effective against major cereal diseases of barley and wheat, such as powdery mildews and leaf spots, at 250 g ai/ha. It is active at lower rates (125 to 187.5 g ai/ha) when mixed with sterol C-14 demethylase inhibiting (DMI) fungicides (e.g., propiconazole or cyproconazole), which add eradicant activity and protection against development of fungicide resistance.

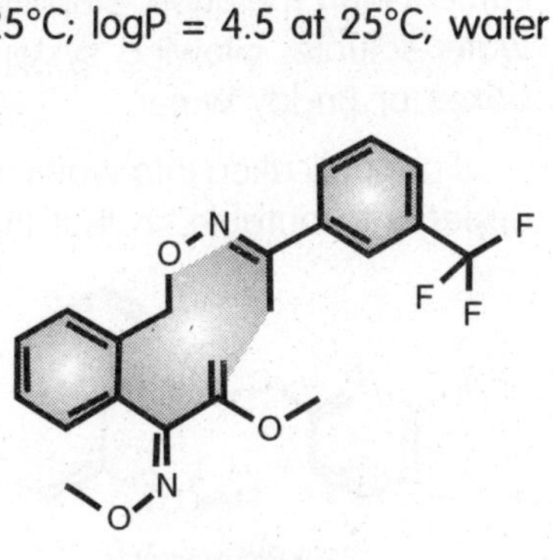

(8) Trifloxystrobin

On grapes, trifloxystrobin controls all major diseases when used at 6.25 to 7.5 g ai/hL and is particularly effective against grape downy mildew when used in combination with cymoxanil at 12.5 + 12 g ai/ha. In apples, scab is controlled at 3.75 to 5.0 g ai/hL, and powdery mildew is controlled at 5 to 7.5 g ai/hL when applied preventatively at 10- to 12-day intervals or 14- to 16-day intervals with larger fruit.

Curative scab activity (3 to 4 days after infection) is possible at the higher rate of 5 g ai/hL. Powdery mildew of cucurbits (*Sphaerotheca*, *Erysiphe*, and *Microsphaera*) can be treated successfully with 6.25 to 12.5 g ai/hL at 10-day intervals. At 70 to 105 g ai/ha, trifloxystrobin is effective against all three foliar diseases of peanut: leaf spot (*Cercospora arachidicola*), rust (*P. arachidis*), and late leaf spot (*Cercosporidium personatum*).

Finally, banana black Sigatoka (*M. fijiensis*) can be controlled at 12-day intervals at 75 to 90 g ai/ha. In all treatments, trifloxystrobin shows a high level of crop safety. Trifloxystrobin is described as "mesostemic" due to its ability to redistribute to untreated plant parts through vapor action, limited but effective cuticular penetration, and translaminar movement by diffusion. It is rainfast by virtue of its high affinity for the waxy cuticular layer.

Trifloxystrobin has an acute oral LD50 > 5,000 and an acute dermal LD_{50} > 2,000 mg/kg in rats. It is not a skin or eye irritant, is nonmutagenic and nonteratogenic. It shows rapid absorption and elimination in the rat. It has no toxicity to birds in acute studies (LD_{50} > 2,000 mg/kg) but has an LC_{50} of 0.015 mg/L in rainbow trout. The bee LD50 = 200 μg/bee.

Environmental fate studies show it to be hydrolytically stable at pH 5, with a DT_{50} of 11.4 weeks at pH 7. It has a photolytic DT_{50} of 31.5 h at pH 7 (25°C) in water, a soil adsorption coefficient (K_{oc}) of 1,642-3,745 ml/g, and a soil DT_{50} of 5.4 days under field conditions.

Metaminostrobin

Metaminostrobin (9; SSF-126; Oribright; (E)-2-methoxymino-N-methyl-2-(2-phenoxyphenyl) acetamide) is a methoxyiminophenylacetamide strobilurin that is being developed primarily for the control of rice blast (*M. grisea*) of both panicle and leaves. It also has activity against rice sheath blight (*R. solani*), apple scab (*V. inaequalis*), and powdery and downy mildews.

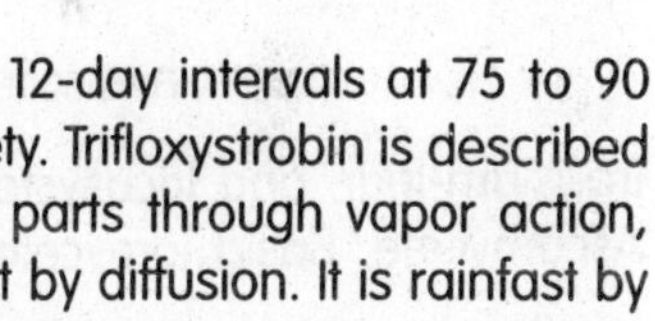

(9) Metaminostrobin

Its pharmacophore consists of an oxime ether amide group, which confers high metabolic stability. Metaminostrobin has a lower logP than azoxystrobin and is more water-soluble, allowing systemic transport through the roots of rice when applied in seedling boxes or paddy water.

For application into water, the compound is formulated as controlled release granules. Toxicity appears favourable, in that the acute oral LD50 is >300 mg/kg, and it is nonmutagenic.

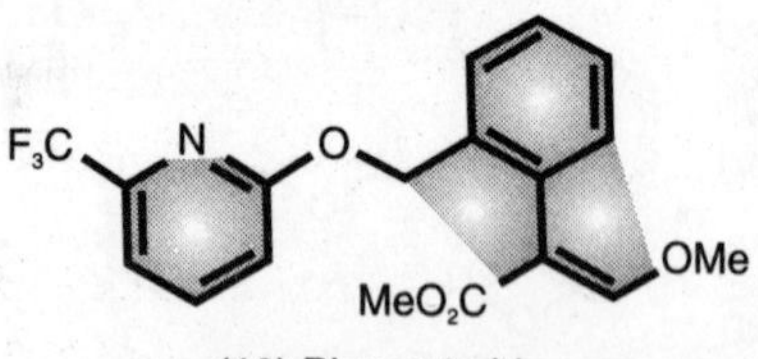

(10) Picoxystrobin

Picoxystrobin

Picoxystrobin **(10;** methyl (E)-3-methoxy-2-[22(6-trifluormethyl-2-pyridyloxymethyl)phenyl]acrylate CAS 1174222-5) from Zeneca Agrochemicals (now Syngenta Crop Protection) was first described by Godwin et al., and is being developed for broad-spectrum control of cereal diseases. The redistribution properties of the molecule, which combine xylem systemicity and vapor phase activity, are considered important in achieving high levels of control of a range of diseases. Picoxystrobin has favourable safety and environmental profiles, and enhances grain yield and quality.

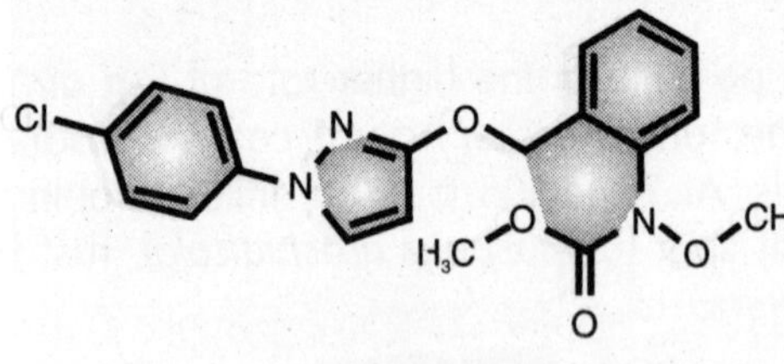

(11) BAS 500F

BAS 500 F

BAS 500 F (11; [2-[[[1-(4-chlorophenyl)-1H-pyrazol-3-yl] oxy]methyl]phenyl]methoxy-, methylester, CAS 175013t18-0, proposed common name pyraclostrobin, is a new broad-spectrum strobilurin fungicide from BASF.

The compound exhibits protectant, curative, translam-inar, and locosystemic properties, and provides control of major plant pathogens from Ascomycete, Basid iomycete, Deuteromycete and Oomycete classes of fungi.

Ammermann et al. present efficacy data on cereals grapevine, tomatoes, potatoes, beans, peanuts, citrus and turf. Pyraclostrobin results in yield enhancement with excellent crop safety. It also possesses favourable toxicological and ecotoxicological profiles. A range of formulations is under development, both as a solo product and in combination with various pre-mix partners.

Famoxadone

Famoxadone **(12;** DPX-JE 874, Famoxate; 5-methyl-5-(4-phenoxyphenyl)-3-phenylamino-2,4-oxazolidinedione; CAS 131807-57-3, MW = 374.4) is an oxazolidinedione fungicide. Although famoxadone is not a strobilurin derivative, it shares the same mechanism of action.

Famoxadone has the following physico-chemical properties: mp = 140.3 to 141.8 °C; vp = 6.4 × 10^{-7} mPa; logP = 4.65; and water solubility = 52 μg L^{-1} at 20 °C. Famoxadone provides control of a broad spectrum of fungi but is particularly effective against downy mildew (*P. viticola*), *late* and early blights (*Phytophthora infestans* and *Alternaria solani*), the *Septoria* complex, and barley net blotch at rates of 50 to 200 g ai/ha.

Target crops include vines, potatoes/tomatoes, cereals, sugar beets, canola, and cucumber. It shows good protectant, translaminar and residual control, with excellent rainfastness and good crop safety. As was seen for trifloxystrobin, it is particularly effective against grape downy mildew when applied in mixture with cymoxanil. The cereal disease spectrum and level of disease control are improved when famoxadone is mixed with triazole fungicides such as flusilazole. Famoxadone has an acute oral LD_{50} > 5,000 mg/kg and an acute percutaneous LD_{50} > 2,000 mg/kg in rats. It is not a skin or eye irritant, is negative in the Ames test, and is nonteratogenic. It shows very little soil degradation, is nonmobile, and has a good ecotoxicological profile.

(12) Famoxadone

Famoxadone inhibits activity of ubiquinol : cytochrome c oxidoreductase at the Q_o site of complex III, the same target protein as the strobilurins. However, studies show a difference in the potency of various strobilurins compared with famoxadone in single amino acid mutants of the cytochrome b protein, suggesting that this compound may interact differently with the target protein.

Fenamidone

Fenamidone (**13;** RPA 407213; (S)-5-methyl-2-methylthi5-phenyl-3-phenylamino-3,4-dihydroimidazol-4-one; CAS 161326-34-7; MW = 311) is also a complex III inhibitor that does not derive from the strobilurins but rather belongs to the imidazolinone chemical class.

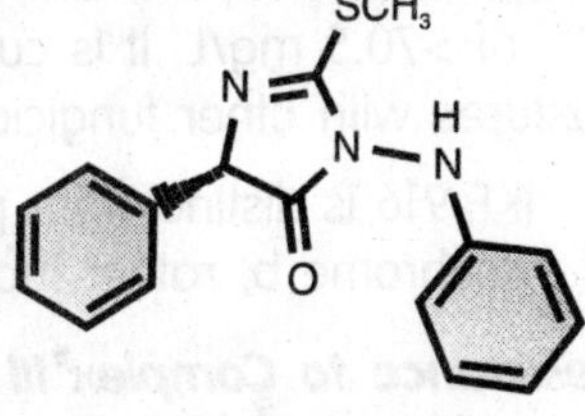

(13) Fenamidone

It was first described by Mercer et al. in 1998. Only the S-enantiomer shows antifungal activity, thus offering a reduction in application rates over the racemic mixture. Fenamidone is a white woolly powder, with the following characteristics: mp = 137°C; water solubility = 7.8 mg/L; logP = 2.8, and vp = 3.4×10^{-7} Pa. Its primary foliar use will be against Oomycete pathogens, although it has activity against other diseases, e.g., *Alternaria* on top fruit and *Mycosphaerella* on bananas.

Oomycete foliar disease control has been demonstrated for *P. viticola* on vine (3 to 6 mg ai/L), *P. infestans* on potato/tomato (25 to 200 mg ai/L), *Pseudoperonospora cubensis* on cucumber (0.2 mg/plant in seedling drench), and *Peronospora spp.* on tobacco (37 mg ai/L) and peas (0.25 to 0.5 mg/plant).

Seed treatments and soil drench applications can be used to control *Pythium* spp. on rice (20 mg/box), maize (12 to 25 g/100 kg seed), and cotton (50 to 100 g/100 kg seed). It has shown protectant, curative, and foliar antisporulant activity as well as translaminar control.

Because of its single-site mode of action, it is recommended that fenamidone be used in mixture with fosetyl-Al or chlorothalonil, which have different modes of action, to manage potential resistance. Fenamidone has an acute oral LD_{50} > 5,000 and an acute dermal LD50 > 2,000 mg/kg in rats. Fenamidone is not a skin or eye irritant, is negative in the Ames test, and is nonteratogenic.

It shows low toxicity to birds in acute and dietary studies but has a 96-h fish LC_{50} of 0.74 mg/L. Fenamidone inhibits ubiquinol : cytochrome c oxidoreductase at the Q_o site, the same target protein as the strobilurins and famoxadone. No evidence of phytotoxicity is seen at dose rates higher than those used for disease control.

H$_3$C SO$_2$N(CH$_3$)$_2$ N CN N Cl

(14) IKF–916

IKF-916

IKF-916 (14; proposed common name: cyamidazosulfamid; Ranman; Docious; Mildicut; 4-chloro-2-cyano-N,Ndimethyl-5-p-tolylimidazole-1-sulfonamide; MW = 324.8) is a cyanoimidazole fungicide.

IKF-916 is an ivory powder with the following characteristics: water solubility = 0.121 mg L^{-1} at pH 5(20°C); mp = 152.7°C; vp = <1.33 × 10^{-5} Pa; logP = 3.2 at 25°C; K_{oc} = 490 to 6,300. IKF-916 is primarily active against Oomycete fungi, including *Phytophthora, Plasmopara, Pseudoperonospora*, and *Pythium* spp., at rates of 60 to 100 g ai/ha.

It is also active against Plasmodiophoromycetes, such as *Plasmodiophora brassicae*. It is under development for preventative foliar use on potatoes, tomatoes, vines, cucurbits, onion, lettuce, and rice. Systemic activity, high residual activity, and rainfastness are claimed.

IKF-916 has an acute oral LD50 > 5,000 and an acute dermal LD_{50} > 2,000 mg/kg in rats. It is not a skin or eye irritant, is negative in the Ames test, and is nonteratogenic. It has a 96-h fish LC_{50} of >70.5 mg/L. It is currently formulated as an SC or a water dispersible granule (WDG) in mixtures with other fungicides.

IKF-916 is distinct from previously described complex III inhibitors in that it binds to the Qi site of cytochrome b, rather than the Q_o site as do the strobilurins and related compounds.

Resistance to Complex III (Q_o) Inhibitors

Laboratory studies have identified a number of mutations in the mitochondrial cytochrome b gene that confer reduced sensitivity to natural Q_o inhibitors. While many of these mutations have been useful in mechanistic studies of the cytochrome bc complex, most also function at a reduced efficiency, thus carrying a fitness penalty which may prevent them from appearing in pathogen populations in the field.

However, one of these mutations, in which the amino acid glycine at position 143 was substituted by an alanine (G143A), has been found in areas in which commercial strobilurins have failed to control diseases as expected. To date, practical resistance due to the G143A mutation has been identified in several pathogens including powdery mildews (*Erysiphe graminis tritici, Erysiphe graminis hordei, Sphaerotheca fulginea*), downy mildews (*Plasmopora viticola* and *Pseudoperonospora cubensis*), apple scab (*Venturia inaequalis*) and black sigatoka (*Mycosphaerella fijiensis*).

Although they are not chemically in the same class as the strobilurins, other Q_o inhibitors like famoxadone and fenamidone are cross resistant with the G143A mutants. A second mechanism of resistance has recently been identified in field isolates of *Venturia inaequalis* which appears to involve enhanced degradation of the compound by the pathogens and not a mutation in the target gene.

While in some pathogens the G143A mutation appears to confer high levels of resistance and is dispersed rapidly throughout the pathogen population, other pathogens show lower levels of resistance and slower spread. In yet other pathogens the G143A mutation and resistance have not yet appeared. Due to the dynamic nature of the situation, the ultimate impact of resistance on the commercial use of the Q_o inhibitors is not clear at present.

Uncouplers

Dinocap

The dinitrophenol derivative dinocap (Karathane, Crotothane, CAS 39300-45-3) is a combination of isomers (2,6-dinitro-4-octylphenyl crotonates and 2,4-dinitrooctylphenyl crotonates where "octyl" denotes a mixture of 1-methylheptyl, 1-ethylhexyl, and 1-propylpentyl groups).

Dinocap is shown in both the 2,4-dinitro and 2,6-dinitro configurations with 1-methylheptyl and 1-ethylhexyl chains, respectively, but may also have a 1-propylpentyl configuration for the "octyl" chain. It is a dark brown liquid with a bp = 138 to 140°C, vp = 5.3×10^{-6} mPa, a K_{ow} LogP = 4.54, and is almost insoluble (<0.1 mg L^{-1}) in water.

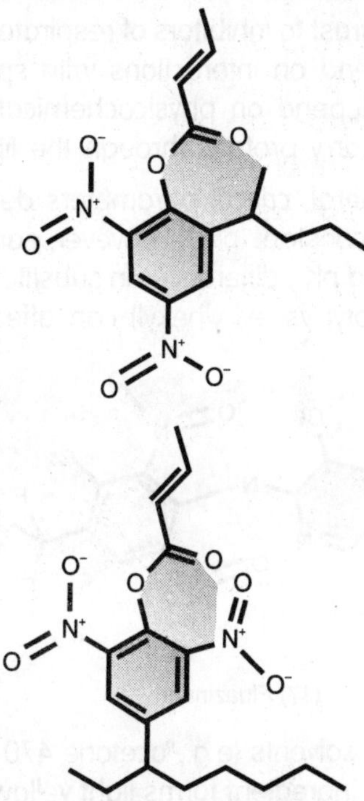

(15, 16) Dinocap

Dinocap is primarily used against powdery mildews in a number of crops, although it shows some secondary utility as a miticide against *Panonychus*, *Tetranychus*, and *Aculus*. Major uses include powdery mildews on vines, apples, ornamentals, cucurbits, and other vegetables under field as well as greenhouse culture.

Early reports indicate that the various isomers can show differentials in mildewcidal, acaricidal, and phytotoxic activities. Although the compound is nonsystemic, it shows protectant and modest curative activity against powdery mildews. Dinocap is most commonly formulated as a WP or EC and may also be used in combinations or coformulations with other fungicides and insecticides.

Dinocap shows moderate acute toxicity to mammals (acute oral LD_{50} in rats of 1,000 to 5,000 mg/kg) with moderate dermal irritation but substantial eye irritation in rabbits (Rohm & Haas MSDS C&P MSDS reference listing; MDL MSDS database listing). Dinocap is teratogenic to mice and hamsters at 10 to 100 mg/kg/day, although no evidence of teratogenic effects were seen in rabbits (Rohm & Haas MSDS C&P MSDS reference listing).

Dinocap is classified as "toxic" to fish but is listed as "nontoxic" to bees. Substituted dinitrophenols like dinocap are classic uncouplers of phosphorylation due to their ability to act as proton ionophores, dissipating the proton gradient across the mitochondrial inner membrane.

A simplified model of this ionophoric activity is shown in Figure elsewhere in this chapter. The uncoupler is dissolved in the membrane due to its lipophilic properties. The anionic form is able to bind a proton on one side of the membrane to form a neutral molecule, which then can diffuse across the membrane.

On the other side of the membrane, the neutral molecule dissociates to reform the anionic form and releases a proton, thus effectively transferring a proton across the membrane. However, because the proton was not transferred through the ATP synthase complex, the energy stored in the transmembrane H+ gradient was not captured and stored in ATP.

In contrast to inhibitors of respiratory electron transport where the structure-activity relationships (SAR) depend on interactions with specific binding sites, the SAR for ionophoric uncouplers like dinocap depend on physicochemical properties affecting the solubility of the molecule and its ability to carry protons through the lipid.

In general, critical parameters determining uncoupling activity are the lipophilicity (LogP) and ionization constant (pK_a). However, early work with Dinocap showed that, within general constraints of LogP and pK_a, differences in substitution patterns (e.g., 2,4-dinitro vs. 2,6-dinitro) and configuration (methylheptyl vs. ethylhexyl) can affect pesticidal activity.

(17) Fluazinam

Fluazinam

Fluazinam (Shirlan, Frowncide, CAS 79622-59-MW = 465.1, Mol form $C_{13}H_4Cl_2F_6N_4O_4$) 3-chloro-N-(chloro-5-trifluoromethyl-2-pyridyl)-α,α,α-trifluoro-2,6-dinitro-p-toluidine shares some structural similarities with dinocap in that it is a dinitro phenyl derivative, although, in this case, a derivative of 2,6-dinitroaniline.

Fluazinam, with a LogP of 3.56, has very low solubility in water (1.7 mg L^{-1}, pH 6.8, 25 °C) but higher solubilities in organic solvents (e.g., acetone 470 g/L, dichloromethane 330 g/L, ethanol 150 g/L, all at 20°C). The active ingredient forms light yellow crystals with an MP of 115 to 117°C and has a vp of 1.5 mPa (25°C). The compound is stable to acid, alkali, and heat.

Fluazinam shows broad-spectrum protectant activity with utility against a range of Ascomycete and Oomycete diseases on a number of crops, although it shows little curative or foliar systemicity. Pathogens controlled include *Botrytis* (grey mold), *Alternaria* (early blight), *Sclerotinia* (white mold), and *Venturia* (apple scab), as well as *Plasmopara* (downy mildew), *Phytophthora* (late blight), and other lower fungi such as *Plasmodiophora* (*clubroot*).

Fluazinam also shows acaricidal activity against citrus red mites and two-spotted spider mites. Fluazinam is formulated as SC, WP, and dustable powder (DP). Despite its high potency as an uncoupler of mammalian mitochondria (see below), fluazinam shows a relatively low mammalian toxicity (acute oral LD_{50} (rat) > 5,000 mg/kg, dermal LD_{50} > 2,000 mg/kg).

The reason for the low mammalian toxicity appears to be related to a metabolic detoxification of the uncoupler by conjugation to cellular glutathione or other sulfhydryl-reactive molecules. Under normal cellular glutathione levels, fluazinam uncoupling is only transient. Where glutathione concentrations in the mitochondrion are decreased, fluazinam activity is enhanced, while additions of exogenous glutathione result in loss of uncoupling by fluazinam.

Guo et al. showed that a fluazinam analog in which the benzyl ring chlorine was substituted with a propoxy group retained good uncoupling activity that was stable over time. From this result they suggested that the loss in fluazinam activity with time was due to glutathione conjugation at the phenyl ring chlorine.

These, and analogous data from Hollingworth and Gadelhak, provide a strong indication that the fungicidal activity and mammalian safety may be related to the levels of glutathione in fungi and mammals and its effect on fluazinam detoxification. Despite its low acute toxicity, fluazinam has other toxicological effects.

Instances of contact dermatitis have been reported for Dutch and Japanese workers from exposure to fluazinam-treated materials, apparently due to the active ingredient and not another formulation component. Bruynzeel et al. noted that the fungicide treated materials were not handled in accordance with the manufacturer's label recommendations and that changes in workplace practices to reduce worker exposure to fluazinam appeared to prevent subsequent cases of dermatitis.

There are potential ecotoxocological hazards associated with the use of fluazinam in that it is toxic to fish (LC50 0.11 mg/L (96 h, rainbow trout).

Fluazinam acts by uncoupling mitochondria through its ability to act as proton ionophore. Detailed considerations of the physicochemical properties of fluazinam and analogs provide an analysis of the properties required for activity. The inhibition of mitochondrial ATP production results in inhibition of spore germination, mycelial growth, appressorial formation, and host penetration in a number of important pathogens.

Triphenyltins

Of the estimated 30,000 tons of organotin compounds produced each year, just over 3% are used in plant protection. There are two main fungicides, triphenyltin hydroxide (fentin hydroxide) and the corresponding acetate (fentin acetate).

Fentins are produced as WP and SC formulations, either as single ai's (e.g., Brestan, Du-Ter) or in combination with other fungicides such as mancozeb or maneb. The fungicidal properties of organotin compounds were first described nearly 50 years ago.

Sn — R

(18) Fentin hydroxide, R= OH;
(19) Fentin acetate, R= OCOCH

Although the triphenyltins exhibit some curative effect, they are used mainly as protectants for control of a range of diseases, examples being early and late blights of potato and leaf spots on sugar beet and peanut. In rice, fentin acetate not only protects against a number of important diseases (blast, sheath blight, brown spot) but also controls algae and snails.

Fentin acetate cannot be used on some crops (e.g., grapevine) because of unacceptable plant injury. Fentin acetate and fentin hydroxide have relatively high octanol : water : partition coefficients, with a logP of 3.43 for both.

They exhibit low water solubilities of approximately 9 mg L^{-1} (20°C, pH 5) and approximately 1 mg L^{-1} (20°C, pH 7) for fentin acetate and fentin hydroxide, respectively.

In aqueous solutions, the acetate is converted to the hydroxide. The triphenyltin fungicides are acutely toxic in mammals, fish, and birds. Toxicity values for fentin hydroxide are as follows: rat: acute oral LD_{50} 140 to 298 mg kg^{-1}; carp: 48-h LC_{50} 0.05 mg L^{-1}; bobwhite quail: 8 day dietary LC_{50} 38.5 mg kg^{-1} diet. The values for both fish and birds place this ai in the EPA's "very highly toxi category.

The lipophilic character of the fentins, reflected in their high logP values, enhances membrane permeability and, thereby, biological activity. However, the same property also increases the potential for undesirable accumulation, e.g., in aquatic invertebrates and fish.

Similarly, strong adsorption to soils (fentin hydroxide has a Koc [adsorption coefficient corrected for organic carbon content] of 23,000) would essentially rule out leaching but might increase losses by run-off with sediment. In view of their hydrophobicity, fentin residues that reach aquatic systems by this or other routes would be expected to bind to the bottom and suspended sediments and to dissolved organic carbon materials (se, for general review of Fate of Pesticides in Aquatic Ecosystems).

As a consequence of their environment and toxicological profiles, organotins in general, including the triphenyltin fungicides, are under regulatory scrutiny in some countries. In the event that uses become more restricted, it will be a challenge to find replacements with comparable efficacy and low cost but with more favourable nontarget effects.

Organotins as a class exhibit biocidal properties and may have toxic effects on both target and nontarget organisms. However, compared with the alkyltins (e.g., tributyltin), triphenyltins are somewhat more selective. Because organotins have effects on several different enzyme systems, they may be considered multisite inhibitors.

Nevertheless, a consistent feature of their activity, irrespective of test organism, is an inhibitory effect on mitochondrial function and energy conservation through uncoupling of oxidative phosphorylation. In chloroplas from algae and higher plants, organotins also inhibit the related process of photosynthetic oxidative phosphorylation.

Although the data used to explain the biochemical basis for the uncoupling have come from studies on the effects of alkyl-substituted analogs on mitochondria from animals and higher plants, sufficient information has been obtained using triphenyltins to indicate that the same principles apply to the activity of these compounds in fungi. Briefly, uncoupling of oxidative phosphorylation is brought about first by inhibition of ATP synthesis by direct binding to a component of ATP synthetase (complex V) and, second, by acting as ionophores and discharging an anion-hydroxide gradient across the inner mitochondrial membrane. For a more detailed account on the mode of action of organotins, the reader is referred to the review by Cooney and Wuertz.

INDIRECT INHIBITION OF RESPIRATION

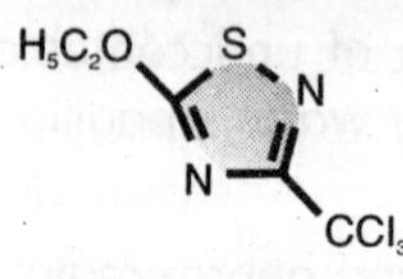

(20) Etridiazole

Etridiazole

Etridiazole was first described in 1969 and is a contact fungicide for control of soilborne diseases caused by *Pythium* and *Phytophthora*. Formulated as a WP (Terrazole), etridiazole may be used in a preventive program for control of *Pythium* blight in turf and of damping-off in ornamentals. It is also formulated in mixture with pentachloronitrobenzene (quintozene) as Terraclor Super X, which is applied in-furrow to cotton for control of seedling diseases (*Rhizoctonia, Pythium, Fusarium*). Application rates in cotton are equivalent to 1 to 1.5 and 0.26 to 0.39 kg ai ha^{-1} for PCNB and etridiazole, respectively.

Physicochemical properties that influence the way in which etridiazole is used include a high vp (1,430 mPa at 25°C) and high photolytic stability in sunlight (<10% decomposition under continuous irradiation for 7 days at 20°C). In uses where the product is incorporated, the high vp in combination with low water solubility (25 mg L^{-1}) should enhance distribution in the soil pore system and, thereby, increase efficacy.

Although the slow rate of photolysis of etridiazole permits the use of surface applications, the high vp would increase the potential for off site losses to the atmosphere through volatilization, with attendant environmental concerns. Etridiazole has low acute toxicity toward mammals (rat: oral LD_{50} 1,100 mg/kg; rabbit: dermal LD50 > 5,000 mg/kg; rat: inhalation LC_{50} (4 h) 5.7 mg/L), and birds (mallard duck: LC_{50} 1,650 mg/kg; bobwhite quail: LC_{50} > 5,000 mg/kg). It is also nonteratogenic.

However, chronic administration to rats was associated with an increased incidence of thyroid follicular cell tumors, and etridiazole (Terrazole) is included by the EPA in its list of Group B2 carcinogens. The only noteworthy ecotoxicological effect concerns the aquatic toxicity of the combination product, Terraclor Super X, which is highly toxic towards both fish and *Daphnia*. However, this can be attributed more to the PCNB component (bluegill sunfish: LC_{50} 0.1 mg/L, EPA designation "highly toxic") than to etridiazole (LC_{50} 3.27 mg/L, "moderately toxic").

Risk in practice would be mitigated by the use pattern of the product involving in-furrow soil application. Early studies by Halos and Huisman (67,68) using mitochondria from *Pythium spp.* and from the insensitive fungus *Fusarium oxysporum f. sp. vasinfectum* suggested that the site of action of etridiazole is the cytochrombc1 complex, identical or close to that of antimycin A (see Complex III).

Contrary to this proposal, later studies by Lyr and coworkers suggested that inhibition of respiration may be a secondary or indirect effect. This conclusion was initially based on observations by electron microscopy of treated *Mucor mucedo* hyphae, which showed profound ultrastructural changes after 2-h exposure to the fungicide at 10 ppm.

The fungicide-induced alterations included lysis of the inner mitochondrial membrane. Deleterious effects on this and other membrane systems were considered due to fungicide-induced lipid peroxidation and activation of a phospholipase.

CONCLUSIONS

Inhibitors of mitochondrial energy metabolism and production of ATP have been and continue to be highly effective and commercially successful fungicides. Disruption of this essential cellular function is achieved both by specific interactions with proteins involved in electron transport functions, as well as nonspecific disruption of the mitochondrial membrane's ability to utilize an electrochemical proton gradient to drive ATP synthesis.

Given the diversity of respiratory targets susceptible to inhibition that have resulted in both broad and narrow-spectrum fungicides, it is very likely that additional effective respiration targets will be identified in the future.

In addition, the recent discovery of diverse chemical classes (strobilurins and oxazolidinediones) that are highly effective respiration inhibitors indicate that novel chemistries could be discovered that act at known target sites. Mitochondrial respiration likely will remain a valuable target for identifying new, safe, and effective means to control fungal diseases.

MELANIN BIOSYUNTHESIS INHIBITORS FUNGICIDES

Melanin biosynthesis inhibitors (MBIs) are used to control the rice blast disease. The disease, caused by the pathogen *Pyricularia oryzae* (teleomorph: *Magnaporthe grisea*), is one of the most serious and damaging in rice in humid and temperate climates.

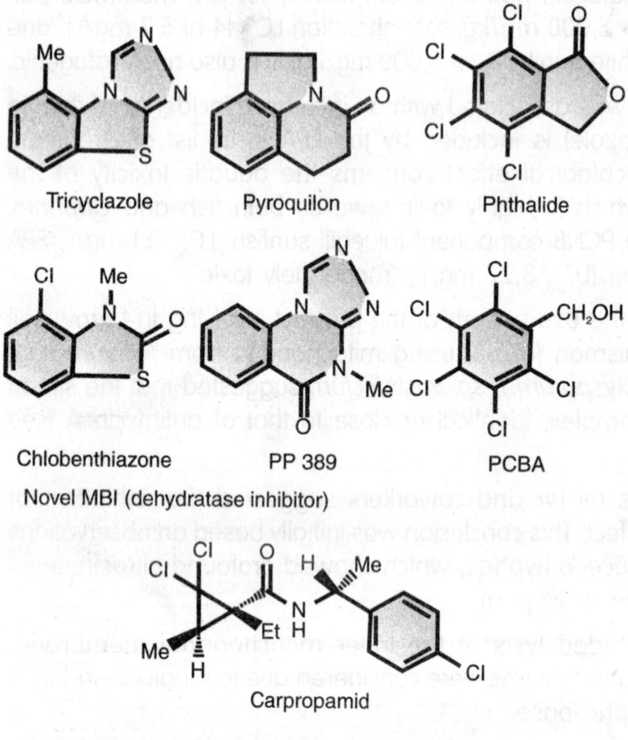

Recent studies proved that nonfungicidal chemicals effectively control rice blast disease, and one of the most potent group of such agents belongs to inhibitors of melanin biosynthesis.

This group includes compounds such as tricyclazole, pyroquilon, phthalide, chlobenthiazone, PP389, PCBA, and carpropamid. Tricyclazole, pyroquilon, phthalide, and carpropamid are currently used for practical control of rice blast disease.

These chemicals are not fungicidal or are only weakly toxic to mycelial growth or spore germination of *P. oryzae*, but possess high protectant activity against the rice blast disease.

These inhibitors show high selectivity for some Ascomycetes and imperfect fungi, and under experimental conditions, they were also shown to control *Colletotrichum lagenarium*, and *C. lindemuthianum*.

Tricyclazole Pyroquilon Phthalide

Chlobenthiazone PP 389 PCBA

Novel MBI (dehydratase inhibitor)

Carpropamid

Figure 13.12: Chemical structures of melanin biosynthesis inhibitors (MBIs). PCBA, pentachlorobenzyl alcohol.

Recent biotechnology facilitated underst-anding of the mode of action of these chemicals at the molecular level, which may provide a design of better and more effective chemicals.

Nomenclature

Tricyclazole

Common name: tricyclazole. Chemical name: 5-methyl-1,2,4-triazolo[3,4-b][1,3]benzothiazole. Chemical abstracts name: 5-methyl-1,2,4-triazolo[3,4b]benzothiazole. Trade names: Beam (DowElanco); Sazole (Sanonda) CAS RN: [41814-78-2] EEC no. 255559-5. Development code EL-291.

Pyroquilon

Common name: pyroquilon. Chemical name: 1,2,5,6-tetrahydropyrrolo[3,2,1-i, j]quinoline-4-one. Chemical abstracts name: 1,2,5,6-tetrahydro-4H-pyrrolo[3,2,1-i,j]quinoline-4-one. Trade names:

Coratop (Novartis); Fongarene (Novartis). Other name: [4lilolidone]. CAS RN: [57369-32-1]. Development code: CGA 49104.

Phthalide

Common name phthalide; fthalide (alternative spelling). Chemical name 4,5,6,7-tetrachlorophthalide. Chemical abstracts name 4,5,6,7-tetrachloro-(3H)-isobenzofuranone. Trade names Rabcide (Kureha); Blasin (Mixture)(Takeda); Kasurabcide (Hokko). Other name TCP. CAS RN [27355-22-2]. Development code KF-32.

Carpropamid

Common name: carpropamid. Chemical name: (1RS,3SR)-2,2-dichloro-N-[1-(4-chlorophenyl)ethyl]-1-ethyl-3-methylcyclopropanecarboxamide. Trade name: Win (Bayer). Test number: 0301. Development code: KTU3616.

Physical Properties

Tricyclazole

MW: 189.2 M.f.: $C_9H_7N_3S$. Form: Crystalline solid mp 187-188°C bp 275°C vp 0.027 mPa (25°C). log K_{ow}: 1.4. Solubility: In water at 25°C, 1.6 g/L. In acetone 10.4, methanol 25, xylene 2.1 (all in g/l, *25°C*). Stability: Stable at 52°C. Relatively stable to ultraviolet light.

Pyroquilon

MW: 173.2 M.f.: $C_{11}H_{11}NO$ Form: White crystals mp 112 °C vp 0.16 mPa (20 °C). log K_{ow}: 1.57. Specific gravity: 1.29 (20 °C). Solubility: In water 4 g/L (20°C). In acetone 125, benzene 200, dichloromethane 580, isopropanol 85, methanol 240 (all in g/L, 20°C). Stability: Stable to hydrolysis, and to temperature up to 320°C.

Phthalide

MW: 271.9 M.f.: $C_8H_2Cl_4O_2$. Form: White crystals mp 209-210 °C vp 3 × 10^{-3} mPa (23 °C). log K_{ow}: 3.01. Solubility: In water 2.5 mg/L (25 °C). In acetone 8.3, benzene 16.8, dioxane 14.1, ethanol 1.1, tetrahydrofuran 240 (all in g/L, 25°C). Stability: Stable for 12 h at pH 2 (2.5 ppm aq. solution); in weak alkali DT_{50} c. 10 d (pH 6.8, 5-10 °C, 2.0 ppm aq. solution); 15% ring opening in 12 h (pH 10, 25°C, 2.5 mg/L aq. solution). Stable to heat and light.

Carpropamid

MW: 334.7 M.f.: $C_{15}H_{18}Cl_3$No Form: Colourless powder. Specific gravity: 1.290 g/cm³ (20 °C) mp 147-149°C (diastereomer A: 161.7°C, diastereomer B: 157.6°C) vp 0.00027 mPa (20°C, OECD). log K_{ow}: 4.23 (22°C). Solubility: In water 3.6 mg/L (A: 3.8, B: 3.0).

In acetone 153, methanol 106, acetonitrile 65, toluene 39, n-hexane 0.9, dichloromethane 350, tetrahydrofurane 345 (all in grams/liter at 20°C), dimethylformamide >200 (25 °C), dimethylsulfoxide about 200 (25°C) g/L. Stability: Stable to heat, acid, base, and light. Formulation: 4GR, 15SC, 0.5DP, mixtures with imidacloprid, etc.

Agricultural Uses

Tricyclazole

Systemic fungicide that controls rice blast in transplanted and direct-seeded rice. Can be applied as a flat drench, transplant root soak, or foliar application. One or two applications by one or more of these methods give a season-long control of the disease.

Pyroquilon

Systemic fungicide giving effective preventive control of the rice blast fungus in rice as foliar spray or seed treatment.

Phthalide

Systemic fungicide giving effective preventive control of the rice blast fungus in rice as foliar spray.

Carpropamid

Systemic fungicide that controls rice blast by nursery box treatment, foliar spray, dusting, seed treatment, and water surface application.

Mode of Action

Primary targets of MBIs are 1,3,8-trihydroxynaphthalene reductase (tricyclazole, pyroquilon, and phthalide) and scytalone dehydratase (carpropamid) in the melanin biosynthetic pathway of the rice blast fungus, P. oryzae. For pathogenic fungi such as Pyricularia spp. and Colletotrichum spp., formation and melanization of an appressorium, the swollen tip of a germ tube or hypha that facilitates attachment and penetration into the host is essential for plant infection.

An appressorium enables these fungi to infect host plants through the hosts' cuticular wax and silica layer, the main obstacle to the entry of pathogens into the plant issue. These appressoria are well-defined, dark, thick walled structures delimited by septa from the germ tubes.

Invasion of host plants is achieved by an infection peg that is formed at the base of an appressorium. Fungal appressoria, melanins, and antipenetrant actiity. Three developmental stages of *P. oryzae appressria* have been defined, i.e., initiation, maturation, and infection peg formation.

The conidial spores, which are disseminated from other diseased plants, land on and attach to the plant surface. They germinate in an infection drop under highly humid conditions at around 25°C. Once the germ tube apex recognizes host surfaces, the germ tube begins to swell.

Subsequently, the swollen apex is delimited by a septum and it starts darkening and excretes mucilaginous substances to securely fasten the pathogen to the plant. The appressoria formed are spherical or ovoid with diameters ranging from 5 to 12 mm, and black pigmentation usually accompanies wall thickening at maturity.

Then penetration pegs (diameter, c_r. 0.1 mm) emerge from the underside of appressoria whe cell walls are thin and mostly not melanized. The infection pegs are visible at around 16 h, and penetration through the cuticle layer into the epidermal cells of rice plant appears complete after 48 h. The pegs succeed in penetrating into the cell lumen, usually of motor cells, and become invasion hyphae with normal diameter (ca. 2.5 mm), which invade adjacent cells by overcoming the plant resistance.

Finally, mature mycelia in plants cause typical blast lesions from which numerous conidia are formed for secondary infections. Mechanical force exerted by appressoria is necessary for successful penetration of the infection peg. It seems likely that vertical pressure of the appressoria on the

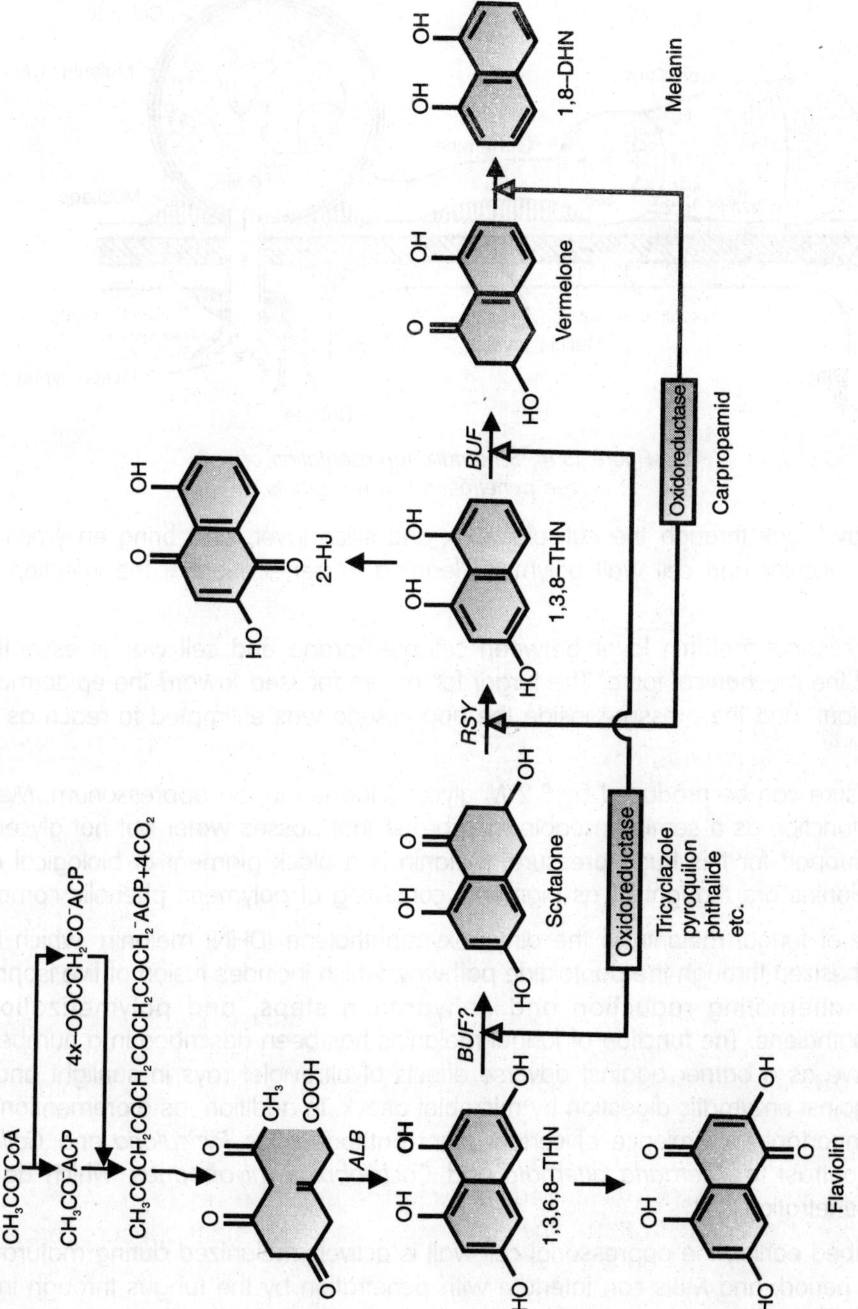

Figure 13.13: The biosynthetic pathway of fungal DHN-melanin. 1,3,6,8-THN, 1,3,6,8-tetrahydroxynaphthalene; 1,3,8-THN, 1,3,8-trihydroxynaphthalene; 1,8-DHN, 1,8-dihydroxynaphthalene; 2-HJ, 2-hydroxyjuglone.

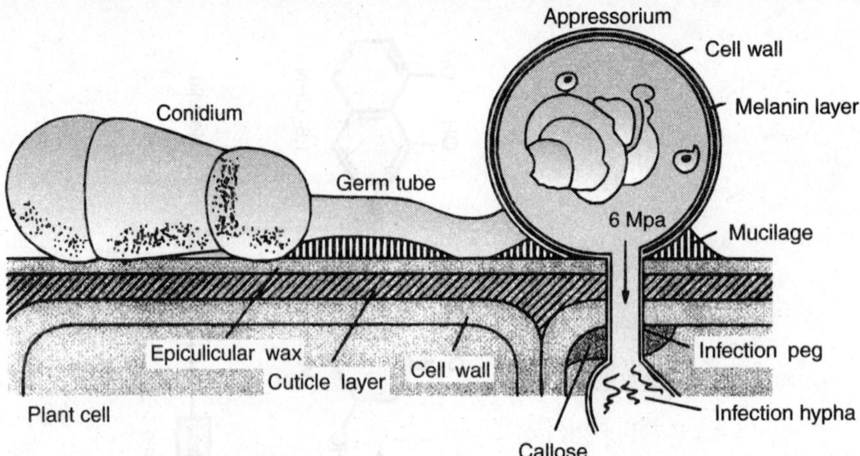

Figure 13.14: Schematic representation of P. oryzae penetration into rice plants.

epidermis may break through the cuticular wax and silica layer, and bring enzymes into easy contact with cuticular and cell wall polymers, leading to penetration of the infection pegs into plant tissue.

The appressorial melanin layer between cell membrane and cell wall is essential for the generation of the mechanical force. The turgor forces are focused toward the epidermal surfaces of the host plant, and the pressure inside the appressoria was estimated to reach as high as 8 MPa.

This pressure can be produced by 3.2 M glycerol formed in the appressorium. Melanin was proposed to function as a semipermeable membrane that passes water but not glycerol and as a structural support for this huge pressure. Melanin is a black pigment of biological origin, but generally melanins are thought of as pigments consisting of polymeric phenolic compounds.

One type of fungal melanin is the dihydroxynaphthalene (DHN) melanin, which is a black pigment synthesized through the pentaktide pathway, which includes fusion of five isoprenyl units, two sets of alternating reduction and dehydration steps, and polymerization of 1,8-dihydroxynaphthalene. The function of fungal melanins has been described in a number of ways; they may serve as a barrier against adverse effects of ultraviolet rays in sunlight and a potent protectant against enzymatic digestion by microbial attack. In addition, as aforementioned, fungal melanin is important for virulence of certain plant pathogens as *Pyricularia* and *Colletotrichum* species, in contrast to *Alternaria alternata* and *Cochliobolus miyabeanus,* which do not need melanin in penetration.

As described earlier, the appressorial cell wall is actively melanized during maturation within a fairly short period, and MBIs can interfere with penetration by the fungus through inhibition of melanin biosynthesis in appressoria, as discussed below in more detail.

Chemicals such as tricyclazole, pyroquilon, phthalide, and carpropamid specifically interfere with this penetration process in the infection cycle of *P. oryzae* and are actually used in practice. These chemicals exhibit a marked protective efficacy but have no curative effects on the disease.

They act only as potent antipenetrants against *P. oryzae*, as indicated by the lack of control if the epidermal layer was punctured or when the application of the chemicals was delayed. The antipenetrant activity was also proven in tests using rice sheaths or cellophane membranes, *Bryophyllum* epidermis or Formvar plastic, and onion epidermis. The fact that their antipenetrant activity in artificial membrane systems parallels their effects on rice plants strongly suggests that the target site(s) of those chemicals resides inside the pathogen but not the host plants.

When *P. oryzae* is cultured on a nutrient agar, mycelia generally show a black colour. The mycelium as well as the culture broth are also markedly melanized in liquid cultures at a later stage of growth. The melanization was specifically inhibited by the aforementioned chemicals with antipenetrant activity.

Among the MBIs, tricyclazole, pyroquilon, chlobenthia zone, PP389, and carpropamid inhibit the black pigmentation of both the mycelia of *P. oryzae* and the culture media, at concentrations that are not inhibitory to conidial germination or the mycelial growth of the pathogen. In contrast, the effect of phthalide and PCBA at nonfungicidal concentrations appears to be weak, leading to browngray mycelia.

Furthermore, the antipenetrant activity of MBIs was shown to be associated with the inhibition of appressorial melanization of *P. oryzae*. Tricyclazole, pyroquilon, and phthalide: Reductase *inhibitors*. The sensitive steps to MBIs are the conversion of 1,3,6,8-THN (1,3,6,8-tetrahydroxynaphthalene) to scytalone and 1,3,8-THN (1,3,8-trihydroxynaphthalene) to vermelone.

The latter step is more sensitive to MBIs than is the former, because 2-HJ (2-hydroxyjuglone) accumulates in fungal cultures more significantly than does flaviolin at low inhibitor concentrations. Both steps are reductive reactions that were, in a cell-free homogenate of *Verticillium dahliae*, catalyzed by reductase(s) requiring NADPH as a cofactor.

The second reductive reaction, from 1,3,8-THN to vermelone, is thought to be more important than the first as a target for the MBIs *in vivo*. Analysis of structural requirements for melanin biosynthesis inhibitors proposed that a benzo-bicyclic ring system, a nitrogen atom at a position alpha to the benzene ring, and small substitution groups at specific positions are important for potent activity of the compounds. From a different point of view, new melanin inhibitors were designed on the assumption that the enzymatic inhibition by known MBIs occurs competitively between the inhibitors and 1,3,8-THN.

The synthesized phthalazine derivatives showed high inhibitory activity on *P. oryzae* melanization as well as on rice blast. These results suggest that the reductase converting 1,3,8-THN to vermelone is the actual target site of MBIs *in vivo*. The 1,3,8-THN reductase (3HNR) genes (cDNAs) were isolated from *P. oryzae* and from some other phytopathogenic fungi.

The 3HNR cDNA of *P. oryzae* codes for a polypeptide of 282 amino acids with a calculated molecular mass of 30 kDa. This protein is a tetramer of four identical 30-kDa subunits. The deduced amino acid sequences of 3HNRs showed about 80% identity to each other. The reductase also shares 56% identity with a putative ketoreductase (Ver-1) involved in aflatoxin biosynthesis in Aspergillus *parasiticus*.

Additional similarities were found with numerous other oxidoreductases. Wheeler et al. have reported the inhibition of aflatoxin B1 production in *Aspergillus flavus* by tricyclazole, PCBA, phthalide, and pyroquilon. In view of the sequence similarity, it was proposed that tricyclazole prevents aflatoxin B1 synthesis by inhibiting the product of *ver-1*. Genetic analysis of melanin-deficient mutants, albino (*alb*⁻), rosy (*rsy*⁻), and buff (*buf*), of *M. grisea* showed that the three mutant phenotypes are due to single gene defects at unlinked loci. Analysis of double mutants between the three classes of mutants revealed epistas relationships.

The data suggested the order of function of *ALB*⁺, *RSY*⁺, and *BUF*⁺ in melanin biosynthesis coding for polyketide synthase, scytalone dehydratase, and 1,3,THN reductase, respectively. However, it has not been clear why mutants at the first reduction step have not been isolated so far. The reason for this phenomenon was solved.

It was found that two naphthol reductases function at the first reduction process in *P. oryzae*. Both 3HNR and a second naphthol reductase (4HNR) specific to 1,3,6,8-THN are responsible for the first reduction step. 4HNR prefers 1,3,6,8-THN as a substrate over 1,3,8-THN by a factor of 310. In contrast, 3HNR prefers 1,3,8-THN as a substrate over 1,3,6,8-THN by a factor of 4.2. As a result, no block mutant at the 1,3,6,8-THN reduction step could be isolated.

In contrast, a mutation in the 3HNR gene may cause a block at the 1,3,8-THN reduction step. Kinetic study revealed that 4HNR has a 200-fold larger K_i for tricyclazole than that of 3HNR, and this accounts for the latter enzyme being the primary physiological target of the fungicide.

Crystal structure of the 3HNR from the rice blast fungus with NADPH and tricyclazole was determined at 2.8-Å resolution. Structural information revealed that four identical subunits of 3HNR form a tetramer and that 3HNR seems to belong to the short-chain dehydrogenase family. The subunit contains a dinucleotide-binding fold that binds the coenzyme, NADPH. Tricyclazole binds at the active site in the proximity of the nicotinamide ring of NADPH.

The N2 and N3 atoms of the inhibitor form hydrogen bonds with the phenolic hydroxyl group of Y178 and the hydroxyl group of S164, respectively. The inhibitor is stacked between the side chain of Y223 on one side and, to a small extent, the nicotinamide ring of NADPH on the other side, with stacking distances of about 3.5 Å.

The structural information indicated that tricyclazole is a competitive inhibitor of 3HNR, the conclusion also drawn from enzyme kinetics data. Tricyclazole and part of NADPH are embedded in the protein by a ^{C}terminal helix-loop-helix region, and their binding and release might be controlled by conformational changes of the enzyme.

Crystal structure analysis of apo-3HNR showed that the helix-loop-helix is extremely flexible in the absence of the inhibitor and NADPH. These data suggest that the helix-loop-helix region may act as a lid that covers the active site upon binding of the substrate. Appressorial melanization of *P. oryzae* in the presence of MBIs was restored by the addition of vermelone or 1,8DHN, both later intermediates in melanin biosynthesis.

However, in contrast to the significant recovery of melanization, the restoration of appressorial penetration by these intermediates was not complete and only partial. Accumulation of toxic shunt metabolites such as 2-HJ in melanin biosynthesis may contribute to the fungicidal effect of MBIs. Alternatively, precisely localized synthesis of melanin in the appressorium may be needed for proper penetration.

Carpropamid: Dehydratase inhibitor. The melanin biosynthetic pathway from 1,3,6,8-THN to

1,8-DHN involves consecutive reduction and dehydration steps, as described earlier. The dehydration reactions were shown to be insensitive to conventional MBIs. Carpropamid was recently developed as a controlling agent against rice blast disease caused by *P. oryzae*.

Application of carpropamid to some plant pathogenic fungi, including *P. oryzae*, remarkably changed the colour of the fungi from black to brown, suggesting that the target of this compound can be the enzyme(s) in the fungal melanin biosynthetic pathway. Then it was shown that metabolism of scytalone, an intermediate in the melanin biosynthetic pathway, is inhibited by carpropamid, resulting in the remarkable accumulation of scytalone in the culture broth of *P. oryzae*.

The efficacy of carpropamid in disease control was nearly parallel to the ability of melanin biosynthesis inhibition. Thus the primary target of carpropamid was presumed to be scytalone dehydratase (SD) that catalyzes two dehydration reactions, from scytalone to 1,3,8-THN and from vermelone to 1,8-DHN. SD genes (cDNAs) are isolated from *P. oryzae, C. lagenarium, Aspergillus fumigatus, and A. alternata. Carpropamid prevents infection of C. lagenarium* by inhibiting SD of this fungus *in vitro*.

The molecular mechanism of carpropamid action on SD of *P. oryzae* was studied by enzyme kinetics and X-ray crystallography with use of a recombinant SD. Inhibition of SD by carpropamid was observed at very low concentrations of the inhibitor, close to the enzyme concentration. Enzyme kinetics data suggested that carpropamid is a tight-binding inhibitor of SD.

Interactions that determine this tight binding were revealed by an X-ray crystallographic study. A structural model of the complex between SD and carpropamid was obtained at 2.1-Å resolution. In the model, three hydrogen bonds and interactions among aromatic rings are considered to be particularly important.

X-ray crystallographic study also showed that carpropamid was embedded in a hydrophobic cavity of SD. This implies that the structural change of SD is needed for inhibitor and substrate binding as well. The most significant part of structural change seems to occur at a C-terminal region (20 amino acids), as this region is extremely flexible and structural change of this region enables the inhibitor to bind to the hydrophobic pocket.

Deletion of the C-terminal of SD remarkably decreased enzyme activity; enzymatic activity was absent when at least 15 amino acids were deleted. The data suggest that the C-terminal portion of SD is important in catalysis or structural integrity. Lundqvist et al. proposed a three-dimensional structure of a complex between SD and a competitive inhibitor, and they suggested that the C-terminal region is involved in the binding of an inhibitor, but not in catalysis.

However, the activity remarkably decreased after C-terminal deletion, suggesting that the C-terminal region may be involved not only in the inhibitor binding, but also in the enzyme reaction. It was also reported that carpropamid inhibits secondary infections. When rice plants were treated with the chemical, the number of air spora dispersed from the lesions decreased significantly. The data suggest that inhibition of the secondary infection by carpropamid is achieved by specifically preventing spore liberation.

Because similar effects were observed when tricyclozole and other MBIs were used, this secondary effect may be common for all MBIs. In addition, carpropamid enhances phytoalexin production after blast infection in rice plants. These data indicate that carpropamid has a secondary effect as a plant activator.

Recently, several patents by Sumitomo Chem. Ind. Co. Ltd., Kumiai Chem. Ind. Co. Ltd., and Rhone Poulenc Co. Ltd. have been applied for SD inhibitors. Some SD inhibitors were also reported by AgrEvo UK Limited and E.I. DuPont de Nemours. Detailed analyses of the action mechanism of carpropamid may shed light on both understanding the inhibition mechanism of this new type of anti-blast agent and designing more efficient molecules.

Metabolism and Environment Fate

Tricyclazole. Animals: Rapid and extensive metabolism. Plants: The principal metabolite is the hydroxymethyl analog. Soil/Environment: Kd 4 (loamy sand, pH 6.5, 1.5% o.m.), 45 (loam, pH 5.7, 3.1% o.m.), 21 (clay loam, pH 7.4, 1.9% o.m.), 22 (silty clay loam, pH 5.7, 4.1% o.m.).

Pyroquilon. Animals: Rapidly metabolized and eliminated via urine and feces. Residues in tissues were generally low, and there was no evidence for accumulation or retention of pyroquilon or its metabolites. Plants: Major metabolites in rice grain were 3,4-dihydro-4-hydroxy-oxoquinoline-8-acetic acid and two other acetic acid derivtives. Soil/Environment: DT50 (silty soil) 2, (sandy loam) 18 w. K_d 1.3–42 μg/g soil, little to moderately mobile. Photolysis in water, DT_{50} 10 d.

Phthalide. Animals: In rats, principal metabolites are 2-hydroxymethyl-3,4,5,6-tetrachlorobenzoic acid and its oxidation products. Plants: 4,7-Dichlorophthalide and 4,6,7-trichlorophthalide are formed in rice. Soil/Environment: Principal metabolites in soil are 2-hydroxymethyl-3,4,5,6 tetrachlorobenzoic acid and its oxidation products.

Carpropamid. Animals and Plants: Carpropamid was metabolized almost in a similar pathway both in animals and in plants. Hydroxylation primarily occurs in the cyclopropane–methyl moiety and in the phenyl ring. Those primary metabolites were further oxidized, conjugated, and then excreted from rat. Soil/Environment: Indirect photodegradation plays a major role in the degradation of carpropamid in natural water, and it was degraded in the similar oxidative pathway in soil, and finally, mineralized to carbon dioxide. Its half-life in the practical paddy field ranged from 60 to 100 days. As the soil adsorption coefficient was high (c. K_{oc}: 1500), the risk for the contamination of surface water and leaching to ground-water was estimated to be low.

TOXICOLOGY AND SAFETY ASPECTS

Tricyclazole. Mammalian toxicology. Oral: Acute oral LD50 for rats 314, mice 245, dogs >50 mg/kg. Skin and eye: Acute percutaneous LD_{50} for rabbits >2000 mg/kg. Slight eye irritant; nonirritating to skin (rabbits). Inhalation: LC_{50} (1 h) for rats 0.146 mg/L air. NOEL: (2 yr) for rats 9.6 mg/kg b.w.; for mice 6.7 mg/kg b.w.; (1 yr) for dogs 5 mg/kg b.w.; 3-generation reproduction for rat 3 mg/kg b.w. ADI: 0.03 mg/kg. Toxicity class: World Health Organization (WHO) (a.i.) II; U.S. Environmental Protection Agency (EPA) (formulation) II RC risk Xn (R22). Ecotoxicology. Birds: Acute oral LD_{50} for mallard ducks and bobwhite quail >100 mg/kg. Fish: LC_{50} for bluegill sunfish 16.0, rainbow trout 7.3, goldfish fingerlings 13.5 mg/l. Daphnia: LC_{50} (48 h) > 20 mg/l; NOEC (21 d) 0.96 mg/L.

Pyroquilon. Mammalian toxicology. Oral: Acute oral LD_{50} for rats 321, mice 581 mg/kg. Skin and eye: Acute percutaneous LD_{50} for rats >3100 mg/kg. Not a skin irritant, minimal eye irritant (rabbits). Not a skin sensitiser (guinea pigs). Inhalation LC50 (4 h) for rats >5.1 mg/l air. NOEL: (2 yr) for rats 22.5, mice 1.5 mg/kg b.w. daily; (1 yr) for dogs 60.5 mg/kg b.w. daily. ADI: 0.015 mg/kg b.w. Other: Not teratogenic, not mutagenic, not oncogenic. Toxicity class: WHO (a.i.) II; EPA (formulation) II RC risk R22. Ecotoxicology. Birds: LC_{50} (8 d) for Japanese quail 794, chickens 431 mg/kg. LC_{50} (8 d) for Japanese quail >10000 mg/kg. Fish LC50 (96 h) for catfish 21, rainbow trout 13, perch 20, guppy 30 mg/l. Bees: Practically nontoxic to honeybees; LD_{50} (oral) > 20, (contact) > 1000 mg/bee. Daphnia LC_{50} (48 h) > 60 mg/L. Algae: No effect on *Scenedemus acutus*.

Phthalide. Mammalian toxicology. Oral: Acute oral LD_{50} for rats and mice >10000 mg/kg. Skin and eye: Acute percutaneous LD_{50} for rats and mice >10000 mg/kg. Not a shaved skin irritant, eye irritant (rabbits). Inhalation LC_{50} (4 h) for rats >4.1 mg/L air. NOEL: (2 yr) for rats 2000, mice 100 mg/kg diet. Other: Acute i.p. LD_{50} for male rats 9780, female rats 15000, mice 10000 mg/kg. Toxicity class: WHO (a.i.) III; EPA (formulation) IV. Ecotoxicology. Birds: No effect on hens fed 1.5 mg/kg for 7 days, and for another 3 days at 15 mg/kg. Fish LC_{50} (48 h) for young carp >320 mg a.i. (as tech. or DP)/L, 135 mg a.i. (as 50% WP)/L. Bees: Nontoxic to bees; LD_{50} (contact) > 0.4 mg/bee. Daphnia: LC_{50} (3 h) > 40 mg/L.

Carpropamid. Mammalian toxicology. Oral: Accute oral LD_{50} for rats >5000, mice >5000 mg/kg. Skin and eye: Acute percutaneous LD50 for rats >2000 mg/kg. Nonirritating to skins and eyes (rabbits). Not a skin sensitiser (guinea pigs). Inhalation LC_{50} for rats >5.063 mg/L air. Other: Not teratogenic, not mutagenic, not carcinogenic. No effect on reproduction. No chronic toxicity. NOEL: for rats (male 24.7, female 34.0), mice (male 13.6, female 20.8), dogs (male 5.90, female 1.43) mg/kg/day. Ecotoxicology. Birds: LD_{50} for Japanese quail >2000 mg/kg. Fish LC_{50} (48 h) for carp 5.6, trout 12.4 mg/L. Daphnia: LC_{50} (3 h) > 20 mg/l. Carprpamid is environmentally harmless, and despite the longlasting property, it does not accumulate residues in rice grain (residue level of carpropamid in hulled rice was 0.019 ppm (allurial soil) and 0.011 ppm (volcanic soil)).

General Comments

Although tricyclazole is known to inhibit melanin biosynthesis in a number of Ascomycetes and imperfect fungi, the control spectrum is actually limited to rice blast disease under practical conditions. This is because melanin biosynthesis is critical for *P.* oryzae to penetrate and invade rice plants armed with rigid epidermis.

The selective activity of MBIs may offer advantages over conventional fungitoxic chemicals that act on biochemical processes common to many nontarget organisms; therefore, it would ordinarily constitute a reduced environmental hazard. Effects of MBIs on other microorganisms have been reported; tricyclazole induces melanin shunt products in fungi, such as *Leptosphaeria maculans* and *Alternaria solani*.

It also inhibits the accumulation of the phytotoxin altersolanol A by *A. solani* and the accumulation of aflatoxin B by *A. flavus*. The latter effects of tricyclazole may imply that MBIs have activities other than inhibition of melanin biosynthesis, such as inhibition of the *ver-1* protein in the aflatoxin biosynthetic pathway, as described earlier.

Inhibition of conidial pigmentation of *Penicillium* and Aspergillus species by reductase inhibitors was also reported (83). This is supported by the cloning of *A. fumigatus* arpl gene, which is needed for the pigmentation. The deduced amino acid sequence showed high similarity to SD of *M. grisea* and *C. lagenarium*, suggesting that the pigments and the DHN-melanins are formed by a similar biosynthetic pathway. Most MBIs have been developed empirically by conventional screening. New MBIs may be developed by novel methods, such as structure-based drug design. This approach needs 3D-structure information as well as understanding of catalytic mechanism.

The X-ray crystal structures of SD (67,68) and 3HNR are the first 3D structures determined for target enzymes of commercial fungicides to control an agronomically important pathogen. The catalytic mechanism of SD was analyzed, and a structure-based approach is in progress.

The melanin biosynthetic enzymes of *A. alternata*, which does not require melanin for infection, could complement the deficiency of mutants in the melanin biosynthetic pathway of *C. lagenarium* and *M. grisea*. Such data may contribute to elucidation of the molecular mechanism of the fungal melanin biosynthesis.

Dehydratase and reductase in the melanin biosynthetic pathway are now major targets of new controlling agents for rice blast. Pentaketide synthases and oxidases in the melanin biosynthetic pathway are also potential targets of novel fungicides. Elucidation of fungal metabolism essential for the expression of pathogenicity should provide promising and unique targets for the development of new disease-control agents.

In spite of a long application of MBIs in fields, appearance of resistant rice blast fungus has rarely been reported. This is a superior characteristic of MBIs. However, the reason for this phenomenon is not known. Recent progress in understanding of the mechanism of melanin biosynthesis may shed light on it.

MULTISITE INHIBITORS—BROAD SPECTRUM SURFACE PROTECTANTS FUNGICIDES

Chemicals first used in agriculture to control plant dieases were inorganic fungicides such as elemental sulfur, lime sulfur, and copper fungicides, including Bordeaux mixture. Those fungicides are primarily applied to cover the surface of the plants to protect them from various plant pathogens. They are, therefore, broad-spectrum surface protectants and have been used since the nineteenth century.

However, their use has been limited because of their lower level of effectiveness, risk of their toxicity to the plants, and lack of compatibility in use with other fungicides and insecticides. Since the 1930s, chemical industries have made remarkable progress, especially in the field of organic synthesis, and some of these products were introduced in agriculture as fungicides.

The organic fungicides developed until the 1960s were mostly simple compounds such as dithiocarbamates and quinones followed by phthalimides, anilazine, guanidines, chlorothalonil, and sulfamides. They are generally more effective than are inorganic fungicides, because organic fungicides more easily penetrate lipid layers of fungal cell membranes than do inorganic fungicides.

These organic fungicides are more or less reactive to functional groups of cellular constituents of fungi, such as sufhydryl (-SH), hydroxyl (—OH), and amino (—NH_2) radicals, and they inhibit various physiological processes in the fungal cells. They are, therefore, called multisite inhibitors and used as broad-spectrum protectants in contrast to site-specific inhibitors, which inhibit a single specific site or function in the fungal cells.

The site-specific inhibitors have been developed since the 1960s as agricultural fungicides and are used now more widely. Toxicity of inhibitors to various organisms may be affected by factors such as penetration of the molecules through cell membranes and detoxification and activation mechanisms within the cells. However, sensitivity of the site of action to the inhibitor is undoubtedly a basic factor defining the toxicity to the target organisms.

Sensitivity of a specific target site to an inhibitor is often variable among various organisms, so that site-specific inhibitors are generally selective in their action, and fungitoxic spectra are more or less restricted. On the contrary, multisite inhibitors act on multiple target sites in organisms, and thus selectivity in toxicity and their fungitoxic spectra are generally broad.

Careful tests for toxicity to beneficial organisms are, therefore, necessary with these multisite inhibitors in using them as agricultural fungicides. To avoid risk of phytotoxicity of multisite inhibitors, the fungicides have been developed for applications only to cover the surface of crop plants and protect them from infection with pathogens.

Multisite inhibitors qualified as agricultural fungicides do not easily penetrate into plant cells and are not toxic to crop plants. If they do penetrate, they are highly phytotoxic because of their reactivity with multisites in the plants. On the other hand, most site specific inhibitor–type fungicides are designed for their chemical structure most fitted to inhibit the target site in the fungal cells but not the other sites in organisms.

Thus, they are much less phytotoxic and suitable as systemic fungicides, which penetrate into plant cells and act effectively on the invading pathogens. This type of site-specific, systemic, and selective fungicide has, in general, advantages over multisite inhibitors in modern agriculture. Multisite inhibiting fungicides are, however, effective to a wider range of diseases and sometimes indispensable for preventing diseases that few or none of the modern selective fungicides can control effectively.

Broad spectrum fungicides are sometimes indispensable as soil and seed fungicides because various pathogens are often the simultaneous targets in controlling soil- and seed-born diseases. Multisite inhibiting fungicides are also important to control fungal populations resistant to specific-site inhibiting fungicides, which are now occurring in the field and are a troublesome problem.

Strategies have been developed that mix or alternate multisite and specific-site inhibitors to delay the onset of fungal resistance. Without the continued availability of broad-spectrum fungicides, the use of systemic fungicides would be in jeopardy.

Anilazine

Anilazine has been developed late as a fungicide in the class of multisite inhibitors, reported in 1955 and introduced into agriculture in the 1960s. Anilazine has chemical structure of chlorinated anilinotriazine. It is solid, sparingly soluble in water, and soluble in ordinary organic solvents. Anilazine is effective primarily by foliar application as a surface protectant for the crop plants. Vapor pressure is low. It is stable under neutral and slightly acidic conditions but unstable under alkaline condition.

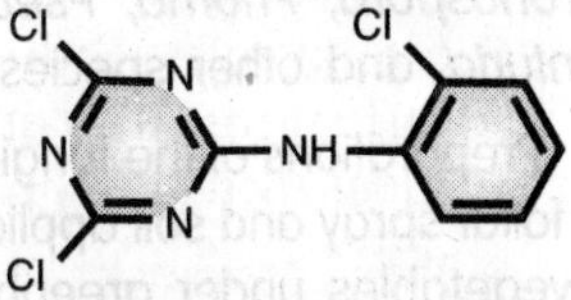

(1)

Figure 13.15: Chemical structure of anilazine.

It is effective in controlling a variety of diseases caused by *Alternaria, Ascochyta, Botrytis, Cercospora, Cladosporium, Colletotrichum, Fusarium, Helminthosprium, Leptosphaeria,*

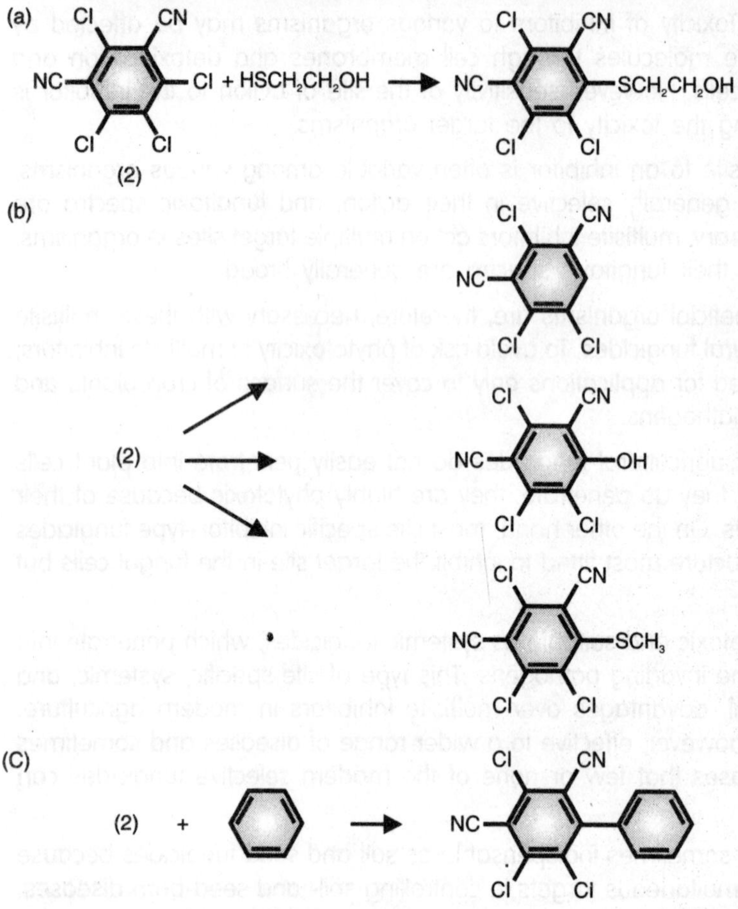

Figure 13.16: Reactions and conversions of chlorothalonil, (a) with 2-mercaptoethanol in methanolic buffer solution, pH 6.8, (b) in soils, and (c) in benzene solution under sunlight.

Mycosphaerella, Peronospora, Phytophthora, Pyrenophora, Puccinia, Rhizoctonia, Septoria, Stemphylium, and other species on a variety of crops, including berry fruit, cereals, coffee, cucurbits, ornamental plants, potatoes, tobacco, turf, vegetables, and so on.

The mechanism of action of anilazine was suggested to involve reactions of the molecule at the site of chlorine substituents on triazine ring with free amino and thiol groups of fungal cell constituents.

Chlorothalonil

Chlorothalonil was introduced in the 1960s. The fungicide has a chlorinated isophthalonitrile structure. It is colourless crystalline almost insoluble in water and soluble in ordinary organic solvents. It has a certain degree of vapor pressure and, hence, some fumigating properties. The fungicide is sometimes used as a soil fungicide to control *Aphanomyces, Phytophthora, Plasmodiophora, Rhizoctonia, Rhizopus*, and other species. Chlorothalonil is effective as a surface-protecting foliar fungicide to control *Alternaria, Botrytis, Cercospora, Cercosporella, Cladosporium, Colletotrichum, Corynespora, Elsinoe, Erysiphe, Helminthosporium*, Monilinia, *Mycosphaerella, Peronospora, Phoma, Pseudoperonospora, Sclerotinia, Septoria, Sphaerotheca, Stemphylium, Venturia*, and other species.

Preparations of the fungicide are usually suspension concentrate, wettable powder, and granule for foliar spray and soil applications. Preparations for fogging are also used, especially for cultivation of vegetables under greenhouse condition.

The crops to be protected by this fungicide are banana, beans, berry fruit, bush and cane fruits, coconut palm, coffee, cucurbits, hop, mango, mushrooms, oil palm, ornamental plants, peanut, pepper, pome fruit, potatoes, rice, rubber, stone fruit, sugar beet, tea, tobacco, turf, vegetables, vines, and wheat.

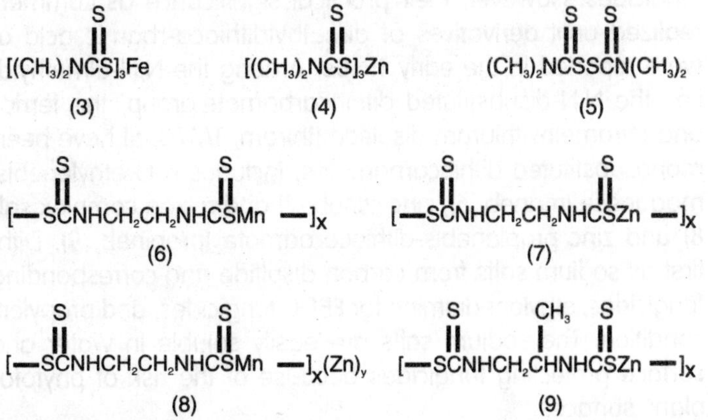

Soil application of this fungicide is effective in controlling seedling blight of rice caused *by Rhizopus* spp., which has been one of the serious diseases since the introduction of transplanting machines into rice cultivation.

Another merit of this fungicide is that it is effective in controlling diseases caused by *Oomycetes* spp. such as *Phytopht-hora, Pero-nospora,* and *Pseudo-peronospora* spp., against wh other fungicides are less effective or where resistance has occurred with modern selective fungicides.

Mechanism of action of this fungic-ide may be attributed to inhibition of physiological activit-ies of fungal cell constituents by binding reaction. The reaction was observed in buffer solution to substitute hydroxyethylthio radic-al(s) of 2-mercaptoethanol for chlorine radical(s) on the benze ring of the fungicide molecule preferably at 4-position (i.e., also 6-) followed by other positions.

$[(CH_3)_2NC(=S)S]_3Fe$ (3) $[(CH_3)_2NC(=S)S]_2Zn$ (4) $(CH_3)_2NC(=S)SSC(=S)N(CH_3)_2$ (5)

$[-SC(=S)NHCH_2CH_2NHC(=S)SMn-]_x$ (6) $[-SC(=S)NHCH_2CH_2NHC(=S)SZn-]_x$ (7)

$[-SC(=S)NHCH_2CH_2NHC(=S)SMn-]_x(Zn)_y$ (8) $[-SC(=S)NHCH_2CH(CH_3)NHC(=S)SZn-]_x$ (9)

Figure 13.17: Chemical structures of dithiocarbamates, ferbam, ziram, thiram (TMTD, 5), maneb, zineb, mancozeb, and propineb. N,N'-Ethylenebi-sdithiocarbamates and N,N'-propylenebisdithiocarbamate are polymeric salts as shown by their polymerization degree x. Mancozeb is a complex salt of maneb with zinc salt(s) (chloride and others) as shown by the amount of the added zinc in the atomic/molecular ratio y.

Similar reactions in fungal cells were observed between the fungicide and glutathione and high molecular weight cell constituents having a sulfhydryl group. The fungicide inhibits activities of thiol-dependent enzymes such as alcohol dehydrogenase, gycerald-ehyde-3-phosphate dehydrogenase, and malate dehydrogenase.

Preliminary addition of gluta-thione or dithiothreitol protects the thiol enzymes from inhibition but later addition does not reverse the enzyme inhibition. Chymotrypsin, a non-thiol enzyme, was not inhibited by this fungicide. Binding of the fungicide to the sulfhydryl group of cell constituents appears to be the primary mode of its action.

Degradation pathways of chlorothalonil in upland and paddy soils and by soil bacteria were studied, and most initial products were identified to be the results of chlorine substitution reactions, by hydrogen (i.e., dechlorination), by hydroxyl, and by methylthio groups.

These reactions took place first at the 4-position of the ring followed by reactions at other positions in the reaction with thiol compounds. Paddy soil degraded the fungicide faster than did upland soil. Chlorine substitution reaction at 4-position of the fungicide molecule was also reported in benzene solution under sunlight, and the phenyl-substituted product was identified. Similar photolysis was observed in other aromatic hydrocarbon solutions but not in acetone, hexane, and ether solutions.

Dithiocarbamates

Dithiocarbamates, patented in the early 1930s, are one of the oldest groups of organic synthetic fungicides. However, their practical significance as commercial fungicides in agriculture was not realized until derivatives of dimethyldithioca-rbamic acid and of ethylenebdithiocar-bamic acid were reported in the early 1940s. Among the N,N-dimethyldithioc-arbamic acid (DDC) derivatives, i.e., the N,N-disubstituted dithiocarbamate group, the ferric salt (ferbam, 3), zinc salt (ziram, 4), and tetramethylthiuram disulfide (thiram, TMTD, 5) have been widely used. The other group the N-monosubstituted dithiocarbamates, includes N,N-ethylenebisdithiocarbamic acid (EBDC) salts with maganese (maneb, 6), zinc (zineb, 7) and mixed complex salt of manganese with zinc (mancozeb, 8) and zinc proplenebis-dithiocarbamate (propineb, 9). Dithioc-arbamates are usually produced first as sodium salts from carbon disulfide and corresponding amines, i.e., dimethylamine for DDC fungicides, ethylenediamine for EBDC fungicides, and propylenediamine for propineb, under alkaline condition. The sodium salts are easily soluble in water and not necessarily suitable for use as surface protecting fungicides because of the risk of phytotoxicity and lack of tenacity to stick on plant surfaces.

Only one member of this group, methyldithiocarbamic acid (metam, 10) as its sodium salt (metam-sodium) is used as a soil fungicide and nematicide. Other dithiocarbamates are converted to less soluble derivatives such as metal salts and disulfide, and used as surface protectants. These metal salts and thiram are solid and stable when they are stored under dry storage conditions at room temperature. They slowly decompose with moisture or in aqueous solution or suspension.

Under acidic conditions, the dithiocarbamates decompose, liberating carbon disulfide and the corresponding amines, reversing the process of production. It means that dithiocarbamic acids are so unstable that their isolation as free acids is impossible. Metal dithiocarbamates and thiram are sparingly or slightly soluble in water. They are slightly soluble or practically insoluble in organic solvents, but thiram is soluble in most organic solvents, except alcohols and aliphatic hydrocarbons.

Vapor pressure of metal dithiocarbamates at room temperature is negligible, but thiram has high vapor pressure, and this may be an advantage of thiram among this group when it is used as seed and soil fungicides. Thiram may be able to reach seed- and soil-borne pathogens by its vapor. Formulation types of dithiocarbamate fungicides are wettable powder, suspension concentrate, and other granule and powder-type preparations for use in foliar spray and seed and soil treatments.

Dithiocarbamates are nonselective fungicides and effective in controlling numerous diseases on various crops. They are old fungicides but still useful for control of diseases that fungitoxic spectra of modern selective fungicides do not cover. Against concurrent incidence of multiple diseases, dithiocarbamates are often used in mixture with modern systemic fungicides. In particular, this kind of mixture is sometimes essential for control of diseases caused by various seed-born and soil-born pathogens at seedling stage of crop plants.

Thiram is an important component of mixed preparations for this type of seed treatment and soil application to expand the fungitoxic spectra and control the concurrent incidence of various diseases. Mixed use or alternative use of dithiocarbamates with modern selective fungicides is a practical and effective method to delay the development of fungal resistance to modern fungicides. An example is mixtures of dithiocarbamates with a class of selective fungicides, acylalanines, and related fungicides, as a countermeasure against emergence of populations of the pathogens resistant to the selective fungicides.

Dithiocarbamates control various pathogens, including *Alternaria, Aphanomyces, Botrytis, Cercospora, Cercosporella, Cladosporium, Colletotrichum, Diaporthe, Fusarium, Gloeosporium, Monilinia, Mycosphaerella, Peronospora, Phoma, Phytophthora, Plasmopara, Pseudoperonospora, Puccinia, Pythium, Septoria, Stemphylium, Uromyces,* and *Venturia* on various crop plants, including banana, berry fruit, cereals, citrus fruit, cucurbits, legumes, ornamentals, pome fruit, potatoes, stone fruit, tobacco, vegetables, and vines.

Although both N-monosubstituted and N,N-disustituted dithiocarbamates are used similarly in control of plant diseases, there seems to be a difference in mode of action between them. The difference was already suggested early in the 1950s shortly after this group of compounds was introduced as agricultural fungicides.

The disubstituted group such as DDC fungicides, including thiram, often showed unusual relations in their dosage response curves (DR curves) in the tests for inhibitory action of the fungicides on fungal spore germination. With increasing dosage, the inhibition increased in the lower concentration region, but it decreased once it reached a certain dosage and increased again in the further higher concentration region.

This type of bimodal curve is called the TMTD curve, and it is often observed with DDC fungicides but never observed with the monosubstituted group such as EBDC fungicides. As an explanation of the bimodal dosage response relation, a suggestion was first proposed that the first peak is due to the action of dithiocarbamic ions, and the action at higher dosage region is due to the undissociated molecules.

This hypothesis, however, did not explain the decreased inhibition in the intermediate concentration region. Research has been carried out on the mode of action of DDC fungicides, particularly on their action in relation to heavy metals in fungal cells or in the test media.

A suggestion was first proposed that DDC deprives fungal cells of essential heavy metals such as copper by chelation and inhibit the fungal growth. This proposal was, however, not confirmed in the tests supplemented with copper salts and with other nontoxic chelating agents such as EDTA in the test media.

Addition of the copper ion did not reverse the toxicity of DDC but increased it, and EDTA markedly depressed the toxicity of DDC. Thus, the copper ion seemed necessary for the fungitoxic action of DDC. After these experiments, a plausible explanation was proposed for the bimodal fungitoxic action of DDC fungicides. DDC and copper forms a 1 : 1 complex DDC-Cu^+, which is toxic to fungal cells.

The toxicity of DDC-Cu^+ is, however, antagonized by further addition of DDC by forming a 2 : 1 complex DDC-Cu-DDC, which is not toxic to fungal cells because of insolubility. After copper in the medium forms the nontoxic 2 : 1 complex, further addition of DDC brings about increased fungitoxicity caused by excess copperfree DDC.

In regard to the biochemical mode of action of DDC at the site(s) of its action, the most probable primary mechanism may be binding of the 1 : 1 complex DDC-Cu^+ to enzymes or other components of fungal cells, which results in inhibition of fungal growth. In fact, the role of the copper ion in coupling dithiocarbamates to proteins has been demonstrated. Besides this mechanism, other possible mechanisms were suggested as follows:

1. Transport of excess heavy metals into fungal cells by DDC resulting in inhibition of fungal growth by the metals.
2. Introduction of irregular metals catalyzed by DDC into metal-containing constituents or their precursors originally essential to fungal growth.
3. Binding of the dithiocarbamic acid (DSH) to sulfhydryl groups of cellular constituents (HSR), forming mixed disulfides DS-SR.
4. Depriving fungal cells of essential heavy metals, which is stated above and unlikely as a primary action of high potency but probable as a mechanism of action of low potency occurring in copper deficient media.

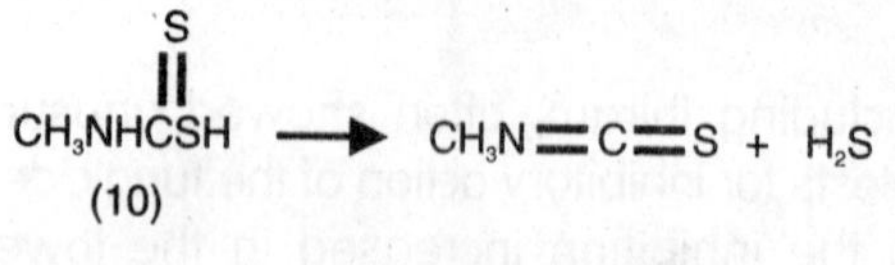

$$\underset{(10)}{CH_3NHC(=S)SH} \longrightarrow CH_3N{=}C{=}S + H_2S$$

Figure 13.18: Conversion of metam to methyl isothiocyanate.

For the above-stated possible mechanisms of action of DDC, as a representative of N,N-disubstituted dithiocarbamate fungicides, the intact structure of the molecule has an important role in the transport of metal ions or binding of the fungicide molecule to essential constituents of fungal cells.

In contrast to this, N-monosubstituted dithiocarbamates seem to be often converted before they act on fungal cells, owing to a reactive hydrogen attached to the nitrogen atom in the

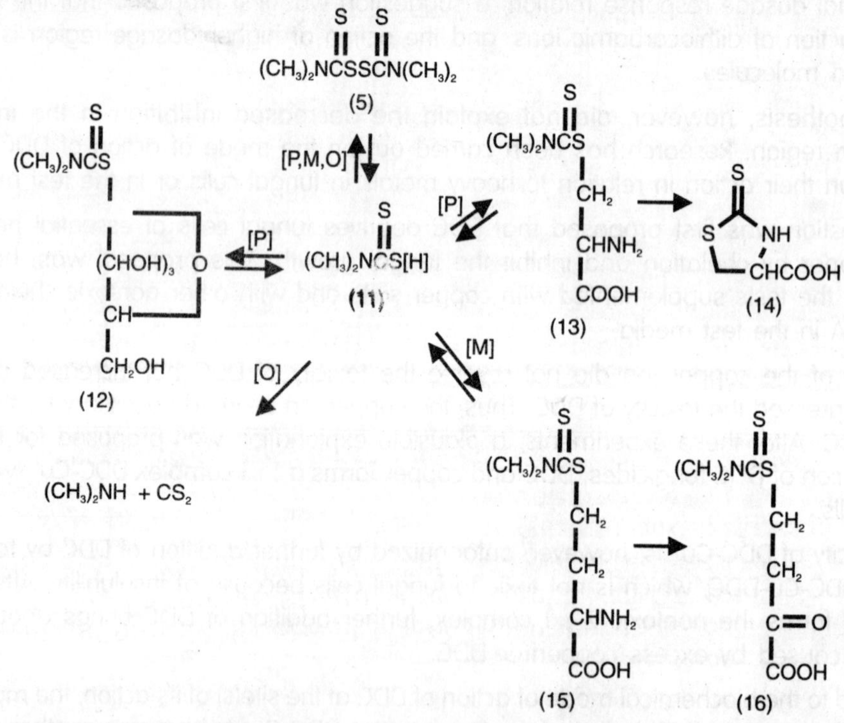

Figure 13.19: Conversions of N,N-dimethyldithiocarbamic acid (DDC) tested as its salts. Most tests were conducted with the sodium salt, i.e., ionic hydrogen [H] in DDC molecule was replaced by sodium. Similar results were expected with other metal salts. Conversions were observed by plants [P], by microbes [M], and under other conditions [O] into thiram , DDC-,-glucoside, ,-DDC-alanine, thiazolidine-2-thione-4-carboxilic acid (TTCA) γ-DDC-α-aminobutyric acid, and γ-DDC-α-ketobutyric acid.

molecules. A typical example was demonstrated by the soil fungicide and nematicide, metam. It is easily converted to methyl isothiocyanate, which is volatile and toxic to the pathogens.

Similar conversion was suggested as a mechanism of action of EBDC fungicides, and the resultant isothiocyanates were thought to bind to enzymes. As another degradation product of metam, carbonyl sulfide (COS) has been detected and suggested as one of the fungitoxic principles.

There has been another suggestion that fungitoxic action of methyl isothiocyanate is different from that of metam. The latter (metam) but not the former (methyl isothiocyanate) caused alteration of hyphal permeability. Fungitoxic action of EBDC seemed due to an alteration of hyphal permeability, which suggests similarity between metham and EBDC.

Evidence available is poor, which demonstrates similarity in fungitoxic mechanisms between N,N-disubstituted and N-monosubstituted dithiocarb-amates. Of course, there may be differences in fungitoxic action between them based on chemical reactivity. However they basically possess a similar chemistry, and EBDC can also form chelate complexes with metals like DDC.

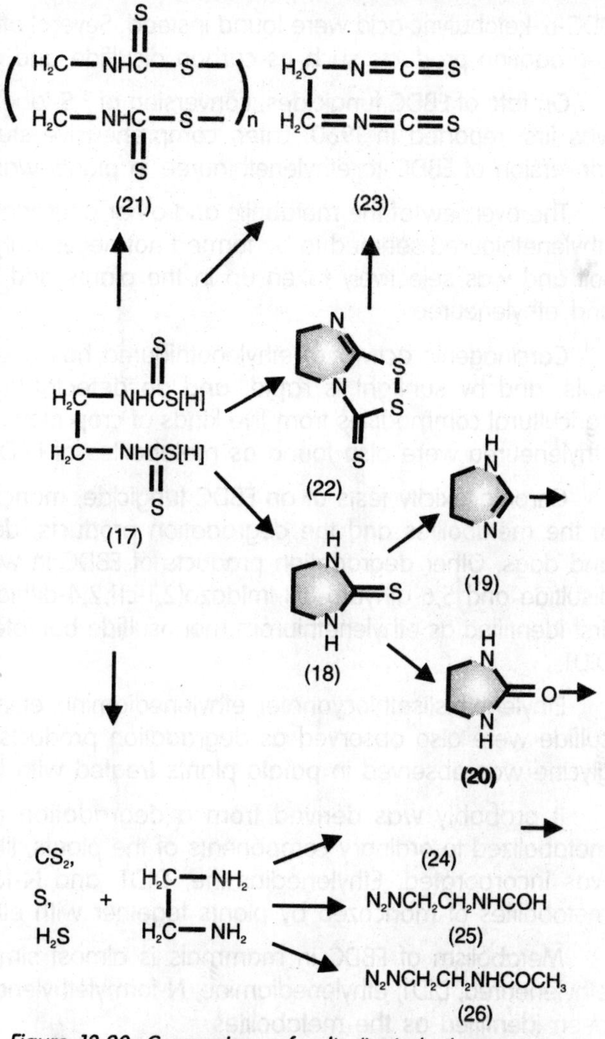

Figure 13.20: Conversions of salts (ionic hydrogen [H] was replaced by sodium and other metals) of ethylenebisdithiocarbamic acid (EBDC, 17) into ethylenethiourea, 2-imidazoline, ethyleneurea, polymeric ethylenthiuram disulfide (n is the integral number representing degree of polymerization), 5,6-dihydro-3H-imidazo[2,1-c]-1,2,4-dithiazole-3-thione (DIDT, 22), ethylene bisisothiocyanate, glycine, N-formylethylenediamine, and N-acetylethylen-ediamine.

There has been a proposal that a mechanism of toxicity analogous to that of disubstituted dithiocarbamates should not be ruled out as a possibility for monosubstituted dithiocarbamates. Conversion of DDC tested as its salts by plants and microbes has been investigated, and the main metabolites have been identified as follows.

Metabolites by plants were *DDC-β-glucoside*, β-dimethylthiocarbamoyl-thioalanine (β-DDC-alanine) and thiazolidine-2-thione-4-carboxylic acid (TTCA). TTCA was supposed to be nonenzymatically converted from β-DDC-alanine. Among metabolites of DDC by fungi and bacteria, β-DDC-alanine was not found but γ-DDC-α-aminobutyric acid and its corresponding keto acid γ-

DDC-α-ketobutyric acid were found instead. Several other unidentified metabolites and reasonable degradation products such as carbon disulfide and dimethylamine were reported.

On fate of EBDC fungicides, conversion of ^{35}S-labeled zineb to ethylenethiourea on crop plants was first reported in 1960. Later, comprehensive studies with EBDC salts were conducted and conversion of EBDC to ethylenethiourea in plants was reconfirmed.

The overview of the metabolic and other degradation pathways of EBDC is shown in Figure 6. Ethylenethiourea seemed to be formed not necessarily only in plants, but also in water and in the soil and was selectively taken up in the plants and further converted rapidly into 2-imidazoline and ethyleneurea.

Carcinogenic activity of ethylenethiourea has been suspected, but its degradation in plants, soils, and by sunlight is rapid, and no detectable amounts of residual ethylenethiourea on agricultural commodities from five kinds of crop plants have been observed. Ethylenethiourea and ethyleneurea were also found as metabolites of EBDC by mammals.

Chronic toxicity tests of an EBDC fungicide, mancozeb, which obviously involve an evaluation of the metabolites and the degradation products, demonstrated no undesirable effects on rats and dogs. Other degradation products of EBDC in water solution are polymeric ethylenethiuram disulfide and 5,6-dihydro-3H-imidazo[2,1-c]1,2,4-dithiazole-3-thione (DIDT). The latter product was first identified as ethylenethiuram monosulfide but later revised to its dehydrogenated compound, DIDT.

Ethylene bis(isothiocyanate), ethylenediamine, elemental sulfur, carbon disulfide, and hydrogen sulfide were also observed as degradation products in water solution. Formation of radioactive glycine was observed in potato plants treated with ^{14}C-labeled mancozeb.

It probably was derived from a degradation product ethylenediamine that was further metabolized to ordinary components of the plants, like glucose and starch, in which radioactivity was incorporated. Ethylenediamine, DIDT, and N-formylethylenediamine were also found as metabolites of mancozeb by plants together with ethylenethiourea and ethyleneurea.

Metabolism of EBDC in mammals is almost similar to that in plants, and ethylenethiourea, ethyleneurea, DIDT, ethylenediamine, N-formylethylenediamine, and N-acetylethylenediamine have been identified as the metabolites.

Guanidines

Two derivatives of guanidine, dodine and iminotadine, are being used as agricultural fungicides. The former was reported in 1957, whereas the latter is new as a member of the multisite inhibitor group of fungicides. Both have one or two guandine group(s) connected to a long chain alkyl or iminodi (polymethylene) group and seem to belong to a kind of cationic surface active agent.

To the structure of iminoctadine, the name "guazatine" was first applied and its use as an agricultural fungicide was reported in 1968, but its technical product was a reaction mixture consisting of the related compounds. The high purity product was prepared later and reported in 1986. The new name "iminoctadine" has been given to the active ingredient. Now the original name "guazatine" refers to the reaction mixture.

Dodine is solid and soluble in water and alcohols but almost insoluble in most other organic solvents. Its vapor pressure is negligible, and its diffusion by vapor is not to be expected. Its

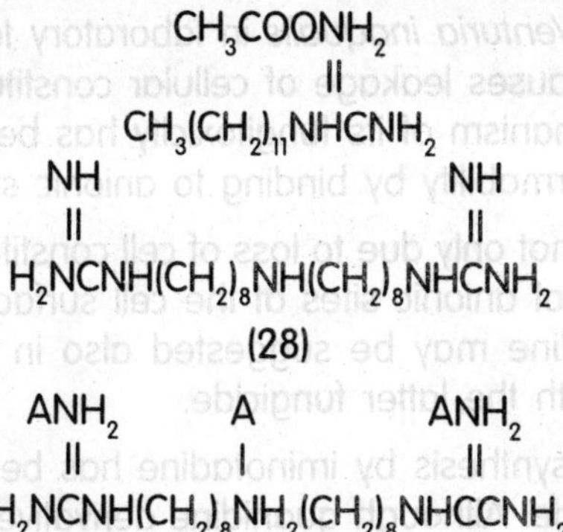

Figure 13.21: Chemical structures of guanidine fungicides, dodine, iminoctadine, and iminoctadine salts, triacetate and trisalbesilate. Acid group in the molecules of the salts is designated as A, which is (CH_3COO—) for triacetate and ($CnH_{2n}+1 - C_6H_4 - SO_2O$–, where n shows carbon number in the alkyl chain) for trisalbesilate.

formulation types are wettable powder, suspension concentrate, and soluble concentrate for spray application.

It is a typical surface protectant for control of foliar diseases of berries, ornamental plants, pome fruit, stone fruit, and vegetables caused by various pathogens such as *Alternaria, Cercospora, Mycosphaerella,* and *Venturia* spp. It has been widely used for control of scab on apples and pears caused by *Venturia spp. Phytotoxicity* is sometimes observed on sensitive crop plants.

Iminoctadine triacetate is colourless crystal easily soluble in water and alcohols. Formulation types of the triacetate are liquid preparations for spray, painting, and drenching. Iminoctadine tris(albesilate) has been developed to improve the phytotoxicity of the triacetate by decreasing the water solubility. The tris(albesilate) is waxy solid and slightly soluble in water, easily soluble in alcohols but almost insoluble in most other common organic solvents.

Formulation type of the tris(albesilate) is wettable powder. Because iminoctadine triacetate has a tendency to cause phytotoxicity, its application has been limited primarily to control summer diseases of apples, storage diseases of citrus, and seed- and soil-born diseases of cereals.

However, it is important in controlling canker of apple caused by *Valsa mali,* which most other fungicides do not control. Iminoctadine tris(albesilate) has a wider range of application and has been used on crop plants sensitive to phytotoxicity such as cucubits, pears, peaches, and vegetables.

Pathogens to be controlled by iminoctadine include *Alternaria, Botrytis, Cercospora, Cladosporium, Colletotrichum, Fusarium, Gloeosporium, Helminthosporium, Monilinia, Penicillium, Pestalotia, Phomopsis, Ramularia, Sclerotinia, Septoria, Sphaerotheca, Stemphylium, Tilletia, Valsa,* and *Venturia spp.*

Fungitoxicity of a homologous series of 1-alkylguanidine (i.e., dodine series) increases with increasing alkyl chain length up to a certain homologue, but the higher homologues beyond that lose the activity, owing to some limitations such as lower solubility in water.

The homologues exhibiting high fungitoxicity were those having alkyl groups of 13–14 carbons against cells of Saccharomyce*s pastorianus* and conidia of *Monilinia fructicola* and those of 12–

15 carbons against conidia of *Venturia inaqualis* in laboratory tests. Like other ordinary cationic surface-active agents, dodine causes leakage of cellular constituents through cell membrane of yeasts and fungi, and the mechanism of its fungitoxicity has been discussed in connection with its action on cell membrane permability by binding to anionic sites on the cell membrane.

But its fungitoxicity seemed not only due to loss of cell constituents by the fungilytic action, but also due to interference with vital anionic sites at the cell surface and possibly with certain vital enzymes. Action similar to dodine may be suggested also in iminoctadine, but few research works have been conducted with the latter fungicide.

Inhibition of fungal lipid biosynthesis by iminotadine has been suggested, presumably as a part of the effects of this fungicide. Although guanidine derivatives are a multisite inhibitor type of wide-spectrum fungicides, they show sometimes a certain selectivity in their action. Laboratory resistance to dodine has been reported in *Hypomyces solani*, and four genes have been observed to be related to the resistance.

Field resistance has been also reported after intensive and exclusive use during more than 10 years of the same fungicide against *Venturia inaequalis* on apples. Level of the field resistance was low (less than three times), and two major genes are suggested to be related to the resistance.

Biochemcal mechanism of the resistance to dodine has not been elucidated, but decreased detoxification was suggested as a mechanism because variance in detoxification was also suggested as the reason for differential sensitivity to dodine among four phytopathogenic fungi. Fate of dodine in fungal cells and in apple trees has been tested with the radioactive compound labeled on the guanidine moiety of the molecule.

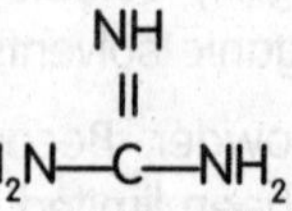

$$H_2N{-}C({=}NH){-}NH_2 \quad (30) \qquad H_2N{-}C({=}NH){-}N(CH_3)CH_2COOH \quad (31)$$

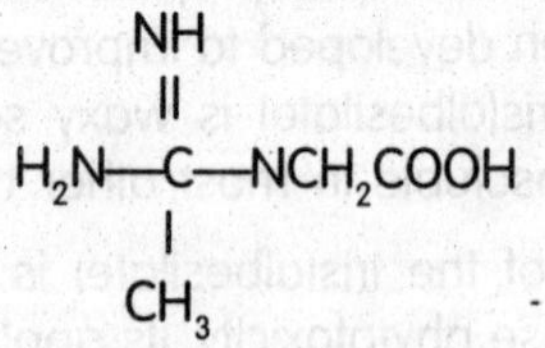

Figure 13.22: Chemical structures of guanidine and [cre]tine. Their derivatives have been found as the metabolites of dodine.

Various metabolites have been detected, among which derivatives of guanidine and creatine have been observed, suggesting transfer and methylation of the guanidine moiety of dodine in its metabolic conversions.

Phthalimides

Three fungicides, captan, folpet, and captafol are N-(chlorinated alkyl)thio substituted derivatives of tetrahydr-ophthalimide or phthalimide. They are solid, sparingly soluble in water, and hydrolyzed slowly at neutral pH but rapidly in alkaline solutions.

Captan and folpet were reported in 1952, but captafol was developed later. They are basically a multisite inhibitor type of wide-spectrum fungicides used as surface protectants. They are also used as soil and seed fungicides.

Although their vapor pressures are negligible, their gaseous degradation products are fungitoxic and may play a fumigating role in soil and seed treatments. Formulation types of this class are powder and granule types such as wettable powder and wettable granule for spray suspensions and for soil and seed treatments.

Pathogens controlled by this class of fungicides include *Alternaria, Botrytis, Cercosp-ora, Cladosporium, Colletotr-ichum, Elsinoe, Gloeosporium,* Guignardia, *Monilinia, Mycosph-aerella, Phoma, Phytophthora, Plasmopara, Pythium, Rhizoctonia, Rhynchosp-orium, Sclerotinia, Septoria, Sphaerotheca, Stemphy-lium,* Taphrina, and *Venturia* on various crops, including cereals, citrus fruit, coffee, cucurbits, ornamentals, pome fruit, potatoes, stone fruit, turf, vegetables, and vines.

The mode of action of captan has been investigated in relation to its reaction with the sulfhydryl group of cellular constituents because it reacts with various sulfhydryl compounds. The reactions have been tested with low molecular weight sulfhydryl compounds, RSH, and the products were observed as summarized in Figure elsewhere in this chapter.

A remarkable reaction is oxidation of the sulfhydryl group shown by formation of disulfides RSSR, such as cystein from cystine and oxidized glutathione from glutathione. The reaction was accompanied by formation of tetrahydropht-halimide and release of hydrogen chloride and thiophosgene $SCCl_2$.

Presumably, the reaction proceeded via unstable intermediates $RSSCCl_3$ formed by transfer of the trichloromethylthio group of cap-tan to the sulfhydryl compounds.

Thiophosgene is a reactive compound, and the release of it means further progress of its reactions with sulfhydryl, amino, hydroxyl, and other groups. Trithiocarbonates RSC(=S)SR are another type of product in the reaction of captan with sulfhydryl compounds, presumably produ-ced by the secondary reaction of thiophosgene

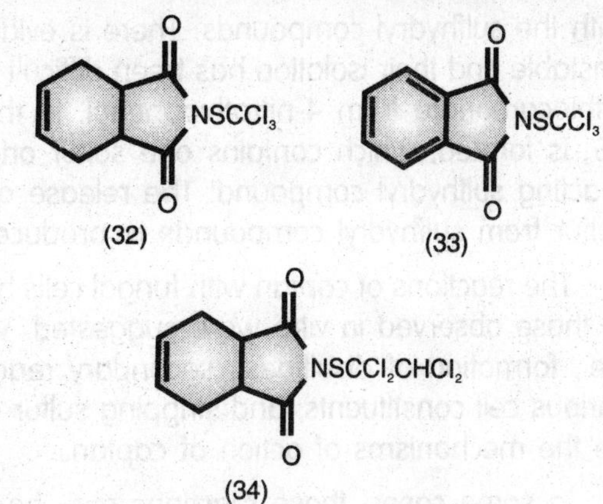

Figure 13.23: Chemical structures of phthalimide fungicides, captan, folpet, and captafol.

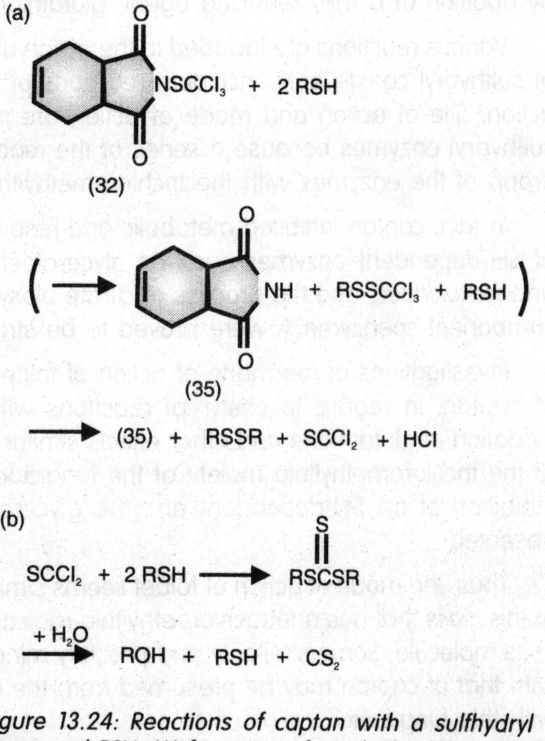

(a)

(32) NSCCl3 + 2 RSH

→ (35) NH + RSSCCl3 + RSH

→ (35) + RSSR + SCCl2 + HCl

(b)

SCCl2 + 2 RSH → RSC(=S)SR

+ H2O → ROH + RSH + CS2

Figure 13.24: Reactions of captan with a sulfhydryl compound RSH, (A) formation of tetrahydrophthalimide, disulfide RSSR and thiophosgene SCCl2 and (B) secondary reaction of thiophosgene with the sulfhydryl compound to form trithiocarbonate RSC(=S)SR followed by its breakdown to form the desthio deriv tive of the sulfhydryl compound ROH and carbon disulfide CS_2.

with the sulfhydryl compounds. There is evidence that this reaction occurs, but the products are unstable and their isolation has been difficult except in the case of formation of bis (4-nitrophenyl) trithiocarbonate from 4-nitrothiophenol. In the breakdown of trithiocarbonates, carbon disulfide CS_2 is formed, which contains one sulfur originating from captan and one originating from the reacting sulfhydryl compound. The release of carbon disulfide is, therefore, a reaction stripping sulfur from sulfhydryl compounds to produce desthio derivatives.

The reactions of captan with fungal cells have been also investigated, and the reactions similar to those observed in vitro were suggested. Various reactions, i.e., oxidation of sulfhydryl groups (i.e., formation of disulfides), secondary reactions of a degradation product thiophosgene with various cell constituents, and stripping sulfur atom from sulfhydryl constituents, are considered to be the mechanisms of action of captan.

In some cases, those reactions may be detoxification of the fungicides when the reactions take place with constituents not vital to the fungus. Sometimes reverse reactions might be possible, as in the case of reduction of oxidized forms of sulfhydryl constituents to the original form, which is suggested by fungistatic action at low concentrations of the fungicide or by recovery from inhibition by addition of a mild reducing agent, glutathione.

Various reactions are included in the action of this fungicide, and reactions other than oxidation of sulfhydryl constituents, including reactions of thiophosgene, may lead to irreversible fungicidal action. Site of action and mode of action are multiple, but this fungicide may be most fatal to sulfhydryl enzymes because a series of the reactions seem to start with binding of the sulfhydryl group of the enzymes with the trichloromethylthio group of the fungicide.

In fact, captan inhibited metabolic and respiration process utilizing glucose, in which activities of SH-dependent enzymes such as glyceraldehyde-3-phosphate *dehydrogenase, carboxylase,* and *hexokinase,* and the process of citrate biosynthesis from acetate mediated by SH-dependent component coenzyme A were proved to be strongly inhibited by captan.

Investigations of the mode of action of folpet have been conducted in a way similar to those of captan. In regard to chemical reactions with low molecular weight sulfhydryl compounds, oxidation of glutathione and other effects similar to those of captan have been observed. Transfer of the trichloromethylthio moiety of the fungicide to fungal cell constituents was suggested, and inhibition of an SH-dependent enzyme glyceraldehyde-3-phosphate dehydrogenase was also observed.

Thus, the mode of action of folpet seems similar to that of captan. Captafol is another fungicide of this class that has a tetrachloroethylthio radical instead of a trichloromethylthio radical of captan in its molecule. Some similarity and possibly minor differences in its mode of action in comparison with that of captan may be presumed from the chemical structure, but little work has been done with this fungicide.

Quinones

Quinones are an old group of organic synthetic fungicides, and their use as agricultural fungicides has been initiated with a chlorinated benzoquinone, chloranil, reported in 1940. A little later, a chlorinated naphthoquinone, dichlone was found to be more effective and has been used not only as an agricultural fungicide but also as preservatives for textile and other industrial products.

Both of the quinones are basically multisite, wide-spectrum fungicides and have been used as foliar fungicides and more widely as seed fungicides. When used as foliar spray, they sometimes show slight phytotoxicity.

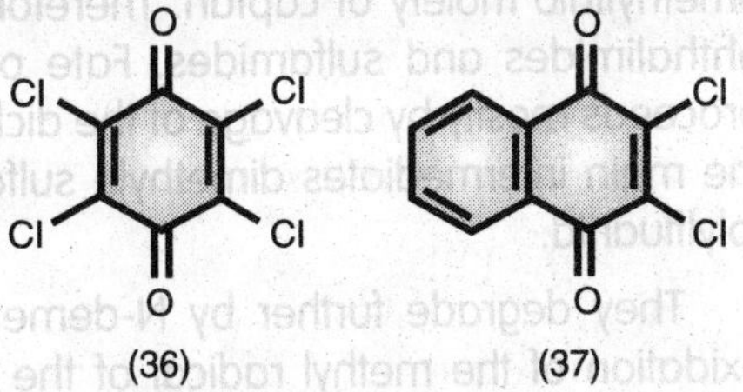

However, because they have suitable vapor pressure for providing fumigating effects, their use for seed treatment is enhanced. Recent development of systemic fungicides has greatly reduced the amount of quinones used for agriculture. Chloranil and dichlone are yellowish crystals. Chloranil is slightly soluble, and dichlone is practically insoluble in water at room temperature.

O Cl Cl Cl Cl O (36) O Cl Cl O (37)

Figure 13.25: Chemical structures of quinone fungicides, chloraniland dichlone.

They are stable in neutral and acidic media but unstable in alkalis. Chlorine atoms bound to carbon atoms adjacent to carbonyl radicals in the molecules are easily substituted by residues of compounds having sulfhydryl, amino, hydroxyl, and other functional groups.

Thus, the mode of action of dichlone was postulated as its action on sulfhydryl and amino groups of enzymes and other cell constituents followed by inhibition of phosphorylation, dehydrogenation, functions of coenzyme A, and other metabolism.

Sufamides

Dichlofluanid and tolylfluanid are N-dichlorofluoromethylthio-substituted sulfamides having N-phenyl and N-p-tolyl substituents, respectively. Activities as broad-spectrum fungicides with protective action have been reported in 1964 for dichlofluanid and in 1967 for tolylfluanid.

Dichlofluanid is solid sparingly soluble in water, soluble in most organic solvents, and decomposes in alkaline media. Tolylfluanid has physical and chemical characteristics almost similar to those of dichlofluanid. These fungicides are used as wettable powder and other water-dispersible formulation types.

Dichlofluanid is also used in formulations for fumigation. Both the fungicides are effective in controlling various pathogens, including *Alternaria, Botrytis, Cladosporium, Peronospora, Phytophthora, Plasmopara, Sphaerotheca, Uncinula,* and *Venturia* on various crop plants, including berry fruit, cucurbits, hops, ornamentals, pome fruit, stone fruit, vegetables, and vines.

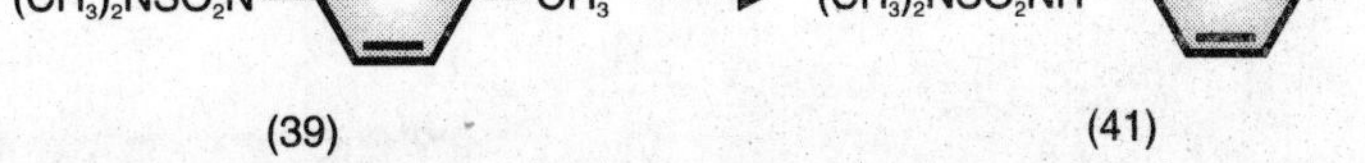

A detailed mode of action of the fungicides has not been elucidated, but he dichlorofluoromethylthio moiety in the molecules of the fungicides is mobile in metabolism by organisms and degradation in the environment, just like the trichlor-

Cl_2FCS $(CH_3)_2NSO_2N$ (38) → $(CH_3)_2NSO_2NH$ (40)

Cl_2FCS $(CH_3)_2NSO_2N$ CH_3 (39) → $(CH_3)_2NSO_2NH$ (41)

Figure 13.26: Degradation of dichlofluanid and tolylfluanid into the main intermediates dimethylsulfanilide and dimethylsulfotoluidide which are converted further as described in the text.

omethylthio moiety of captan. Therefore, some similarities in mode of action might exist between phthalimides and sulfamides. Fate of the fungicides in the environment and organisms thus proceeds mostly by cleavage of the dichlorofluoromethylthio moiety from the molecules and leaving the main intermediates dimethyle sulfanilide from dichlofluanid and dimethylsulfotoluis,dide from tolylfluanid.

They degrade further by N-demethylation, hydroxylation, hydrolysis of the sulfamide bond, oxidation of the methyl radical of the tolylmoiety to the carboxyl radical, and conjugation.

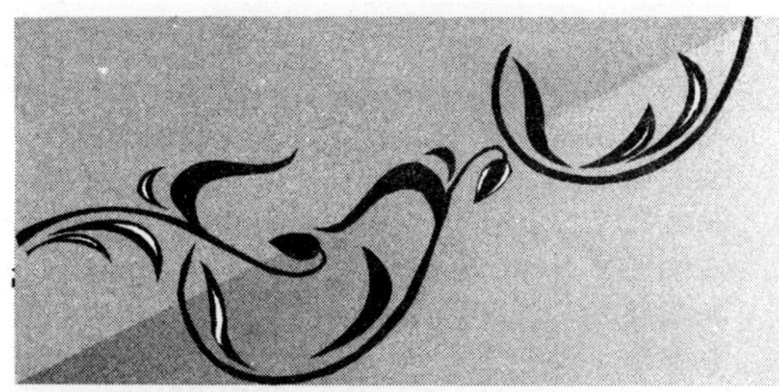

INDEX

A

B

C

D

E

F

G

H

I

J

K

L

M

N

O

P

Q

R

S

T

U

V

W

X

Y

Z